MW00343842

"As the U.S. military refocuses on great-power
essary the hoary canons of the past. Marvin
in his comprehensive, timely, and highly re
ciples of war."
—Capt. Peter M. Swartz, U.S. Navy (Ret.), senior CNA strategy analyst and former U.S. Navy maritime strategist and counterinsurgency advisor and instructor

"Pokrant makes an excellent case for a new set of principles of war. Everyone that
deals in the world of war plans needs this in their reference library. Rich in historical examples to support his case."
—Adm. Stan Arthur, U.S. Navy (Ret.)

"An important book for all who seek to understand how to improve the odds of
winning our nation's wars. Marvin Pokrant develops his new, more useful principles of war through the effective application of history, treating history as data
to test existing and proposed principles against. I have incorporated both his
principles and his methods into my course on wargaming."
—Col. Matthew B. Caffrey Jr., U.S. Air Force (Ret.), author of *On Wargaming: How Wargames Have Shaped History and How They May Shape the Future*

"All active-duty military officers should read this book. It would be perfect as a
text at military war colleges. People dealing with strategic studies and national
security will find this book valuable due to its suggestions of new principles to
guide American military efforts. Readers of military history will find this work
interesting because of the numerous historical examples."
—Phil E. DePoy, PhD, former president of the Center for Naval Analyses and
founding director of the Wayne E. Meyer Institute for Systems Engineering,
Naval Postgraduate School

"Marvin Pokrant's *New Principles of War* is an excellent examination of the development of the principles of war throughout history and how they have differed
over time and among countries. . . . If you are a student of military thought, this
work is a must-read and will be a welcome addition to your library."
—Michael A. Palmer, PhD, author of *Command at Sea: Naval Command and
Control since the Sixteenth Century*

"Marvin Pokrant is a brilliant military analyst. *New Principles of War* is a must-read for any student of the art of war. In fact, Pokrant's concise and straightforward approach will also benefit non-military business executives as they pursue the study of the art of business!"
—Vice. Adm. Cutler Dawson, U.S. Navy (Ret.), and former CEO of the Navy Federal Credit Union

"*New Principles of War* is an example of the power that can result from combining the skills of military analysis with the skills—and tasks—more often associated with the military historian. Both communities will benefit from the insights the author derived from his research and deep thinking."
—Peter Perla, PhD, author of *The Art of Wargaming*

NEW PRINCIPLES OF WAR

NEW PRINCIPLES OF WAR

Enduring Truths with Timeless Examples

MARVIN POKRANT

Potomac Books

AN IMPRINT OF THE UNIVERSITY OF NEBRASKA PRESS

Library of Congress Cataloging-in-Publication Data
Names: Pokrant, Marvin, 1943–, author.
Title: New principles of war: enduring truths with timeless
examples / Marvin Pokrant.
Description: [Lincoln, Nebraska]: Potomac Books,
An imprint of the University of Nebraska Press, [2021] |
Includes bibliographical references and index.
Identifiers: LCCN 2020038246
ISBN 9781640122222 (paperback)
ISBN 9781640124592 (epub)
ISBN 9781640124608 (mobi)
ISBN 9781640124615 (pdf)
Subjects: LCSH: War (Philosophy) | War—History. |
Military history, Modern.
Classification: LCC U21.2 .P565 2021 | DDC 355.0201—dc23
LC record available at https://lccn.loc.gov/2020038246

Set in Garamond Premier Pro by Mikala R. Kolander.

It is a very difficult task to construct a scientific theory for the art of war, and so many attempts have failed that most people say it is impossible, since it deals with matters that no permanent law can provide for. One would agree, and abandon the attempt, were it not for the obvious fact that a whole range of propositions can be demonstrated without difficulty.

—CLAUSEWITZ, *On War*

CONTENTS

TABLES

Fools never learn. Sensible people learn from their mistakes. Wise people
learn from the mistakes of others. The wisest of all learn from
both the successes and failures of other people.

—ANONYMOUS

....

HOW DOES A POTENTIAL MILITARY LEADER LEARN THE ART
of war? No one is born with the skills and knowledge to be a great military leader. People have to learn somehow. We can divide the way people
learn to wage war into five categories:

Their own experience in war.

Mentors: that is, the experience of a few living people.

Military history: that is, the experience of many people, living and dead.

Exercises, simulations, wargames, and analyses.

Indirect means such as studying psychology or playing games of strategy.

One's own experience in war is the most vivid teacher. For instance,
many American officers who served in the Vietnam War as lower-level
officers saw many mistakes made and vowed never to repeat those mistakes when they achieved high rank. These men led the U.S. Army, Air
Force, Marine Corps, and Navy in the first Gulf War, Operation Desert
Storm, and changed the U.S. military into the high-morale, effective fighting force that won that war.[1]

Although one's own experiences are extremely influential, that is not
necessarily the best way to learn. People generally learn more from failure than they learn from success. From a personal growth point of view,
learning from failure is good, but from a results perspective, it is better
to learn *before* failing.

The process of learning from mentors draws on the experiences of multiple people and is the second most vivid way to gain knowledge. Learning from the experiences of a few people is certainly better than learning from the experiences of a single person. However, we can do even better. Reading military history greatly expands the list of "mentors" from whom we can learn. How else could we ever have the opportunity to learn from Alexander the Great, Hannibal, Scipio Africanus, Julius Caesar, Napoleon Bonaparte, Horatio Nelson, and Robert E. Lee? Napoleon claimed that the only way to master the art of war was to study the campaigns of the great generals of the past.[2]

Exercises, simulations, wargames, and analyses also have a role to play, but they are imperfect teachers. Conditions are often unrealistic and fail to consider the impact of physical danger on performance. The value of these teaching tools is that military leaders can consider situations that they would never otherwise encounter, especially when dealing with cutting-edge technology.

Finally, the study of psychology, either formally in a classroom or informally in dealing with people throughout one's life, provides background that helps in understanding one's friends as well as one's enemies. Similarly, playing games of strategy such as chess provides background for devising strategies that work in a simplified situation. However, this at best is a supplement to other sources of knowledge and is not a substitute for other ways to learn.

We thus conclude that the study of military history is the area in which there is the most to gain. However, can we learn from history? Perhaps warfare changes too much over time or is too complex for any historic lessons to be valid. Some military writers have argued that because each situation is unique, principles of war are useless. Let us address the effect of changes in warfare over time and the uniqueness of each situation due to warfare's complexity.

Over the past several thousand years, the weapons carried by individual soldiers have gone from clubs to swords to muskets to machine guns to shoulder-fired missiles. On another scale, the technology of war has given us precision-guided munitions, stealth aircraft, cruise missiles, drones, nuclear-powered submarines, aircraft carriers, nuclear weapons, satellites, and network-centric warfare. Why do we think that we can find useful lessons that are valid across the millennia? We certainly would never study the ancient Greeks or Romans for guidance on building a super-computer or a rocket that could take us to Mars.

However, while technology and social conditions have changed, human nature has not. Remarkably, the human brain today has the same capacity as that of our long-ago ancestors. The wiring of the brain with respect to human emotions and psychological factors has not changed. We still have similar hopes, fears, and cognitive biases today as we did ages ago.

One other thing that has not changed is logic. Two plus two still equals four. Euclid's geometric proofs are still valid today.

Any guide that depends on constantly changing factors, such as technology or social conditions, is not going to be valid across the ages. However, guidance that depends on logic and the behavior of human beings might still be valid.

Warfare might be so complex that we will not be able to find useful guides. Every battle throughout history had unique circumstances. Nevertheless, although history never repeats in exactly the same way, as Mark Twain and many others are reputed to have noted, "it does rhyme."

Our task is to ascertain the ways in which military history "rhymes." We will try to distill important recurring lessons from military history. Because warfare involves human beings, each of whom is unique, we should not expect to find exact laws of nature similar to those in the physical sciences. However, finding recurring themes would still be useful.

If we find useful guidance, what is the best way to present it? Marshall Falwell noted that humans have a psychological drive to make sense of the world. This gives them an innate belief in cause and effect. If people are not provided sound guidance, they will devise their own rules based on their relatively narrow personal experience.[3] To provide guidance, most militaries of the world have institutionalized a list of the principles of war. These principles usually consist of a title of a single word or phrase, a short definition of one or two sentences, and a longer definition of a few paragraphs. Encapsulation of guidance in the form of principles has gained wide, though not universal, acceptance. At the very least, principles form a good framework for studying history. As Patrick MacDougall observed, if someone first learns general principles of warfare, then one will find the details of battles and campaigns more interesting. In addition, one will learn more quickly when one can consider the details in relation to the framework of general principles, rather than as a collection of random facts.[4]

If we express guidance in the form of principles of war, we will need to check whether each candidate principle of war remains valid across a variety of times, places, and circumstances. Because any principles that are valid across the ages will not incorporate ever-changing technology or social conditions, they *cannot* form a complete theory of warfare that accounts for such conditions. Nevertheless, even though they cannot form a complete theory of warfare, principles of war still could be quite useful.

The term "principle" is not universal. Writers have used terms such as precept, tenet, axiom, guide, rule, law, ultimate truth, fundamental truth, essence of warfare, maxim, lesson, aphorism, consideration, judgment, observation, imperative, fundamental, and commandment. The connotations of these terms run the gamut from mere suggestion to a law of nature that one must always follow. Most writers have used the various terms inconsistently. Some writers have differentiated them in excruciating detail. We will not get into arguments about the exact meaning of each term. We prefer to concentrate on the concepts, whatever their originators called them. In the first three chapters, we use the term "principle" in the broadest sense to include all of these variations. Later, in chapter 4, we define what we mean by a "principle of war."

In this book, we first look briefly at the history of the principles of war, the status of such principles in major military powers, and the principles endorsed by various American services. Then we present the criteria that we believe a principle of war should satisfy. We examine the degree to which each of the current American and British principles of war satisfy those criteria. In many cases, we conclude that they do not. The core of this book is our proposed new principles of war.

In this book, we illustrate various points by appealing to examples from history. Rather than undertaking an in-depth analysis of a few cases, we chose to present numerous historical examples; while the brevity of the descriptions necessarily leaves out many details and nuances, throughout, we did our best to ensure that the brief accounts do not mislead the reader about the lessons to be learned.

Of course, no selection of examples can ever prove or disprove the validity of a candidate principle of war because one can always cherry-pick examples from history to support almost any point. We present these examples solely to *illustrate* our ideas and do not claim they prove anything.

ACKNOWLEDGMENTS

FIRST, I WOULD LIKE TO ACKNOWLEDGE THE SUPPORT AND great patience shown by my wife, Gail, during the research and preparation of this manuscript, as well as for her proofreading.

This book would not exist without the enthusiastic support and encouragement of my agent Fritz Heinzen. He made numerous excellent suggestions and criticisms over the course of this project. No one ever had a better agent.

I thank my editor Tom Swanson and the other people at the University of Nebraska Press for their support and professionalism. I would especially like to thank Abigail Stryker and Taylor Rothgeb for their cheerful willingness to answer all my questions. My copyeditor, Irina du Quenoy, made many suggestions to clarify the text and make the writing smoother. She also went beyond the usual role of a copyeditor to make astute suggestions concerning facts and interpretation that improved the book. Of course, I am responsible for any remaining errors.

I would like to thank Capt. Peter Swartz, USN (Ret.), who read the draft manuscript and made helpful suggestions about the discussions concerning the U.S. maritime strategy, of which he was one of the developers. I would also like to thank Dr. Michael Palmer, who read and critiqued the manuscript, pointing out several places where he disagreed with my initial interpretations. The remaining errors are mine alone.

I would like to thank all my colleagues in the military analytic community for the exchange of ideas over the course of my career. I am grateful for the acceptance and friendship from the many military people from all services with whom I have worked during my career. I learned so much from them.

PROLOGUE

Pickett's Charge

IN JULY 1863 GEN. ROBERT E. LEE AND THE ARMY OF NORTHERN
Virginia fought Maj. Gen. George Meade and the Army of the Potomac
at the Battle of Gettysburg. In this prologue, let us consider how the nine
currently recognized American principles of war apply to the climactic
part of that battle from the Confederate point of view.

Objective

The primary political objective of the Confederacy was independence.
If Lee defeated the Union Army in a major battle on northern soil, that
would encourage the northern peace movement and might persuade Great
Britain and France to recognize and perhaps even support the Confederacy. Thus, the military objective in the summer of 1863 was to invade
the North and defeat the Union Army on northern territory.[1] The operational objective was to engage the Army of the Potomac under favorable
conditions and defeat it. On the third day of the battle the tactical objective was to break the Union line in the center where Lee thought it was
weakest. In summary, each objective in the hierarchy of objectives from
political to tactical seemed well defined and logical. Based on a series of
brilliant victories won by the Army of Northern Virginia, Lee believed
these objectives were attainable. Thus the Confederates complied with
the principle of the objective in choosing their objectives.

Mass

The principle of mass calls for concentrating combat effects at the critical
place and time. Both at the time and in retrospect, the assault on Cemetery Ridge on the afternoon of 3 July 1863 was the critical place and time,
not just of the battle, but also of the entire Civil War. Operationally, in
recognition of the importance of this theater, Lee's Army had the largest portion of Confederate forces. It comprised eighty thousand battle-

hardened veterans with a history of success. At this crucial moment Lee intended to devote most of his artillery to the attack and fifteen thousand men to the charge itself. Lee's concentration of artillery was said to be the largest ever on the North American continent. Although neither the artillery nor the infantry were as well concentrated and coordinated as Lee intended, in general the Confederacy and Lee concentrated their forces at the crucial point and time. The Confederates reasonably followed the principle of mass.

Offensive

The military strategy was offensive. Operationally, the invasion of the North was offensive and so was the engagement of the Union Army at Gettysburg. During the three days of the Battle of Gettysburg the Confederates were almost always on the tactical offensive, while the Union Army was mostly on the defensive. Finally, Pickett's Charge was clearly offensive. Thus, the Confederates definitely obeyed the principle of the offensive.

Unity of Command

The chain of command went from Lee to Lt. Gen. James Longstreet to the three division commanders: Maj. Gen. George Pickett, Brig. Gen. J. Johnston Pettigrew, and Maj. Gen. Isaac R. Trimble. The Confederates adhered to the principle of unity of command.

Simplicity

The plan was simple. First, the Confederate artillery would soften up the Union defenses at the point of attack, especially by silencing some of the Union artillery in that sector. Then three divisions would march straight to the Union line and overwhelm them. The plan and the orders were simple. Lee's plan respected the principle of simplicity. He rejected Longstreet's suggestion of a more complex plan to turn the Union's flank.

Economy of Force

The Confederates economized in other theaters to give Lee as large a force as possible for the invasion of the North. Lee moved artillery from other areas to the critical sector for the attack. The Confederates abided by the principle of economy of force.

Maneuver

Simply marching toward the enemy lines did not involve sophisticated use of maneuver.

Security

Lee's other sectors were secure, as was his line of communication. The Confederates heeded the principle of security.

Surprise

There was little surprise in Pickett's Charge. While Lee repositioned most of his artillery at night so the movement would not tip off the Union Army and Pickett concealed his division, the extended artillery bombardment and then the long march across the valley gave the Union defenders plenty of warning. Thus, the attack was not in accordance with the principle of surprise.

Summary

Pickett's Charge adhered to most of the nine currently recognized principles of war, exceedingly well for the most part. In addition, veteran troops of an extremely capable army executed Pickett's Charge with exemplary courage and fortitude. If Pickett's Charge complied with nearly all of the principles of war and brave, disciplined men with high morale executed the attack, why did it fail?

The current principles of war do not explain the failure of Pickett's Charge because they are flawed. This book will examine the current principles of war and show that most do not really reflect the essence of the excellent ideas that led to their articulation. Therefore, we will suggest new principles of war that will better incorporate the lessons of military history.

NEW PRINCIPLES OF WAR

ONE

History of the Principles of War

BEFORE CRITICIZING, REPLACING, OR ENDORSING THE CUR-
rent principles of war, we should understand their genesis and evolution.
For more than two millennia, people have written about how to fight and
win wars. Many of their ideas involved details of the then-current state of
technology and the prevalent type of warfare. Although these ideas may
have been good advice at the time, often the concepts were too specific
to the author's context to be applicable throughout the ages. However,
some of the guidance was sufficiently general to be useful even today. As
the theory of warfare grew, many writers found principles useful to vary-
ing degrees and advocated a variety of uses for them.

In this chapter, we will briefly discuss some of the history of concepts
related to what we now call principles of war. Many people will be surprised to learn that the current form of the principles of war is only about
a hundred years old. Furthermore, the lists of principles and their definitions have changed over the last hundred years.

Ancient Concepts of War

Sun Tzu

Nearly twenty-five hundred years after Sun Tzu wrote *The Art of War*, his
teaching remains relevant. Certainly, much of his advice was quite specific
to his time; for example, he wrote, "Generally, operations of war require
one thousand fast four-horse chariots, one thousand four-horse wagons
covered in leather, and one hundred thousand mailed troops."[1] However,
some of Sun Tzu's guidance is more generalized and still useful. We can
even recognize some elements of current principles of war.

Sun Tzu claimed that generals had to master the five factors of moral
influence, weather, terrain, command, and organization. He claimed that
he could predict which side would be victorious by comparing one's own
situation with the enemy's situation regarding seven elements: the ruler's

moral influence, the commander's ability, taking advantage of terrain, carrying out instructions, strength of troops, training, and rewarding and punishing in an enlightened manner.[2]

Sun Tzu advised avoiding battle until one had a relative advantage over the enemy. A bad general fights a battle hoping for victory; a good general wins before fighting a battle. The ultimate goal is to attain victory without battle: "To subdue the enemy without fighting is the acme of skill."[3]

Sun Tzu stressed the importance of deception, spies, maneuver, and surprise to gain an advantage. This advantage should then be used to attack undefended places or portions of the enemy force. The reader will recognize elements of the modern principles of maneuver, surprise, and mass in these aphorisms:

> Speed is the essence of war. Take advantage of the enemy's unpreparedness; travel by unexpected routes and strike him where he has taken no precautions.[4]

> Attack where he is unprepared.[5]

> If I am able to determine the enemy's dispositions while at the same time I conceal my own then I can concentrate and he must divide. And if I concentrate while he divides, I can use my entire strength to attack a fraction of his.[6]

He emphasized gaining knowledge of the enemy, oneself, and the environment.

> Know the enemy, know yourself; your victory will never be endangered. Know the ground, know the weather; your victory will then be total.[7]

> Determine the enemy's plans and you will know which strategy will be successful and which will not.[8]

Sun Tzu also gave advice on how to discipline troops and maintain their morale. He recognized that only a well-disciplined force could effectively feign disorder to deceive an enemy. Although he gave much advice that looked like rules, Sun Tzu stated that there were no constant conditions. Hence, one should not mindlessly repeat successful tactics. Instead, one should modify tactics according to each specific situation.[9]

There is much wisdom in Sun Tzu that one could use to formulate a set of principles of war. However, the Western world did not learn of his work until the first translation appeared in Paris in 1772. Therefore, military thought in the Western world evolved independently, rather than building on Sun Tzu.[10]

Kautilya

The *Arthashastra* of Kautilya is an ancient Indian treatise, parts of which date back at least two millennia.[11] Scholars thought it had been lost until its rediscovery in 1905. The *Arthashastra* reveals much about ancient Indian culture, law, economics, politics, diplomacy, and warfare. The parts related to warfare give advice on when to invade a country, when to attack an enemy army, and the best marching formations under various circumstances. It also discusses the best ways to employ infantry, cavalry, chariots, and elephants. Much of the advice contains excruciating detail. For example, in addition to specifying that camps should have four gates, six roads, nine divisions, ditches, parapets, and watch towers, the *Arthashastra* even specifies the harem should be located west of the king's quarters.[12]

The *Arthashastra* gives prominence to spies, deceit, and intrigue. Some of the advice is Machiavellian: "When the country is full of local enemies, they may be got rid of by making them drink poisonous (liquids); an obstinate (clever) enemy may be destroyed by spies or by means of (poisoned) flesh given to him in good faith."[13] One can find here elements of the modern principles of concentration, security, surprise, and morale, though the *Arthashastra* does not state any in a manner resembling modern principles. For example, concerning the maintenance of morale, the treatise reads, "Astrologers and other followers of the king should infuse spirit into his army."[14]

The *Arthashastra* also advises knowing oneself, the enemy, and the terrain: "The conqueror should know the comparative strength and weakness of himself and of his enemy."[15] It provides meticulously detailed recommendations on choosing the terrain for a battle. One example of the advice on when to attack is to "strike the enemy's men during the afternoon when they are tired by making preparations during the forenoon; or . . . strike the whole of the enemy's army when it is facing the sun."[16]

The Romans

Sextus Julius Frontinus was a successful Roman general, three-time consul, and supervisor of the aqueducts during the first century CE. His honesty and modest nature are evident in his book describing the Roman aqueduct system. Therefore, we take him seriously when he claims to be the only person to have reduced the rules of military science to a system. He did so in the *Art of War*. Unfortunately, this work has not survived except insofar as Vegetius may have incorporated some material three centuries later. Frontinus followed the *Art of War* with the *Strategemata*, which he claimed illustrated the rules of war. It contains many one-paragraph descriptions of various stratagems taken from the then-current literature, but nothing resembling rules of war.[17]

In late Roman times, Flavius Vegetius Renatus, commonly known as Vegetius, wrote *The Military Institutions of the Romans*, which described the practice of war and the fundamentals essential to victory. Writing about 390 CE during the decline of Roman power, Vegetius's goal was to restore Roman military prowess by showing how Romans had operated during their greatest days. He believed that the Roman army of his time had forgotten the things that had made the Roman army of the past so successful. Vegetius drew on material from Augustus, Trajan, Hadrian, Cato the Elder, Cornelius Celsus, Paternus, and Frontinus.[18]

The Military Institutions of the Romans consists of five books. The first book concerns the discipline and training of recruits. Vegetius emphasizes the importance of skill and discipline. Recruits had to learn to march carrying sixty pounds in addition to their arms. They had to be able to march twenty-four miles in five "summer hours." Recruits had to learn it was more lethal to thrust with the gladius (the short sword used by the Romans) than to slice with it, the reasons for which Vegetius explained in detail.[19]

Vegetius's second book described the organization of a legion, its officers, and the legion's formation for battle. It also discussed promotions, legionary music, and tools of the legion.[20] His fourth book described the attack and defense of fortified positions. His fifth book discussed naval operations.[21]

Vegetius's third book, describing tactics and strategy, was his most influential. He introduced the concept of a reserve and emphasized the importance of knowing an enemy's capabilities, tendencies, commanders' dispositions, and morale. He also emphasized the importance of knowing the terrain in detail. He insisted on making troops practice activities repeatedly. In a telling passage, he lamented that legions in his time neglected to fortify their camps and, as a result, were often surprised by the enemy. Vegetius concluded his third book with a list of general maxims. Most of these are pithy summaries of topics discussed at length earlier in the book. About half of these maxims are specific to warfare of that age, but some are sufficiently general to have applicability today.[22] Vegetius's maxims include the following:

> Those designs are best which the enemy are entirely ignorant of till the moment of execution. . . .
>
> Valor is superior to numbers.
>
> The nature of the ground is often of more consequence than courage.
>
> Few men are born brave; many become so through care and force of discipline. . . .
>
> Novelty and surprise throw an enemy into consternation; but common incidents have no effect. . . .
>
> An army unsupplied with grain and other necessary provisions will be vanquished without striking a blow.[23]

Although the Romans of his time made little use of his books, military thinkers and practitioners relied heavily on Vegetius *for more than a thousand years*. Charlemagne's commanders considered Vegetius a necessity. Henry II and Richard the Lion-Hearted carried their copies everywhere. Niccolò Machiavelli relied on it. Field Marshal Prince de Ligne of Austria called it a golden book in 1770.[24]

The Renaissance

Sixteenth-century writers such as Niccolò Machiavelli found that some aspects of war had not changed in the millennium after Vegetius wrote of them. At that time, Europeans thought that classical Rome had been the peak of civilization. Machiavelli believed that conducting war the way the Romans had done would be a step forward, asserting that "the ancients did everything better and with more prudence than we."[25] Like Vegetius, Machiavelli emphasized discipline and training. He argued that the introduction of firearms and artillery should not change many of the lessons from the Romans, but he did acknowledge that it affected a few things. He also advocated that a single commander be in charge.[26]

Machiavelli gave much advice that he clearly copied from Vegetius nearly word for word. For example, "No proceeding is better than that which you have concealed from the enemy until the time you have executed it."[27]

For the next three centuries, military thinkers leaned on writings from the classical age of Greece and Rome. Most writers still mixed some broad fundamental concepts with many details. About 1644, Henry, the duke of Rohan, listed seven maxims. As with most other writers of his time, these maxims were reasonable advice but were not comparable to current-day principles.[28] For example,

> Never allow yourself to be forced to combat against your will. . . .
>
> Have good leaders at the head of each principle corps. . . .
>
> Put the best troops on the wings and attack with the wing that is the strongest.[29]

The Age of Reason

In 1687 Isaac Newton published his masterpiece, *Mathematical Principles of Natural Philosophy (Principia)*, which presented Newton's three laws of motion and the law of universal gravitation. These four laws explained planetary orbits, the motion of the moon, the paths of comets, tides, and the precession of the equinoxes, as well as the motion of objects on earth. In particular, Newton's Laws enabled the calculation of artillery trajectories. This and other scientific progress, along with technological progress in warfare, inspired military thinkers to seek similar laws that governed the conduct of warfare. As a result, writers often used the language of science and sometimes the terms "principle" or "rule."

In the early eighteenth century, Antoine Manassès de Pas, the marquis de Feuquières, wrote *Memoirs on War*. This book presented Feuquières's advice on various aspects of military operations. He criticized French commanders in the wars in which he had participated, drew lessons from those wars, and claimed those lessons were rules with universal applicability, and indeed that he had collected all the rules that generals needed.[30]

Maurice de Saxe was a successful soldier of the mid-eighteenth century. In the text *My Reveries,* published posthumously in 1757, he presented minutely detailed instructions on the organization, discipline, armament, and motivation of armies. In common with Renaissance writers, he thought highly of the Romans: "We should have a better opinion of the ancients and of the Romans, who are our masters and who should be."[31] He advocated reintroducing the use of cadence in marching, which the Romans had used but more recent militaries had abandoned. He emphasized the role of morale in war and discussed the psychology of motivating men. He claimed that the decisive factor in all battles is the panic men feel when something unexpected happens, especially when it threatens their flanks or rear, asserting, "Human weaknesses cannot be managed except by allowing for them."[32] De Saxe stated that in contrast to the sciences, war had no principles. He also wrote that establishing any system required knowledge of the underlying principles. However, by principles, he meant the details of organization, discipline, and motivation of armies, rather than anything that we would recognize as a principle today.[33]

The Napoleonic Age

Napoleon Bonaparte had unprecedented success as a general. He fought more battles than any previous general and won almost all of them. Of course, people studied his campaigns and his words. During the first part of the nineteenth century, Napoleon influenced many military theorists who wrote about the factors that lead to success in war.

Napoleon

Napoleon claimed that he could write down simple principles of war so clear that any military man could understand them, but he never explicitly did so. Napoleon also said that the proper application of the principles depended on the particular circumstances and "No rule of war is so absolute as to allow no exceptions."[34]

Although Napoleon never actually wrote down a compact list of principles, others extracted maxims from his memoirs and correspondence. General Burnod collected 115 numbered maxims. Here are a few that seem relevant to present-day warfare and the currently accepted principles of war.

> (8) A general should say to himself many times a day: If the hostile army were to make its appearance in front, on my right, or on my left, what should I do? And if he is embarrassed, his arrangements are bad; there is something wrong; he must rectify his mistake. . . .

> (14) . . . Even in offensive war, the merit lies in having only defensive conflicts and obliging the enemy to become the assailant. . . .

(16) A well-established maxim of war is not to do anything which your
enemy wishes—and for the single reason that he does so wish. . . .

(29) When you have it in contemplation to give battle, it is a general rule
to collect all your strength and to leave none unemployed. . . .

(64) Nothing is more important in war than unity in command. . . .

(77) . . . Generalship is acquired only by experience and the study of the
campaigns of all great captains. Gustavus Adolphus, Turenne and Frederic, as also Alexander, Hannibal and Caesar have all acted on the same
principles. To keep your forces united, to be vulnerable at no point, to
bear down with rapidity upon important points—these are the principles which ensure victory.[35]

The reader may recognize elements of the principles of security, concentration (or mass), and unity of command.

Clausewitz

The two most influential writers who came after Napoleon were Antoine-
Henri Jomini and Carl von Clausewitz. Whereas Jomini analyzed military history and then tried to extract principles from his analysis, his
contemporary Clausewitz was more of a philosophical military theorist
who tried to devise a broad theory of war. At the same time, Clausewitz
did not ignore history. He tested each conclusion against military history and rejected those contradicted by it.[36] His unfinished masterpiece,
On War, eventually proved to be more influential than Jomini's writings.

Clausewitz introduced many concepts that remain relevant today. One
of his models was the paradoxical trinity of war. He contended that, under
various circumstances, war displays characteristics of primordial violence,
chance, and reason. Each of these attributes primarily concerns a different group. The blind natural forces of violence and hatred mainly affect
the people. Chance and the creative spirit involve the armed forces and
its leaders. As an instrument of policy, war is also subject to reason, and
it is the government that sets the political objectives. Thus, we have a secondary social trinity of the popular base, the military and its leaders, and
the government.[37]

Clausewitz repeatedly emphasized the role of chance in warfare, how
unpredictable random events can have outsized effects. Indeed, today

we recognize that war is what mathematicians term a chaotic system, in which minute changes in the initial conditions can have enormous consequences. In addition, more than other military theorists, Clausewitz accentuated the psychological (moral) aspects of war. He also highlighted the important role of friction in war, as well as the fog of war.

With regard to the principles of war, Clausewitz believed that every situation a commander faced was unique. Therefore, the application of any general principle would be different in each case. However, like many other writers who said similar things, Clausewitz also presented various principles or maxims in his writings.

King Frederick William III of Prussia appointed Clausewitz to tutor his teenage son Crown Prince Frederick William on warfare. When the king formed an alliance with Napoleon in 1812, Clausewitz resigned his position with the Prussian Army in protest and joined the Russian army. He left a final piece of instruction for the young crown prince, *The Most Important Principles for the Conduct of War to Complete My Course of Instruction of His Royal Highness the Crown Prince*. This source contains a discussion of principles of war and several lists of principles.[38]

Clausewitz's *Principles for the Conduct of War* begins with a section titled "Principles of War in General," in which he wrote, "The theory of warfare tries to discover how we may gain a preponderance of physical forces and material advantages at the decisive point."[39] In this work, Clausewitz calls it a "fundamental principle" that one should never remain completely passive on defense.[40] Defense must set the stage for offense. However, even under the best conditions, a commander on the offensive must consider the possibility of great disaster and take appropriate precautions. He also states that the principles are derived from the study of history and that the maxim "Pursue one great decisive aim with force and determination" is the most important cause of victory.[41]

Clausewitz advocates sending the primary attack against just a single point, which is that part of the enemy force whose defeat will result in a decisive advantage. He also advises concentration of forces, but strongly advises against sending all of one's forces into battle at once. Instead, he urges commanders always to follow the principle of conserving "a decisive mass for the critical moment" after the enemy has been fatigued.[42] He notes that surprise is one of the best weapons.[43] Clausewitz claimed that pursuit of a beaten enemy is extremely important, second only to victory itself, and advocated combining multiple types of arms to increase the strength.[44]

Under strategy, Clausewitz stated that the primary objective should be the enemy army's main body. To achieve this, he claimed there were only a few principles, which one could summarize briefly. His first and most important rule was to use one's entire force with great energy. His second rule was to concentrate forces for the primary attack, even at the expense of weakening positions elsewhere. Clausewitz's third rule was to never waste time. The fourth rule was to pursue a beaten enemy vigorously.[45]

Thus, we can detect in this manuscript elements of the modern principles of the objective, concentration, economy of force, unity of effort, offensive, and surprise. However, Clausewitz cautioned the crown prince that the purpose of the principles was not to provide complete instructions but rather to stimulate thought and to guide his study of military history.[46]

Clausewitz undoubtedly revised many of his ideas after writing his guidance for the crown prince, but his widow thought *Principles for the Conduct of War* provided insight into his early thinking.[47] Nevertheless, most current-day analysts downplay the importance of that document relative to his later writings. In contrast, Clausewitz's *On War*, though unfinished, contains his later ideas and is far more complex than his memorandum for the crown prince.[48]

Clausewitz believed that one should use theory as a guide for learning from books. Principles that emerge from the study of history provide a frame of reference for actions, rather than a guide that specifies exactly the action to take in the midst of battle.[49] Bernard Brodie wrote, "Clausewitz would have been appalled" at many of the current uses of principles of war "and not surprised at some of the terrible blunders that have been perpetrated in the name of those 'principles.'"[50]

Clausewitz repeatedly stated in various ways that concepts such as principles could not be "dogmatically applied to every situation, but a commander must always bear them in mind so as not to lose the benefit of the truth they contain in cases where they do apply."[51] He gives examples of "rules" such as "cooking in the enemy camp at unusual times suggests that he is about to move" and "the intentional exposure of troops in combat indicates a feint."[52]

With regard to concentration, Clausewitz asserted,

History of the Principles of War

There is no higher and simpler law of strategy than that of *keeping one's forces concentrated*. No force should ever be detached from the main body unless the need is definite and *urgent*. We hold fast to this principle, and regard it as a reliable guide. (emphasis in original)[53]

This principle also contains at least part of the currently recognized principle of economy of force. Clausewitz later elaborates on this, writing, "Make sure that all forces are involved—always to ensure that no part of the whole force is idle."[54]

To summarize, Clausewitz believed that principles of war existed, but he repeatedly cautioned against applying them without thought. He believed they were useful as guides both in studying military history and in battle, but in the latter case, the commander had to judge each situation as to whether a principle was relevant and, if so, how to apply it. Some of the principles he mentioned clearly have much in common with today's prevailing principles. However, many others were specific to warfare of his age.

Initially, Clausewitz had limited influence, much less than Jomini. Later, as his work gained in stature, some claimed to detect in his writings nearly all of the modern principles of war, although others dissented. For example, although some military historians have credited him as "a prime contributor to the modern principles of war," John Alger claimed that he was not.[55] However, Alaric Searle disagreed, especially with regard to the development of the British principles of war.[56] Broadly speaking, Clausewitz's greatest influence during the nineteenth century was in Germany.

Jomini

Although Clausewitz is better known today, Jomini was the one who most popularized the idea that there are just a few general principles that could lead to success in war. The evolution of these ideas may be seen in several books he authored.[57]

Jomini believed that war as a whole was an art, not a science. Although he believed that fixed laws similar to those of science governed strategy, he stated unambiguously that this was not true of war viewed as a whole. Nevertheless, he rejected the idea that there were no tactical rules. He claimed that if skillful generals had repeatedly applied certain rules that were successful many times, it would be absurd to reject those rules because they failed occasionally. Jomini allowed that rules could not tell generals what to do in every situation. However, he claimed that it was "certain" that the rules would "always" identify errors to be avoided. He even claimed that these rules thus become, in the hands of skillful generals commanding brave troops, means of almost certain success.[58] Although Jomini's *The Art of War* often describes things he calls principles and contains many lists, none of them resembles what we would recognize today as a list of principles of war. However, we can compile some guidance based on his writing.

Jomini stated that the science of strategy involves choosing the theater of war and determining the enemy's strategy. He describes the art of war as the employment of troops in the chosen theater of war and claimed this should be guided by two fundamental principles, which he also referred to as "fundamental truths." First, use quick maneuvers to concentrate the mass of your troops against a portion of the enemy's troops. Second, strike in the direction in which the enemy's defeat would be most disastrous for the enemy, but an enemy victory would be minimally advantageous.[59]

In other places in *The Art of War*, Jomini dropped the second "fundamental truth" and asserted:

There is one great principle underlying all the operations of war—a principle which must be followed in all good combinations. It is embraced in the following maxims:

1. To throw by strategic movements the mass of an army, successively, upon the decisive points of a theater of war, and also upon the communications of the enemy as much as possible without compromising one's own.

2. To maneuver to engage fractions of the hostile army with the bulk of one's forces.

3. On the battlefield, to throw the mass of the forces upon the decisive point, or upon that portion of the hostile line which it is of the first importance to overthrow.

4. To so arrange that these masses shall not only be thrown upon the decisive point, but that they shall engage at the proper times and with energy.[60]

The modern reader will recognize in this list elements of the principle of mass or concentration.

We can also find other familiar principles in *The Art of War*. We see the principle of the objective when Jomini says: "The choice of objective is by far the most important thing in a plan of operations."[61] Jomini offered a forerunner of the principle of economy of force when he wrote that it would be a grave fault to send out large detachments unnecessarily.[62] Furthermore, he repeatedly emphasized the advantages of the offensive. An enemy on defense must prepare for many possibilities, whereas an attacker can strike at a chosen point and use a large force against a fraction of the defender's force. However, in a limited area (what Jomini terms a tactical situation), the defender can see the attacker's moves and meet them with reserve forces.[63]

The Art of War dominated military thinking for two generations, both in Europe and in the United States. West Point instructor Dennis Hart Mahan and his student Henry W. Halleck both promoted Jomini. The latter translated Jomini and, using ideas from that work, wrote his own book, *Elements of Military Art and Science*, in 1846. He became general in chief of the Union armies in 1862. Generals on both sides of the American Civil War had learned Jominian strategy. Rumors said every successful general carried a copy of Jomini's book. However, Ulysses S. Grant most likely did not. He had little use for military theory and had almost no books in his military library.[64]

MacDougall

Patrick Leonard MacDougall was influential in spreading the concept of principles of war to the British military. MacDougall, appointed superintendent of studies at the Royal Military College in 1854, wrote *The Theory of War* in 1856. In large part, he followed Jomini and accepted the existence of principles. What was new was the explicit listing of those principles and distinguishing between principles and maxims. His three leading principles, the first being most important, were "place masses of your army in contact with fractions of your enemy. . . . operate as much as possible on the communications of your enemy without exposing your own. . . . operate always on interior lines."[65]

MacDougall also presented thirty maxims that he deduced from his principles, with each maxim being subordinate to the principles. On the one hand, most of these maxims addressed details specific to the then-current type of warfare. On the other hand, one can see the wisdom of the ninth maxim in the American leapfrogging campaign in the Pacific during World War II: "To besiege a fortified place whose possession would be useless to yourself, and which gives the enemy no power of annoyance, is to waste time and means."[66] The thirtieth maxim contained guidance that seems timeless: "Every disadvantage may be removed by skill or fortune, except *Time*. If a general has Time against him he must fail. And conversely, *Time* is the best ally" (emphasis in original).[67]

Late Nineteenth Century

After the American Civil War, military schools in the United States, United Kingdom, and France promoted the idea that a few guiding principles govern the conduct of war. However, debate continued between a focus on principles and the study of individual cases. In the war academy in Berlin, the balance tilted toward the study of specific cases. The Prussian approach emphasized unique solutions to each particular situation.[68]

As chief of the general staff of Prussia (and later united Germany) for thirty years starting in 1857, Helmuth von Moltke (the Elder) greatly influenced German military thought and education. He described himself as a disciple of Clausewitz and rejected the role of principles because the great uncertainty in war precluded firm rules. He believed that no plan could survive first contact with the enemy. After that, commanders had to grasp the situation, despite many uncertainties. Moltke believed that solutions to problems required innovative thought, not rigid rules. Training and studying military history fostered the ability of commanders to find those solutions. Moltke championed the concept of *Auftragstaktik*, which is usually translated as mission tactics but perhaps is better described as a cultural philosophy. This emphasized that officers should use judgment and exercise individual initiative within the design of the overall plan.[69] Under Moltke's leadership, Prussia won three stunning successes in wars with Denmark in 1864, Austria in 1866, and France in 1870–71. These Prussian successes forced military professionals everywhere to take notice of the pragmatic approach of the German method.[70]

In France, Ferdinand Foch, building on previous work by Henri Bonnal, presented principles that incorporated a blend of moral and material factors. The titles of his four principles were economy of forces, freedom of action, free disposal of forces, and "security, etc." Foch did not explain what he meant by "etc."[71] Some writers claim Foch's principles were the first modern principles in form because he gave them a title and presented them in a brief list. Although Alger said that it was not clear whether Foch's principles had any enduring influence, Searle claimed that the 1918 English translation of Foch's book contributed attention to the subject of principles of war in Britain.[72]

Alfred Thayer Mahan's masterpiece, *The Influence of Sea Power upon History 1660–1783*, and other writings convinced countries all over the world of the importance of controlling the sea by building a strong navy.[73] Mahan admired Jomini and shared the belief in the existence of a few principles that could guide a commander in understanding war. He applied this idea to naval warfare. Rather than listing principles and presenting historical instances in which commanders did or did not follow the principles, Mahan chose to present a narrative from which a student could deduce the principles. This method forced the reader to use his mind actively to assimilate knowledge of the principles, rather than giving him digested knowledge. Mahan did not present a list of his principles, but scattered them throughout the text. He emphasized that the primary principle of naval strategy and tactics was that of concentration. Famously, he successfully urged Pres. Theodore Roosevelt not to divide the American battle fleet between the Atlantic and Pacific Oceans.[74]

Mahan argued that history showed that the primary objective of a navy is to gain control of the sea. It can do that only by *destroying* the enemy battle fleet. This is usually interpreted as bringing about, and winning, a *decisive* battle. To do this, one must concentrate one's fleet and ignore calls to disperse parts of it to defend various locales. In particular, one should never divide the battle fleet. A navy must always be on the offensive; otherwise, the enemy will have the initiative and may attack where and when they will do the most damage. An inferior fleet might have to avoid battle unless conditions are favorable, but being on the defensive for a long time causes sailors to lose their edge. Well-trained and well-led personnel are important attributes. Mahan and his idea of bringing about a decisive battle had great influence on Japanese naval strategy in World War II.[75]

Twentieth Century

In the years prior to World War I, military writers generally acknowledged the existence of principles of war, but there was still little agreement on their number. Nor was there agreement on their degree of permanency.[76]

In first decade of the twentieth century, both British and American army documents asserted that the principles of war were few and not complicated, but did not list any. However, they seemed to be warming to the idea that actual enumeration of principles of war was both possible and desirable.[77]

During World War I, orders and directives sometimes contained lists of principles. They often described these as applying to strategy or tactics, or to a particular branch of the army, or to the defensive. They started to resemble the modern form of principles, with a title composed of a single word or phrase, followed by a pithy aphorism. Many "principles" were still quite narrow conceptions compared to current principles. However, some were beginning to resemble the modern principles of objective, concentration, surprise, offensive, and unity of command. The principles were sometimes claimed to be immutable and other times offered as a guide whose application would vary with circumstances.[78]

British Principles of War

The 1909 version of the British Army's *Field Service Regulations* stated that the fundamental principles of war were few in number, though their application was difficult. There was no list of principles, but nearly every section began with a discussion of "general principles." These sections covered numerous subjects, such as marches, movement by rail, quarters, billets, information, outposts, attack, defense, the encounter battle, and warfare against an uncivilized enemy. The regulations stated that all ranks should regard these principles as "authoritative" and that violating these principles has caused mishaps, even disaster, in the past. Whenever a commander in the field had to make a decision, he was instructed to give these principles "full weight."[79]

Although many of these discussions of principles concerned minutiae, some contained elements of the modern principles of war. A few sentences, many in bold print, would be recognizable to a modern reader: "Every commander is responsible for the protection of his command against surprise." "Timely information regarding the enemy's dispositions and the topographical features of the theatre of operations is an essential factor of success in war." "Decisive success in battle can be gained only by a vigorous offensive." "The defensive attitude must be assumed only in order to obtain or create a favourable opportunity for decisive offensive action."

"The objective of the decisive attack should be struck unexpectedly and in the greatest possible strength."[80]

J. F. C. FULLER WAS THE SINGLE MOST INFLUENTIAL INDIVIDual in promoting a move toward the modern version of the principles of war. Fuller claimed that when he read in the 1909 British Army *Field Service Regulations* that the principles of war were few in number, he searched the document but could not find any principles. (However, see previous paragraphs.) Therefore, in 1912 he studied Napoleon to devise six principles. He titled these objective, mass, offensive, security, surprise, and movement. The explanation of each resembled modern explanations. One elaboration was that the true objective should be the "the point at which the enemy may be most decisively defeated," which would usually be the line of least resistance. In 1916 he added economy of force and cooperation to his list of principles of war.[81]

In 1919 the British Army assigned a committee led by Col. John Dill to revise the *Field Service Regulations*. Fuller urged the committee to define and list the principles of war. In 1920 the British Army issued a provisional version of the regulations, which was the first official publication to present a brief list identified as "the principles of war." It identified each of the eight principles by a title and a brief explanation. These principles clearly followed Fuller's list, though with different titles:[82]

Maintenance of the objective—In every operation of war an objective is essential; without it there can be no definite plan or coordination of effort. The ultimate military objective in war is the destruction of the enemy's forces on the battlefield, and this objective must always be held in view.

Offensive action—Victory can only be won as a result of offensive action.

Surprise—Surprise is the most effective and powerful weapon in war. Whether in attack or defence the first thought of a commander must be to outwit his adversary. All measures should therefore be taken, and every means employed to attain this end.

Concentration—Concentration of superior force, moral, and material, at the decisive time and place, and its ruthless employment in the battle are essential for the achievement of success.

Economy of force—To economize strength while compelling a dissipation of that of the enemy must be the constant aim of every commander. This involves the correct distribution and employment of all resources in order to develop their striking power to the utmost.

Security—The security of a force and of its communications is the first responsibility of a commander. To guard against surprise; to prevent the enemy from obtaining information; to dispose his covering troops so as to allow his main forces to move and rest undisturbed; these are the considerations which must govern his actions in obtaining security. A force adequately protected retains its liberty of action and preserves its fighting efficiency against the day of battle.

Mobility—Mobility implies flexibility and the power to maneuver and act with rapidity, and is the chief means of inflicting surprise. Rapidity of movement for the battle should, therefore, be limited only by physical endurance and the means of transportation available.

Cooperation—Only by effective cooperation can the component parts of the fighting forces of a nation develop fully their inherent power, and act efficiently towards success.[83]

This was the birth of the modern principles of war in official documents. They were principles of *war*, not principles of strategy or principles of tactics or principles of infantry. This edition of the British *Field Service Regulations* did note that, because each situation was unique, the application of the principles would vary, leaving room for judgment and genius. In contrast, Fuller claimed these principles were eternal and applied to every battle.[84]

The Royal Air Force and the Royal Navy adopted the army's principles of war with almost no changes in 1922 and 1925, respectively. The debate among the services about the principles contributed to a common language that was useful in future joint operations.[85]

In his 1926 book *The Foundations of the Science of War*, J.F.C. Fuller claimed that one could reduce war to a science, though an inexact science. He elevated economy of force to a *law* and expanded his list of principles to nine, grouped in three sets of three.[86] However, these ideas did not gain traction.

The British Army's list of principles periodically changed, as did its position on the immutability of the principles. For example, the British dropped the principle of maintenance of the objective in 1929. The 1935 edition of the *Field Service Regulations* retained the principles but deemphasized them.[87]

After World War II, Field Marshal Bernard L. Montgomery, as chief of the Imperial General Staff, revised the list of principles of war and promulgated a list of ten: selection and maintenance of the aim, flexibility, economy of effort, concentration of force, offensive action, surprise, security, cooperation, administration, and maintenance of morale. This is similar to the 1920 list, but deleted the principle of mobility and added the principles of flexibility, administration, and maintenance of morale.[88]

The British still use nine of these ten principles today. The exception is that the principle of sustainability has replaced the principle of administration.[89] Table 1 shows the evolution of the British principles of war:

TABLE 1. Evolution of British principles of war

J. F. C. Fuller 1916	1920	1929	Post–World War II	2014
Object	Maintenance of the objective	—	Selection and maintenance of the aim	Selection and maintenance of the aim
Offensive	Offensive action	Offensive action	Offensive action	Offensive action
Mass	Concentration	Concentration	Concentration of force	Concentration of force
Economy of force	Economy of force	Economy of force	Economy of effort	Economy of effort
Movement	Mobility	Mobility	Flexibility	Flexibility
Surprise	Surprise	Surprise	Surprise	Surprise
Security	Security	Security	Security	Security
Cooperation	Cooperation	Cooperation	Cooperation	Cooperation
			Administration	Sustainability
			Maintenance of morale	Maintenance of morale

Sources: Fuller, *Foundations of the Science of War*; Alger, *Quest for Victory*; UK Ministry of Defence, *Joint Doctrine Publication 0–01*, 2014.

In 1921 the United States War Department issued *Training Regulations 10–5, Doctrines, Principles, and Methods*. It enumerated nine principles of war: objective, offensive, mass, economy of force, movement, security, surprise, simplicity, and cooperation. This list was similar to Fuller's, but with the addition of the principle of simplicity. It also resembled the British list of 1920, but with slightly different titles for four of the principles. This American document listed just the titles, with no explanations. The publication asserted the principles were immutable, though it also said the application of the principles would vary with the situation.[90]

The 1923 version of the U.S. Army's *Field Service Regulations* discussed general principles of combat, but neither listed nor titled the principles. The discussion clearly covered seven of the nine principles from 1921. The publication did not discuss any principle of movement, but did mention the role of movement in overcoming numerical inferiority. In addition, the regulations did not mention either unity of command or cooperation under general principles of combat. However, in the section on command, the document stated, "unity of command is essential for success." The discussion of the objective was limited to declaring, "The ultimate objective of all military operations is the destruction of the enemy's armed forces by battle." The discussion of surprise mentioned the moral effect of surprise and the need to exploit successful surprise.[91]

In 1949 the principles of war, including both titles and one-paragraph explanations, reappeared in U.S. Army doctrine in *Field Manual 100–5, Field Service Regulation—Operations*. The nine principles of war were objective, simplicity, unity of command, offensive, maneuver, mass, economy of forces, surprise, and security.[92] The 1949 list differs from the 1921 list in that the principles of maneuver and unity of command replaced movement and cooperation, respectively. Revisions of this field manual up to 1968 contained the same nine principles with no significant change in titles or explanations. However, the 1976 revision no longer listed principles of war. This absence was short-lived. In 1978, *Field Manual 100–1, The Army* listed principles of war nearly identical with the previous version.[93] The current list differs from the original 1921 list with just the minor change from movement to maneuver and the significant change from cooperation to unity of command.

The explanations varied since 1949, but the core of the explanations remained the same.

American Principles of Joint Operations

In 1991 the U.S. Joint Chiefs of Staff published a list of nine principles of war that was identical to the U.S. Army's list. They then presented four "concepts" that resulted from applying the principles of war to the context of joint warfare. These concepts were unity of effort, concentration, initiative, and agility. Unity of effort was *not* a substitute for unity of command. The explanation of the concept of unity of effort was that national authority set the objectives and assigned a combatant command to achieve those objectives, but other combatant commanders and all component commands were expected to support the effort. Concentration meant the use of overwhelming force to win quickly with few casualties. They claimed it was the American tradition to seize and maintain the initiative to keep the enemy off balance. Agility referred to the ability to act faster than the enemy could react.[94]

In 1993 the U.S. Army introduced a second set of principles—principles of operations other than war, which included both potentially hostile operations and operations intended to be peaceful. This set of six principles included three taken from the principles of war—objective, security, and unity of effort—with slightly different applications. Unity of effort *replaced* unity of command because operations other than war often preclude unity of command and therefore require cooperation. Note that unity of effort here is different from the concept of unity of effort in the joint doctrine of 1991. The three additional principles—legitimacy, perseverance, and restraint—were a recognition that peace operations required actions different from those of combat:[95]

Legitimacy. Sustain the willing acceptance by the people of the right of the government to govern or of a group or agency to make and carry out decisions.

Perseverance. Prepare for the measured, protracted application of military capability in support of strategic aims.

Restraint. Apply appropriate military capability prudently.[96]

The first two of these principles came from Just War doctrine. The third came from the International Law of Armed Conflict and the Uniform Code of Military Justice.[97] The Joint Chiefs of Staff adopted the same six principles in 1995.[98]

The initial draft of the U.S. Army's 1998 revision to its operations manual combined the two lists, with some changes, into a single list of principles of operations. However, the final version dropped this.[99]

The 2006 revision of *Joint Publication 3–0: Joint Operations (Joint Publication 3-0)* supplemented the nine previously recognized principles of war with three "other principles" to form twelve "principles of joint operations." The three new principles were restraint, perseverance, and legitimacy. The publication claimed that the three new principles might apply across the range of military operations.[100] The 2018 revision of *Joint Publication 3–0* listed the same twelve principles of joint operations but now stated that most of them applied to combat operations.[101] We will present the full explanation of the current American principles in the chapters that discuss each of them.

Other Principles of War

Many writers have presented alternative lists of principles or "axioms" of war. Two authors, in particular, have presented principles of war that are out of the mainstream, but noteworthy. First, the British writer Basil H. Liddell Hart was the apostle of indirect attack. He advocated that all attacks should be indirect not only physically but also, more importantly, psychologically. He criticized modern principles of war as meaning different things to different people and stated that such immutable principles are a mirage. However, he claimed that military history gives us "a few truths of experience which seem so universal, and so fundamental, as to be termed axioms." He hastened to add that these were not abstract principles but rather were practical guides. Furthermore, he reduced all principles of war into "concentration of strength against weakness." Following this principle should prevent the error of allowing your enemy the "freedom and time to concentrate to meet your concentration."[102]

Robert R. Leonhard presented an innovative set of principles in *The Principles of War for the Information Age*. He argued that the currently recognized principles of war were hopelessly outdated in the information age, and regarded his principles as "arguments" that taught the commander what to think about, as opposed to what to think.[103]

IN THIS CHAPTER WE HAVE GIVEN A BRIEF HISTORY OF THE principles of war. Many writers have addressed the history of these principles. Three works are especially useful. John Alger's *The Quest for Victory* is a thorough history of the principles of war up to 1982.[104] Soviet colonel Vasiliy Savkin gives a briefer history from a culturally different point of view. He includes many principles devised by Russian and Soviet writers that are likely to be unfamiliar to the typical Western reader. Savkin criticizes many Western military theorists from a Marxist-Leninist point of view.[105] Finally, Wayne Hughes provides cogent discussions of the virtues and drawbacks of principles of war from a naval point of view. His books on fleet tactics contain many lists of principles.[106]

Although principles of war in various forms date back more than two thousand years, no one presented principles in the modern form until early in the twentieth century. After the British military adopted formal principles of war in 1920 that bear a strong resemblance to the currently recognized principles of war, the Americans soon followed. The British and American principles varied from time to time, but have generally settled into their current form after World War II. In this chapter we presented the principles of war currently espoused by the British and American military. In the next chapter we will see how other major militaries have approached the principles of war and will compare them with the British and American principles.

TWO

Principles of War in Various Countries

CHAPTER 1 COVERED THE HISTORY OF BRITISH AND AMERI-
can principles of war through the twentieth century. In this chapter we
review the status of the principles of war in selected countries. Although
English-speaking countries other than the United States mostly follow
the British principles of war, other countries have followed unique paths
that we summarize in this chapter. The chapter concludes with a comparison of the principles of war espoused by the nations considered here.

France

In the early 1920s, Marshal Henri Philippe Pétain led a board that considered and then rejected the idea of listing general principles of war in
French doctrine. Board members felt that any principles that attempted
to span circumstances that varied so greatly would result in vague formulas that were useless in practice. Nevertheless, French military theorists continued to write about principles. For example, in 1925, Charles de
Gaulle accepted the existence of principles, provided their application was
adapted to circumstances. In 1936 French military doctrine endorsed the
idea of principles of war but provided no clear list of principles.[1]

After World War II, French doctrine moved slowly toward listing principles but did not distinguish between laws, principles, and rules. In the
1950s instructors at the French École supérieure de guerre presented six
"fundamental laws of war and of strategy." These were the laws of movement, force, offensive, protection, friction, and the unforeseen. For tactics, they listed six fundamental laws of movement, fire, shock, protection,
friction, and the unforeseen. Four of the "laws" in each list were the same.
In 1973 a French general directive on land forces listed three permanent
principles: concentration of efforts, freedom of action, and economy of
forces. It also listed five rules derived from these principles that "guaranteed success": initiative, surprise, aggressiveness, continuity of action, and
simplicity and flexibility.[2]

The 2013 edition of the French *Capstone Concept for Military Operations* listed just three "principles of military action" at the strategic level that it claimed were enduring. Although observance of these principles "enables success," commanders were expected to understand them in a modern context. Further on in the text, the doctrine stated that the principles were a guide rather than a dogma. They were a "basis for reflection and guide for action" and their value depended "solely" on how a commander "adapts them to circumstances."[3] These three principles, the same as those listed in 1973, were:

> freedom of action, which means retaining both room for initiative and control of one's options;

> concentration of efforts, which is not confined to a concentration of forces, but means combining complementary efforts; and

> economy of force, which must not lead to excessive caution, but to the allocation of the appropriate resources to the situation, for the best possible efficiency.[4]

French doctrine also listed five factors for operational success that bore little resemblance to the 1973 rules: agility, combination of technological superiority and operational expertise, mastery of information, ability to act in unconventional conditions, and the ability to cope with complex and changing operational environments.[5] The first, fourth, and fifth factors are similar to some aspects of the British principle of flexibility.

Although overall French *Capstone* doctrine only claimed that the three principles of military action applied at the strategic level, in 2010 a French Army publication, *General Tactics*, called three similar concepts—freedom of action, unity of effort, and economy of means—warfighting principles that were "to be applied to the land forces at the tactical level." Furthermore, commanders should apply each of these principles using surprise-oriented methods. Finally, *General Tactics* quoted Gen. Charles de Gaulle as saying that the most important warfighting principle was "the capability to adapt to circumstances."[6]

After promulgation of the 2013 French *Capstone* doctrine, French army doctrine referred to its definition of the three principles of military action, but without noting the restriction to the strategic level. French army doctrine also listed seven factors for success that included unity of command, agility, morale, and logistical self-sufficiency.[7]

Germany

Although some German military theorists wrote favorably about principles of war after World War I, German military doctrine did not include lists of principles of war. Those writing the doctrine followed the teachings of von Moltke the Elder, who rejected the constraints of unchanging rules in favor of a unique solution to each concrete problem.[8]

German doctrine after World War II still did not endorse principles of war, but official publications did discuss principles. One 1962 document presented "basic operational principles," along with warnings that principles were neither immutable nor definitive and that there were no formulas for success in war. Nevertheless, commanders should use clear principles as a guide. Twenty-two paragraphs described principles, with key words emphasized that often coincided with principles used by other militaries: for example, freedom of action, mobility, speed, and simplicity. However, German doctrine held to the German tradition of not referring to principles by title alone and did not present an explicit list of principles.[9]

A 1973 document presented "tactical principles" that were similar to principles of war used by other countries. However, the document emphasized von Moltke's belief in the uniqueness of every situation, while claiming that the principles presented were a "decisive prerequisite." The document presented a very long list of paragraphs containing advice that in some cases resembled the principles of war of other nations.[10]

Although German doctrine does not endorse a list of principle of war similar to other nations, Germany is part of NATO, which does recognize principles of war.

Israel

The Israeli Defense Force defines the principles of war as rules that form a "permanent basis for combat doctrine." It says the principles remain the same even though the list may change and application of the principles will vary according to circumstances.[11]

The ten Israeli principles of war, with commentary by Yaakov Amidror in parentheses, are:

1. Mission and Aim—Adherence to the mission by being guided by the aim (understanding the force's mission within the framework of the aim—and acting accordingly)

2. Optimal utilization of forces (achieving the maximum with what is available while correctly combining capabilities)

3. Initiative and offensive (the commander in the field determines action; he must aim for contact and engagement with the enemy)

4. Stratagem (achieving surprise, but more importantly, identifying, targeting, and exploiting weak points of the enemy)

5. Concentration of efforts (every effort, action, and effect are made to attain the principal mission and aim)

6. Continuity of action (unswerving pressure to prevent the enemy from reorganizing; exploiting our forces' successes)

7. Depth and reserves (to distance threats in order to enable continuity of action in crises)

8. Security (to avoid exposure of the flanks and weakness following a concerted effort)

9. Maintenance of morale and fighting spirit (impels the soldier forward and preserves the unit's vitality under pressure; essential for a small army to compensate for its materiel weakness)

10. Simplicity (each element of the stratagem must be simple to execute even if the broader plan and mission are complicated).[12]

The Israeli principles of war have some unique features. The principle of depth and reserves is included because potential enemies greatly outnumber Israel, a country that has very little geographic area for defense in depth. The principle of continuity of action emphasizes the importance of the initiative in the sense of keeping the enemy off balance. Israeli doctrine expands the principle of the offensive to the principle of initiative and offense. Here "initiative" refers to resourcefulness by lower level officers in carrying out the intent of their mission. The Israelis did not include the principle of maneuver because they thought the concept was obvious. The list does not include the principle of unity of command, nor the British principle of cooperation. Amidror suggests that this is because of the structure of Israel's chain of command.[13]

Another unique feature is the Israeli use of the principle of stratagem rather than surprise. There is a distinct difference between the two concepts. The principle of stratagem involves exploiting surprise to achieve decisive results. Surprise is the means, not the end. One should use the effects of achieving surprise to attack a weak point and destroy the enemy's center of gravity. Amidror claims, "Adopting stratagem in every move must be at the heart of military thought."[14]

Great Britain

The 2014 version of Great Britain's principles of war are intended to "inform and guide" commanders, rather than be rigidly followed. The United Kingdom's *Defence Doctrine* lists ten principles but notes that this list is not exhaustive. Although the British doctrine claims the principles are enduring, it cautions that their application depends on the context.[15]

The titles of these principles differ from those introduced by Montgomery only in that the principle of sustainability has replaced that of administration. Because many countries follow the lead of the British, we discuss these principles in greater depth later in this book. Here, we provide only a brief summary given in the British publication:

> Selection and maintenance of the aim is regarded as the master principle of war. A single, unambiguous aim is key to successful military operations.
>
> Maintenance of morale enables a positive state of mind derived from inspired political and military leadership, a shared sense of purpose and values, wellbeing, feeling of worth and group cohesion.
>
> Offensive action is the practical way in which a commander seeks to gain advantage, sustain momentum, and seize the initiative.
>
> Security is providing and maintaining an operating environment that gives freedom of action, when and where required, to achieve objectives.
>
> Surprise is the consequence of confusion induced by deliberately or incidentally introducing the unexpected.
>
> Concentration of force involves decisively synchronising [and] applying superior fighting power (physical, intellectual, and moral) to realise intended effects, when and where required.

Economy of effort is judiciously exploiting manpower, materiel, and time in relation to the achieving objectives.

Flexibility is the ability to change readily to meet new circumstances—it comprises agility, responsiveness, resilience, and adaptability.

Cooperation incorporates teamwork and a sharing of dangers, burdens, risks, and opportunities in every aspect of warfare.

Sustainability requires generating the means by which fighting power and freedom of action are maintained.[16]

Canada

Canadian doctrine presents ten principles of war as guiding principles that form the fundamentals of military operations. They are not rigid laws and occasionally individual principles may conflict. However, in the view of the Canadian military, violating the principles increases the risk of failure.

The titles of the Canadian principles closely follow the British model, except that Canadian doctrine retains the older British principle of administration, which includes both administration and logistics, rather than replacing it with the principle of sustainability. Although the words of the Canadian explanations of the principles differ from those of the British, the substance is generally similar. The explanation of the principle of cooperation states that aims must be unified, but does not mention anything resembling unity of command. Canadian doctrine emphasizes the primacy of the principle of selection and maintenance of the aim.[17]

Australia

Australian military doctrine states that the principles of war are "time proven and fundamentally important to success," as well as being "relevant irrespective of changes over time." Australian Defence Force principles of war closely follow the British model, with two trivial exceptions of slightly altered titles. The doctrine emphasizes that selection and maintenance of the aim is the "overriding" principle of war. It recognizes that problems might arise because of multiple political goals, subsidiary objectives, conflicting aims, and changing objectives. These factors reinforce the necessity of clearly stating the aim of military action. In addition to the principles of war, Australian doctrine also lists six principles of command, the first of which is unity of command.[18]

The Australian Army's principles of war are the same as those of the Australian Defence Force. Australian Army doctrine explains that principles of war are ideas based on experience—factors that successful commanders have found necessary to consider. It emphasizes the principles are not like the laws of natural science that produce predictable results, nor are they like the rules of a game that carry a penalty for breaking them. Rather, commanders must use judgment in applying the principles.[19]

Australian Air Force doctrine follows the same principles as the Australian Defence Force and Australian Army, but adds its own explanations. It emphasizes the primacy of the principle of selection and maintenance of the aim. Its explanations of the principles are similar to those given by the U.S. Air Force (as described in chapter 3 of this book). In particular, it emphasizes that all the air power should be under the command of an airman reporting directly to the joint force commander. It claims that only an airman has the depth of knowledge necessary to command air power effectively.[20]

Australian maritime doctrine lists the same ten principles of war as those in Australian Defence Force doctrine.[21]

New Zealand

The titles of New Zealand's principles of war are identical to the British principles, with generally similar explanations. New Zealand Defence Force doctrine asserts that the principles provide guidance, but they are not "rigid law" and require judgment in their application. However, it warns, "blatant disregard of them involves increased risk and a proportionate increase in the likelihood of failure." It emphasizes that selection and maintenance of the aim is the "master principle." It acknowledges that in complex situations there may be multiple, and possibly conflicting, aims. In its explanation of the principle of surprise, New Zealand doctrine notes that surprise can have an "immense" psychological effect on the enemy. It does not mention unity of command in the explanation of the principles. However, in another section, it states that if there is not unity of command, there should be unity of purpose.[22]

India

The Indian Army's eleven principles of war closely follow the British model. One exception is that the Indian Army uses the principle of administration instead of the principle of sustainability. Both primarily concern logistics. In addition, in 2004, the Indian Army added the principle of intelligence because of its expected importance in future wars. Indian Army doctrine regards the principles of war as broad guidelines based on experience. The doctrine warns, "They are not rules; yet disregarding them involves risk and could result in failure."[23]

Indian Air Force doctrine has a different set of principles of war. Seven of the principles are the same as those of the Indian Army. However, Indian Air Force doctrine replaces the principle of surprise with that of deception and surprise. It supplants flexibility with flexibility and managing change. It changes cooperation to synergy, synchronization, and cooperation. Finally, it adds the principle of generation and sustenance of favorable asymmetry. Because of its uniqueness, we quote the explanation of this principle:

> The increase in battle space transparency may reduce the impact of surprise at all levels of wars. Therefore, the emphasis now has to be, not only on catching the enemy off guard, but also on keeping him off balance. This would require generating asymmetry at the desired time and place, for wresting significant combat advantage. The methods employed include generating surprise in terms of time, space, and force, a favourable differential in technology and weapon systems at the decisive point, exploiting sound operational art, formulating effective and synergistic strategy and maintaining information superiority.[24]

The emphasis on "keeping the enemy off balance" is one part of the new principle of the sustained initiative that we will introduce in chapter 7. "Generating asymmetry at the desired time and place" is one component of the new principle of relative advantage that we will introduce in chapter 6.

NATO

NATO's Allied Joint Doctrine lists twelve principles of joint and multi-national operations: unity of effort, concentration of force, economy of effort, freedom of action, definition of objectives, flexibility, initiative, offensive spirit, surprise, security, simplicity, and maintenance of morale. According to NATO, freedom of action means to empower commanders and minimize restrictions on them. Initiative means to recognize and seize opportunities. (This is not the same as the new principle of the sustained initiative that we will propose later.) The list seems to be a compromise between the British and American principles. This is most obvious in the explanation of unity of effort, which states that it is achieved mainly through unity of command.[25]

Japan

In the late 1960s the Japanese Board for Study of the Principles of War conducted a systematic study of principles for land warfare.[26] The resulting report became required reading for Japanese Ground Self-Defense Force officers. It states that one can regard the principles as "rules for victory" and that they are "a standard for observing military history." However, they continually evolve with time. One should use the principles only after studying history with them as a guide and then improving one's judgment through exercises and wargames. In any case, one should not follow the principles blindly.

The report states that the most basic principle of war is that "the superior wins and the inferior loses."[27] The concrete principles seek methods for improving combat power relative to that of the enemy to obtain such superiority. The study does not present an explicit list of principles of war in one place but refers to many concepts as principles of war throughout the report. We have extracted some things the document describes as being principles of war:

Surprise

Concentration (of combat power at decisive time and place)

Economic use of combat power (economy of force to concentrate maximum force at the decisive point)

Centripitalism (concentrating combat power by multiple forces situated on the circumference of a circle attacking the center)

Eccentricity (use of interior lines to move repeatedly from a central position to attack elements of the enemy force on the circumference of a circle one at a time)

Defeating piecemeal (taking advantage of the separation of an enemy's combat power by attacking one part at a time)

Reverse use of momentum (lure the enemy force into a battle under circumstances in which you have a relative advantage in combat power)

Never allowing one's lines of communications to be threatened

Changing lines of communications only if necessary for the security of the force.

The report discusses other concepts that it does not label as principles of war, but which appear to be similar to principles of war. For example, the study claims that "unity of command is an absolute prerequisite" for organizing combat power.[28] In addition, the primary principle of command is flexible organization and thinking. The study claims that to concentrate combat power, commanders should consider the following necessary conditions:

Establishment of objective, and a consistent plan . . .

Positive initiative . . .

Selection of the critical time and place . . .

Integrated application of combat power . . .

Economic use of combat power . . .

Application of tactical mobility.[29]

China

For countries such as China and (formerly) the Soviet Union, cultural, language, and secrecy issues make it difficult to form a list of principles of war similar to those of the English-speaking countries. The pervasive influence of Marxist-Leninism makes many of the concepts seem alien to a Western reader.

The writings of Sun Tzu still influence Chinese (and Japanese) military thinking.[30] Ancient Chinese writers generally emphasize deception and secrecy far more than most Western military writers.[31] For instance, the Thirty-Six Stratagems, a collection of wisdom from various classical Chinese sources, primarily involve cunning and deception.[32]

Mao Tse-Tung enumerated six "considerations" in 1936 that are quite different from principles in the Western world. Each of these starts with the phrase "giving proper consideration to." For example, the first two are "giving proper consideration to the relation between the enemy and ourselves" and "giving proper consideration to the relation between various campaigns or between various operational stages." The sixth consideration includes thirty-three paired items to which commanders should give proper consideration. Later, Mao developed and periodically revised ten principles of war. He warned that, on the one hand, principles merited study because they summarized past experience of war, but, on the other hand, commanders should test the validity of these concepts using their own experiences. In the 1950s his list of principles included: aims, mobile concentration, annihilation, fighting on the move, offensive, surprise attack, continuous attack, autonomy, unity, and military spirit.[33]

Identifying current Chinese principles is difficult because the Chinese keep their keystone doctrine documents classified. Analysts try to estimate doctrine by studying open-source documents that they believe are based on classified doctrine.[34]

As of 1999 the Chinese divided strategy into the three spheres of major war, peacetime, and local war. Authoritative sources indicate that Chinese doctrine at this point had six principles for planning and guiding local wars:

> Prepare in many directions to respond flexibly to contingencies. . . .
>
> Combine the offense and defense to seize the initiative. . . .
>
> [Conduct] integrated operations for key point attacks. . . .
>
> Strive for quick decisions, but prepare for protracted conflict. . . .
>
> [Prioritize] overall planning to win victory through coordinated efforts. . . .
>
> Unified leadership, centralized command.[35]

The principle concerning key point attacks includes the idea of concentrating one's best forces to gain local superiority and attacking the enemy's most important points. The principle regarding overall planning indicates that one should use centralized command to employ all aspects of national power, not just military means, to wage local war.[36]

Based on open Chinese sources, several studies have identified strategic principles of warfighting. Three studies that we examined list from seven to nine principles, but all three sources contain roughly the same total information. These strategic principles are not *general* principles of war. They specifically address the challenges of defeating a militarily superior enemy. Here is our amalgamation of the collections of principles in three sources:[37]

Seizing the initiative early. The most emphasized Chinese principle is seizing the operational initiative. The weaker party must be offensive from the start. Passively awaiting the stronger party's attack would be disastrous. The Chinese view is that during the first Gulf War, Iraq's passivity in the face of the American buildup squandered a window of opportunity.

Surprise. Surprise is required to seize the initiative against a powerful foe.

Preemption. Preemptive attacks are desirable to disrupt the enemy's deployment and to seize the initiative against a militarily superior enemy. Preemption improves chances of achieving surprise and seizing the initiative.

Avoidance of direct confrontation. Instead of force-on-force battles, exploit enemy vulnerabilities.

Key-point strikes. Even militarily superior enemies have vulnerabilities. Identifying and attacking enemy centers of gravity can paralyze the enemy. Vital targets include command, information, weapon, and logistic systems, as well as links between them.

Concentrated attack. The most advanced platforms and best units should attack key targets in a limited time and space.

Information superiority. Destroying enemy information systems can paralyze the enemy. This can include cyberattacks, jamming, and physical destruction of vital facilities.

Raising the cost of conflict. China must prevent the enemy from fighting the type of war the enemy wants to fight. By inflicting casualties on an enemy that is sensitive to casualties, China can undermine the enemy's resolve and cause a loss of will to continue the war.

Limited strategic aims. The relative imbalance in capabilities precludes gaining total victory over a militarily superior adversary. The weaker party should seek an initial military victory that raises the costs to its adversary, then negotiate to attain its political objectives. This may require simultaneous fighting and negotiating to maintain military pressure during negotiations. Gains from an initial offensive may present the superior power with a situation where reversing the fait accompli is not worth the cost of doing so.[38]

Tsarist Russia

Prior to the 1917 revolution, Russian doctrine accepted principles due in part to the influence of Jomini.[39] G. A. Leyer, one of the Russian Empire's leading military theoreticians of the late nineteenth century, proposed a list of twelve basic principles of military art. He believed these principles were "eternal, unchangeable, and unconditional." However, they had to be applied flexibly depending on the specific situation.[40] A U.S. Army publication claims the tsarist general staff taught these principles as summarized in the following list:

> Extreme exertion of force at the very beginning of a war
>
> Simultaneity of actions
>
> Economy of forces
>
> Concentration
>
> Chief objective—the enemy's army
>
> Surprise
>
> Unity of action
>
> Preparation
>
> Energetic pursuit
>
> Security
>
> Initiative and dominance over the enemy's will
>
> Strength where the enemy is weak.[41]

If this is accurate, then tsarist Russia was ahead of the United States and Britain in formulating principles of war.

Soviet Union

After the 1917 revolution, the communists debated whether "bourgeois" principles were valid. Trotsky asserted that war could have no laws. On the other hand, M. V. Frunze proposed "proletarian" principles of military art that included maneuverability, activeness, and the offensive. Although Soviet army doctrine did not list principles of war during this period, the 1936 *Field Service Regulations* contained content that resembled British and American principles.[42]

Joseph Stalin promulgated five "permanently operating factors." These were not what we consider principles of war; they were factors that determined the outcome of wars. After Stalin's death, the Soviets reexamined the permanently operating factors and replaced them with laws of war.[43]

The Soviets believed Marxism-Leninism provided a scientific basis for discovering objective laws of war; accordingly, several Soviet writers proposed various such laws. In 1972 Col. Vasiliy Savkin suggested four laws of war, plus two laws of armed conflict.[44] By 1977 the Soviets officially approved six laws of war. The first law was that the essence of war depended on the political goals. The next four laws stated that the "course and outcome of war" depended on the relative strength of the warring states in four areas. These four areas were the economy, scientific potential, moral-political strengths, and military force potentials. The sixth law asserted that historically the side with the more progressive social and economic order wins.[45]

The Marxist-Leninist system leads to a rigid hierarchy of terms and an array of lists. Although this system has an internal logic, it appears bewilderingly complex when one first encounters it. Furthermore, translated Russian terms do not necessarily mean the same as the usual English usage. Let us try to make some sense of it. Communist party leaders at the top determine what they call *military doctrine*. This sets national estimates for expected enemies, types of wars to prepare for, necessary forces, military aims, means of warfare to be used, and preparation for war. Military doctrine is guidance *for* the military, not *by* the military. It is not subject to debate by the military. The professional military controls *military science*, which is the study of everything related to combat. It includes technology, organization, training, military history, and military art. The goal of military science is to develop practical ways to achieve victory in war.

Military art (not to be confused with operational art) is a subdivision of military science. It deals with the methods of combat and includes strategy, operational art, and tactics. The principles of military art are the basic ideas for conducting warfare.[46] They serve approximately the same purpose as Western principles of war.

In 1978 the Soviet Military Encyclopedia identified the eleven most important principles of military art. The presentation of these principles is rather verbose, especially when compared to the single word or phrase used by most Western countries. Here is our condensation of the eleven Soviet *principles of military art* listed by Baxter:[47]

Preparedness. Maintain a high state of readiness, even in peacetime.

Surprise. This includes seizing and retaining the initiative.

Full use of all assets. This implies efficient coordination of all forces. (This is the opposite of graduated response.)

Cooperation. Unity of command is necessary to ensure close coordination in joint operations.

Concentration. There should be a decisive concentration of essential force at the critical time on the main axis.

Destruction of enemy to full depth. One should use rapid maneuver and intensive combat for quick, simultaneous destruction of the enemy to the entire depth of his deployment.

Exploit moral-political factors. This includes propaganda to motivate troops and psychological warfare to demoralize enemies.

Strict and uninterrupted leadership. This means continuous detailed supervision of subordinates and precludes mission-type orders.

Steadfastness. Subordinates must be determined and decisive in fulfilling both the spirit and the letter of assigned missions. They must be energetic in solving emerging problems. According to Baxter, "To the Soviets, initiative means finding ways to execute the plan as written in spite of difficulties. It does not include the concept of revising intermediate steps to meet changed circumstances."[48]

Security. Secrecy is an important part of planning.

Restoration of reserves and combat capability. This means rapid resupply and reconstitution to replace losses in high-tempo operations.

Principles of strategy are constructed from the principles of military art. In turn, the principles of operational art are derived from the principles of strategy and are nearly identical to them.[49] Finally, principles of tactics are based on the principles of operational art.

Savkin listed seven basic principles of operational art and tactics, writing that they were current in the 1950s. However, the American editor of his book claimed that this time frame was a literary device and that the principles were current at the time of publication (1972).[50] This list included some principles similar to Western principles, but with different phrasing, and several principles that are quite different in nature. Here is the list of seven *basic principles of operational art and tactics* with our condensations of Savkin's titles and explanations.[51]

Mobility and high tempo of combat operations. Mobility is more than just movement over terrain. Mobility and high tempo are required for other principles such as concentration, surprise, combat activeness, and preservation of effectiveness. During offensive operations, the basic measure of tempo is the rate of advance.

Concentration of efforts. To achieve victory, one must concentrate the main efforts at the most important place and at the right time to create superiority over the enemy.

Surprise. Use of surprise brings success. The goal of surprise is to dumbfound the enemy and paralyze his will to resist. The duration of the effect of surprise depends on the training, discipline, and morale of those surprised.

Combat activeness. Success goes to the side that is more active, seizes the initiative, and maintains it. Activeness generally means a bold and energetic offensive. (Most of the explanation of this principle is similar to American or British explanations of the principle of the offensive.)

Preservation of combat effectiveness. This concept was elevated to an independent principle because of the possibility that an initial nuclear strike could destroy a large portion of the combat force.

Conformity. The goal and plan of the operation must conform to the actual situation. One must not use subjective judgments to decide the attainability of objectives. Instead, one must make thorough estimates of the situation, accounting for factors of time and space as well as the relative strength of forces compared with those of the enemy.

Interworking. Success requires the close coordination of all forces from all branches of the services and the fullest use of their capabilities, including nuclear strikes. (Much of the explanation of this principle describes combined arms warfare and unity of effort, without using those words.)

Gen. V. G. Reznichenko refers to principles of tactics as principles of combined arms combat. We base our condensation of the titles and explanations of these ten *principles of combined arms combat* on the discussion by Reznichenko, with some input from Baxter:[52]

Readiness. All units should maintain continuous high combat readiness in both peace and in battle. This includes political training.

Aggressiveness. Success in combat, all things being equal, goes to the side that operates more aggressively. This principle calls for continuous intense offensive combat, bold actions, decisiveness, persistence, tenacity, resoluteness, and courage. Decisiveness means destruction of the enemy's capability and will. Aggressiveness implies not just the offensive but also a resolute high-tempo offensive that ensures destruction of the enemy. Continuity means exerting constant pressure on the enemy, seizing and retaining the initiative.

Surprise. Surprise has always been the most important principle of war. It should form the basis of all combat actions. Unexpected actions can cause panic and paralyze the enemy's will to resist. Secrecy, deception, and detailed understanding of the enemy helps achieve effective surprise. To prevent the enemy from recovering from the effects of an initial surprise, one should maximize exploitation of the initial surprise and achieve new surprises. Nuclear weapons increase the importance of surprise.

Coordination. Continuous coordinated employment of all combat arms and services is a prerequisite for victory. Effective combined arms combat requires unity of command, unity of purpose, and unity of action.

Concentration. Concentrating combat power in the main effort at the necessary time is essential for victory. Concentration has been a decisive factor in most past battles. However, one must disperse troops when the enemy makes nuclear strikes.

Maneuver. This applies to forces, resources, and nuclear and conventional strikes. Maneuver enhances flexibility. It is the method of concentrating combat power and seizing and maintaining the initiative.

Moral, political, and psychological factors. People are the most important force in combat. One should consider moral-political and psychological factors in all decisions. Indoctrination of troops enhances cohesion. One should study the enemy with the goal of exploiting moral-political and psychological weaknesses.

Support. Comprehensive support involves maintaining combat readiness. It includes technical and logistic support. Combat support includes reconnaissance, nuclear-biologic-chemical warfare defense, electronic warfare, camouflage, and security.

Maintenance and timely restitution of combat forces. Commanders should plan for heavy casualties and damage.

Firm and continuous troop control. Experience shows that victory always depends on the quality of troop control, which must be centralized, firm, and continuous. Objectives should be attainable considering the actual combat situation, which includes the quality of the troops. Orders "must clearly define the objective of the engagement *and the methods by which it is to be achieved*" (emphasis added).[53]

The last sentence is astounding. We cannot imagine any Western military endorsing the idea that orders must define the methods a subordinate must use to achieve an objective—that is, *how* to carry out the mission.

The various sets of principles obviously have a fair amount of overlap. Reasons for the differences are generally not obvious. Discussions of all the lists of Soviet principles repeatedly mention the use of nuclear weapons throughout in a rather matter-of-fact way. All three lists contain the principles of surprise and concentration. The Soviets consistently call surprise the most important principle, an emphasis likely due in large part to the traumatic German invasion in 1941. The principles of readiness, aggressiveness (or activeness), mobility (or maneuver), and morale each appear in two of the lists. In addition, the descriptions of the principles of cooperation and coordination proclaim that unity of command is a necessity.

A 1984 U.S. Army publication on Soviet operations and tactics claims that when the Soviets add modern ideas, they do not discard earlier concepts. Combining then-recent writings with classical Russian principles resulted in a list of thirty-one Soviet principles of operational art and tactics. We can identify versions of the following familiar principles in this list: offensive, maneuver and mobility, coordination and cooperation, unity of command, surprise, concentration of fires, security, logistics, and flexibility. Some of these principles appear in multiple variants in the list. In addition, the list contains principles that we might term as initiative (in the sense of keeping the enemy off balance), relentless attack, deep attack, and attack weak points. Some of the principles are rather narrow. For example, "conserve fighting strength through the use of combat vehicles with collective NBC [nuclear, biological, and chemical] protection."[54]

In the Soviet system, all sets of principles apply equally to each service. However, the Soviets recognized that the application of the principles varies according to the service. Admiral V. Chernavin claimed that the five most important principles of naval art were readiness, surprise, coordination, maneuver, and mass. Mass refers to the massing of fires, which is the naval equivalent of concentration. Coordination refers to combined action.[55]

Although we were unable to locate any definitive statements of principles dating after the demise of the Soviet Union, it is reasonable to assume that the Russian Federation military would use principles somewhat similar to those from the Soviet Union.

Summary

Most countries with major military forces, other than Germany, recognize principles of war. However, they do not agree on which specific principles to include. In some cases, various countries use different titles for principles that are essentially the same; for example, mass instead of concentration of force. In other cases, there are subtle differences. Still other times, the differences are substantial.

The following summary generally excludes Japan, China, Russia, and the USSR from the discussion because cultural and language differences make comparisons problematic. Most other major military organizations seem to agree on the following aspects of the principles of war:

Because warfare is an art, not a science, the principles of war are not laws like those in the physical sciences.

The principles of war are enduring but not immutable.

The principles of war are a useful guide.

The application of the principles of war varies with conditions and may change as warfare changes.

Commanders must use judgment in applying the principles to each particular situation.

The doctrine of some, but not all, of these countries mentions each of the following points.

The principles of war form a framework that is useful when studying history.

TABLE 2. Comparison of principles of war in English-speaking countries

UK	Canada	Australia	New Zealand	India	NATO	United States[1]
Selection and maintenance of aim	Selection and maintenance of aim	Selection and maintenance of aim	Selection and maintenance of aim	Selection and maintenance of aim	Definition of objectives	Objective
Offensive action	Offensive action	Offensive action	Offensive action	Offensive action	Offensive spirit	Offensive
Concentration of force	Concentration of force	Concentration of effort	Concentration of force	Concentration of force	Concentration of force	Mass
Economy of effort	Economy of effort	Economy of effort	Economy of effort	Economy of force	Economy of effort	Economy of force
Cooperation	Cooperation	Cooperation	Cooperation	Cooperation	Unity of effort	Unity of command
Security	Security	Security	Security	Security	Security	Security
Surprise	Surprise	Surprise	Surprise	Surprise	Surprise	Surprise
Maintenance of morale	Maintenance of morale	Morale	Maintenance of morale	Maintenance of morale	Maintenance of morale	
Flexibility	Flexibility	Flexibility	Flexibility	Flexibility	Flexibility	
Sustainability	Administration	Sustainment	Sustainability	Administration		
					Simplicity	Simplicity
					Initiative	
					Freedom of action	
				Intelligence		
						Maneuver

Sources: UK Ministry of Defence, *Joint Doctrine Publication 0–01*, 2014; Canadian Forces Experimentation Centre, *Canadian Military Doctrine*; Australian Department of Defence, *Foundations of Australian Military Doctrine*; New Zealand Defence Force, *New Zealand Defence Doctrine*; Indian Army Training Command, *Indian Army Doctrine*; NATO Standardization Office, *Allied Joint Doctrine*; Joint Chiefs of Staff, *Joint Publication 3–0*, 2018.

1. Principles of joint operations includes the nine principles of war shown, plus three principles (legitimacy, perseverance, and restraint) not shown that were previously called principles of operations other than war.

The principles of war are concepts based on experience.

The principles of war can help educate military personnel.

The principles of war are aids to assist in organizing thoughts.

The principles of war are tools for planning and executing operations.

The principle of the objective or selection and maintenance of the aim stands above all other principles in importance.

The relative importance of each principle depends on the circumstances.

Not all principles are applicable to each situation.

Ignoring the principles of war increases risk and may result in failure.

English-speaking militaries generally follow the British model in forming the lists of principles of war, with some modifications. Table 2 compares the titles of the principles of war of those countries. Table 3 contrasts the British and American principles with those of various countries that use quite different lists.

Almost all countries examined list some variation of eight principles of war: selection and maintenance of aim, offensive action, concentration of force, cooperation, economy of effort, security, surprise, and morale. Of the eight principles in this list, the U.S. military does not include morale or cooperation. Germany is unique in not recognizing any principles of war, and France identifies only three principles of war.

The United States is alone in using unity of command as a principle of war, though it recognizes that unity of command may not always be possible and, in that case, cooperation is important. Britain's corresponding principle is cooperation, but it urges "clearly divided responsibilities" and a common aim, or at least unity of purpose. NATO's corresponding principle is unity of effort, which the organization states is primarily achieved through unity of command. The United States is alone in designating maneuver as a principle of war and only NATO and Israel join it in recognizing simplicity.

Only the British Commonwealth and India designate administration or sustainability as a principle of war. Only those countries and NATO recognize the principle of flexibility.

TABLE 3. Comparison of principles of war in selected countries

UK	U.S.[1]	Israel	France	USSR[2]
Selection and maintenance of aim	Objective	Mission and aim		
Offensive action	Offensive	Initiative and offensive		Aggressiveness
Concentration of force	Mass	Concentration of efforts	Concentration of efforts	Concentration of power
Economy of effort	Economy of force	Optimum use of force	Economy of force	
Cooperation	Unity of command			Coordination (unity of command)
Security	Security	Security		
Surprise	Surprise	Stratagem		Surprise
Maintenance of morale		Maintenance of morale and fighting spirit		Morale
Flexibility				
Sustainability				Support
	Maneuver			Maneuver
	Simplicity	Simplicity		
			Freedom of action	
		Continuity of action		
		Depth and reserves		Maintenance and restoration of forces
				Readiness

Sources: UK Ministry of Defence, *Joint Doctrine Publication 0–01*, 2014; Joint Chiefs of Staff, *Joint Publication 3–0*, 2018; Amidror, *Winning Counterinsurgency War*; French Defence Staff, *Capstone Concept for Military Operations*; Reznichenko, *Tactics: A Soviet View*; Baxter, *Soviet AirLand Battle Tactics*.

1. Principles of joint operations includes the nine principles of war shown, plus three principles (legitimacy, perseverance, and restraint) not shown that were previously called principles of operations other than war.

2. Principles of combined arms combat tactics.

THREE

Principles of War in the U.S. Services

U.S. MILITARY SERVICES ARE REQUIRED TO ACKNOWLEDGE and adopt joint doctrine but are free to supplement it with service-specific doctrine, provided it is consistent with joint doctrine. Service doctrine includes the foundational philosophy of the service and the methods the service uses to organize and support a joint force commander. Some of the services modify the joint explanation of the principles of war to draw attention to features of special importance to that service. Some also augment the joint principles with additional principles, perhaps called by a different name, that apply to that service.

American naval doctrine discusses each of the nine principles of war in conjunction with the phase of the campaign or operation in which they are most likely to be important. It does not offer any additions to the joint principles of war but points out aspects of special interest to the naval forces.[1] Generally, the U.S. Navy has shown little recent interest in the principles of war.[2] This is unsurprising, because historically navies have not shown as much interest as armies in the principles of war.[3]

U.S. Army

Because the twelve joint principles of operations were taken from those of the U.S. Army, it is not surprising that the Army does not offer any significant changes to the joint definitions. The U.S. Army's *Doctrine Primer* states that a principle is "a comprehensive and fundamental rule or an assumption of central importance . . . to capture broad and enduring guidelines."[4] It says that while the principles are not a checklist, they "have proven to have a positive effect on the outcome of operations" and provide a guide to future operations. Specifically, "they summarize characteristics of successful operations. Their greatest value lies in educating the military professional. Applied to the study of past operations, the principles of joint operations are powerful tools that can assist commanders in analyzing pending operations."[5]

Army doctrine also presents six Principles of Unified Land Operations that are "comprehensive and fundamental" and "of central importance." These six principles are mission command, development of the situation through action, combined arms, the law of war, security, and creation of multiple dilemmas for the enemy.[6]

The Army also presents four tenets of unified land operations, which it describes as "desirable attributes that should be built into all plans and operations." They are simultaneity, depth (of operations), synchronization, and flexibility.[7]

U.S. Marine Corps

Gen. Al Gray transformed the U.S. Marine Corps when he became commandant in 1987. He had absorbed the ideas of former U.S. Air Force fighter pilot John Boyd and became a champion of maneuver warfare. There had been a debate within the U.S. Marine Corps about the merits of maneuver warfare. Gray settled the debate by making maneuver warfare the official doctrine of the Marine Corps, and the 1989 *Warfighting* manual was his guide to it.[8] The manual contained an unusually personal cover letter from Gray. It began, "This book describes my philosophy on warfighting." He also stated, "I expect every officer to read and reread this book, understand it, and take its message to heart." It both provides guidance for fighting and instills a way of thinking about war.[9] The manual is a brilliant discourse on how to fight a war, presenting concepts clearly and concisely.

More specifically, the text presents just two concepts as principles of war: concentration and speed. The concentration should be in both space and time. The explanation of concentration is similar to that of the principle of mass. Because concentration could make one vulnerable to enemy fire, one might consider a pattern of alternating dispersion and concentration. Superior speed is understood here as a weapon that allows one to seize the initiative. The *Warfighting* manual claimed that combining concentration and speed gave momentum, which produced "shock."[10]

The only other reference to the principles of war in the manual was a quotation from Field Marshal Sir William Slim:

> Many years ago, as a cadet hoping some day to be an officer, I was poring over the "Principles of War," listed in the old Field Service Regulations,

when the Sergeant-Major came up to me. He surveyed me with kindly amusement. "Don't bother your head about all them things, me lad," he said. "There's only one principle of war and that's this. Hit the other fellow, as quick as you can, and as hard as you can, where it hurts him most, when he ain't lookin'!"[11]

THE 1997 VERSION OF THE *WARFIGHTING* MANUAL CONTINued the philosophy found in the 1989 manual. General Gray wrote the preface. The new manual hardly mentioned any principles of war, designating speed and focus as two "concepts of universal significance." The concept of focus was quite close to that of concentration. In the explanation of focus, the manual states that use of excessive force in secondary sectors violated the *principle* of focus. The description of speed and focus used many of the same words used to describe the "principles" of speed and concentration in the manual's 1989 version.[12]

In the 2001 *Marine Corps Operations* manual, an appendix listed the nine American principles of war with explanations close to those of joint doctrine. It described the principles as aids to assist a commander in organizing his thinking and tools for planning, executing, and assessing operations. It cautioned that the principles were not prescriptive and commanders had to use their judgment in applying them.[13]

This document also listed six principles that were the basis for Operational Maneuver from the Sea. One of these was to "generate overwhelming tempo and momentum." Another was to "pit friendly strength against enemy weakness . . . rather than attacking his center of gravity when it is strong."[14] The publication also listed sixteen interrelated "fundamentals of offensive action," describing these as "rules" derived from the principles of war.[15]

The 2011 revision of *Marine Corps Operations* listed the twelve principles of joint operations. The description generally followed joint doctrine. One exception was an addition to the explanation of the principle of maneuver: "The Marine Corps maneuver warfare philosophy expands the concept of maneuver to include taking action in *any* dimension, whether temporal, psychological, or technological, to gain an advantage."[16]

The 2011 doctrine listed six enduring U.S. Marine Corps principles that followed from the core values of honor, courage, and commitment.

These principles define "the cultural identity and beliefs of Marines." The first principle is:

> *Every Marine is a rifleman.* Every Marine, regardless of military occupational specialty, is first and foremost a rifleman. (emphasis in original)[17]

This means that the Marine Corps aviator flying close air support trained as a rifleman before he or she became an aviator. As a result, the ground forces have great confidence that Marine Corps aviators will do whatever they can to support them.

Another of the enduring Marine Corps principles is:

> *The Marine Corps is a combined arms organization* ... Marine Air-Ground Task Forces ... contain organic air, ground, and logistic elements under a single command element, making them integrated and self-sustaining (emphasis in original).[18]

This embodies the Marine Corps belief in how to implement unity of command: every type of force in a geographic area should be under the command of one person. As we will see, this conflicts with the Air Force approach toward this same question.

The 2011 revision of *Marine Corps Operations* also presented fifteen fundamentals of offensive action that were identical to those of the 2001 revision, except it dropped one rule concerning concentration. Furthermore, the publication listed Gen. Anthony Zinni's twenty-two principles of peace operations.[19]

U.S. Air Force

U.S. Air Force basic doctrine offers much supplementary material. It significantly alters the definitions of the joint principles to draw attention to aspects important to airmen. It also adds seven "Tenets of Air Power" that are, in effect, additional principles for the application of airpower.

Whereas joint doctrine recognizes twelve principles of joint operations—nine principles of war plus three additional principles—the Air Force refers to thirteen principles of joint operations, the twelve from joint doctrine plus unity of effort. The idea appears to be that while unity of command is primary, in some circumstances such as international coalitions, unity of command is impossible and therefore unity of effort is imperative.[20] This is certainly consistent with the spirit of the principles in joint doctrine.

U.S. Air Force doctrine defines the principles of war as aspects of warfare that are "universally true and relevant" and have "tended to produce military victory" in the past. The explanation of each principle emphasizes aspects that particularly apply to the Air Force. In what follows, we use brief quotes from the Air Force's *Basic Doctrine* to illustrate some of the unique pieces of the Air Force's explanation of each of the nine principles of war.[21]

Unity of command is vital in employing airpower. Airpower is the product of multiple capabilities, and centralized control is essential to effectively fuse these capabilities and provide unity of command. . . .

From an Airman's perspective . . . the principle of objective shapes priorities to allow airpower to concentrate on theater or campaign priorities and seeks to avoid the siphoning of force elements to fragmented objectives. . . .

[The principle of the offensive] is particularly significant to airpower because it is best used as an offensive weapon. . . .

[With respect to the principle of mass] Airpower is singularly able to launch an attack from widely dispersed locations and mass combat power at the objective, whether that objective is a single physical location or a widely dispersed enemy system or systems. . . .

Maneuver places the enemy in a position of disadvantage through the flexible application of combat power in a multidimensional combat space. Airpower's ability to conduct maneuver is not only a product of its speed and range, but also flows from its flexibility and versatility during the planning and execution of operations. . . .

Economy of force . . . calls for the rational use of force by selecting the best mix of air, space, and cyberspace capabilities. To ensure overwhelming combat power is available, maximum effort should be devoted to primary objectives. At the operational level of war, commanders ensure that any effort made towards secondary objectives does not degrade achievement of the larger operational or strategic objectives. . . .

. . . Security also may be obtained by staying beyond the enemy's reach,

Principles of War in U.S.

> physically and virtually. Airpower is uniquely suited to capitalize on this
> through its ability to operate over the horizon....
>
> ... The speed and range of air, space, and cyberspace capabilities, coupled
> with their flexibility and versatility, allow air forces to achieve surprise
> more readily than other forces. The final choice of timing and tactics rests
> with the air component commander because terrain and distance are not
> inhibiting factors....
>
> ... Simple guidance allows subordinate commanders the freedom to oper-
> ate creatively within their portion of the operational environment, sup-
> porting the concept of decentralized execution.[22]

The reason for many of these explanations is the Air Force leadership's firm belief in the proper use of airpower and a desire to fend off potential encroachments on that doctrine. Air Force leaders believe that airpower is uniquely versatile in its ability to attack a wide variety of targets and to shift targets quickly. Airpower can be concentrated on campaign-level center-of-gravity targets one day and then devoted to tactical ground support the next day. Often, aircraft can be redirected even after takeoff. If much of the airpower in a theater were spread out under control of individual ground units, then the ability to concentrate that airpower on a particular vital sector would be lost.[23]

The U.S. Air Force also lists seven "Tenets of Air Power" that are "fundamental guiding truths" that complement the principles of joint operations and "refine" the application of airpower. The tenets are more specific to the use of airpower than the general guidance of the principles of war: centralized control and decentralized execution, flexibility and versatility, synergistic effects, persistence, concentration, priority, and balance.[24]

The explanations of these tenets contain arguments for the way in which Air Force leaders think airpower should be controlled in a joint force. Part of the description of the tenet of centralized control and decentralized execution declares that "control of a valuable yet scarce resource (airpower) should be commanded by a single Airman, not parceled out and hardwired to subordinate surface echelons as it was prior to 1943." The elucidation of the tenet of priority states that "effective priorities for the use of airpower flow from an informed dialogue between the [Joint

Force Commander] and the air component commander." The explanation of the tenet of balance asserts, "An Airman is uniquely—and best—suited to determine the proper theater-wide balance between offensive and defensive air operations, and among strategic, operational, and tactical applications."[25]

THE MOST NOTABLE DIFFERENCE IN THE PRINCIPLES OF THE various U.S. services is the divergent approaches that the U.S. Air Force and the U.S. Marine Corps advocate for the command of air assets in a theater. Both services invoke the principle of unity of command to support their position. As we will discuss in chapter 7, this is the result of conflicting views on how best to implement unity of command. However, before examining particular principles, we must define what we mean by a principle of war.

FOUR

Criteria for Principles of War

IN THE FIRST THREE CHAPTERS OF THIS BOOK, WE SAW THAT most military organizations believe there exist principles of war that provide useful guidance for commanders. The principles generally are derived from a study of military history. However, there is considerable disagreement about the identity of those principles. Before we can criticize or endorse particular principles of war, we must be clear about what we think constitutes a principle. In this chapter, we present the criteria we will apply to each candidate principle of war in subsequent chapters.

War is arguably the most complex activity in which human beings engage. It involves technology, psychology, politics, and a dozen other fields of study. It has elements of games of strategy such as chess, as well as games like poker that involve probabilities and "reading" people. Above all, war is a high-stakes activity that involves danger, hardship, and courage. Thus, we should not expect the conduct of war to have rigid laws like physics or abstract truths like mathematics.

The challenge is to discern order in the apparent chaos of military history and develop practical guidance for making decisions. We seek principles that can help commanders apply the recurring lessons of military history to their unique problems by summarizing those aspects of operations from the past that made a difference between victory and defeat. We should expect that any principle of war will give general guidance and will never be a substitute for study and good judgment. One should examine how great military leaders of the past applied the concepts in these principles to specific situations.

Because warfare changes so much over time, we must be careful to select principles that have endured over time and are likely to remain useful in the future. In general, because the technology of war changes rapidly, any candidate principle of war that depends on a particular technology is not going to be durable. Similarly, a candidate principle will not be useful if it depends on details of the conduct of war in a particular era. On the other hand, human nature has not changed appreciably over the past millennia. Therefore, candidate principles of war that depend on human nature and our cognitive biases might be enduring. Of course, we must check whether each candidate principle of war is valid across a variety of times, places, and circumstances. If we find such principles, we can take advantage of what we have learned over the millennia about how to fight wars. We hope to find essential elements of warfare that span the ages.

Here are the criteria that we think a concept should satisfy to qualify as a principle of war:

A principle must be a fundamental element of successful warfare.

A principle must be enduring.

A principle must give practical guidance.

A principle should be broadly applicable to various levels and types of warfare.

A principle must be a concept that has made a *difference* in the outcome throughout history. We must show that whether or not commanders followed the concept frequently determined whether they succeeded or failed. Thus, a candidate principle must pass a four-part discrimination test:

Adhering to the concept often contributed to victory.

Adhering to the concept rarely contributed to defeat.

Violating the concept often contributed to defeat.

Violating the concept rarely contributed to victory.

Commanders should have violated the principle sufficiently often in military history that it is worth paying attention to it.

A principle of war must be applicable to *war*, as opposed to operations other than war in which the military may participate.

Fundamental Element

To be considered a fundamental element of warfare, a principle of war must be a concept that is a basic and essential component of warfare. It should be broad in conception, rather than concerned with details.

Sun Tzu, writing twenty-five hundred years ago, gave us much wisdom that is still applicable today. He also offered advice that is not very useful at all. For example, "When light chariots first go out and take position on the flanks the enemy is forming for battle."[1] This advice does not help us today because it depended on the technology and tactics of his day. Better would be something like "study your enemies and learn what they do when they are about to attack." This is still narrow, but at least it is useful advice for us today. "Know your enemy," which covers the previous concept and a myriad of similar issues, would have been useful in Sun Tzu's day and is likely to be so in the future.

Several of the principles of war commonly accepted today confuse the goal with the *means* of achieving the goal. They describe tools for achieving desired effects, not the effects themselves. This misplaces the emphasis. An explanation of a current principle typically begins, "*The purpose of an offensive action is to seize, retain, and exploit the initiative.*" If the goal is gaining the initiative, then that might be the more fundamental concept. If so, the principle should emphasize the initiative rather than the offensive, which is merely a means for achieving the initiative. Ideas that are tools are often not essential and not fundamental, and thus may not have the staying power required to qualify as enduring. It is amazing that the U.S. military, which universally proclaims the value of mission-type orders—tell subordinates the desired end state but do not dictate *how* to achieve it—endorses principles that tell *how* to achieve a desired end. Principles should help a commander determine what the goal should be, rather than *how* to achieve that goal.

This does not mean that a "tool" can never be a principle of war. If a tool is *both necessary and sufficient* to achieve a desired effect, then it may be a principle of war.

Enduring Concept

A principle of war must be an enduring concept that does not require frequent revision. We should be skeptical of any concept that has been revised repeatedly. Ideally, we would like a candidate principle to be able to pass a ten-thousand-year test. That is, it should be valid for battle among cavemen fighting with clubs, battles using cruise missiles and stealth fighters dropping smart bombs, as well as for robotic "Star Wars" battles far in the future and nth generation warfare. We should be able to find examples in ancient history in which adherence to the concept paved the way to success but violation of the concept led to defeat. We should also be able to locate examples from recent history.

To enhance longevity, a principle should not be overly specific. It should not depend on a particular technology or on details of the conduct of war in a certain era. "Thou shalt not steal" has endured because it is direct and simple, yet general—it does not list burglary, bunco, armed robbery, embezzlement, fraud, shoplifting, Ponzi schemes, tax cheating, identity theft, computerized phishing, and so forth.

If we relaxed the requirement for a principle to be enduring and just required it to be valid recently, then we could identify more concepts as principles, and they might better address today's concerns, such as information warfare. However, this would be at the cost of a weaker historic foundation and thus such principles would be more likely to become obsolete in the future.

Practical Guidance

A principle of war must give practical guidance to be useful. Vague generalities such as "make good decisions" or "use good judgment" are not helpful. Wayne Hughes says that "to be useful a principle must be current and prescriptive, clearly implying what actions must be taken to put it into practice."[2]

In addition, a principle of war should not be a Rorschach inkblot in which everyone sees what he or she wants to see. Referring to a principle by a single word or short phrase results in a tendency for people to stretch the meaning far beyond its original meaning. To quote Wayne Hughes again, "A word is not a principle. A principle is a statement of general truth."[3]

Several of today's commonly accepted principles of war embody factors that are subject to trade-offs. A factor subject to trade-off is just one element that a commander must consider in conjunction with other competing factors. Like many complex endeavors, successful military operations require balancing many competing factors. Each factor is desirable, but a commander will always have to decide how to balance conflicting factors in his or her particular situation.

We prefer, but do not require, that principles have the feature that "more is nearly always better." Having a well-defined and achievable objective is always good, having it even better defined and more easily achievable is better if that is practical. More security at secondary points is not always better because it comes at the expense of economy of force and vice versa. The commonly accepted principles of war in effect advise exercising economy of force, but not too much economy, and provide security, but not overly much security. A "principle" that tells you to use your judgment in making such a trade-off is not a useful principle. Good judgment is always desirable. Thus, we prefer that a principle of war advise doing things that cannot be overdone or at least give guidance for how to make the trade-offs.

To be practical, principles must be concise, rather than burdened by many qualifications and limitations that might not be enduring. Therefore, commanders must use their judgment in applying principles to each specific situation. Principles are not a substitute for using one's brain.

Applicability

We prefer that a principle be applicable to the strategic, operational, and tactical levels of warfare. This is not an absolute requirement, but the broader the applicability of a principle, the more useful it is.

Ideally, a principle should be applicable to land, sea, and air warfare, and preferably also space warfare, nuclear war, guerrilla warfare, terrorism, and "fourth generation warfare." This is another requirement that is not absolute but certainly desirable.

Discrimination Test

A principle of war must pass the test of history as being a concept that often made a difference by discriminating between victory and defeat. That is, use or neglect of the principle must have been an *important* determinant of victory and defeat in many battles throughout history. Adhering to the concept must have often contributed to victory *as a result of* adhering to the concept, but rarely contributed to defeat. Use of the concept need not have been a necessary condition for victory. Nor must use of the concept have been a factor in every victory, just a frequently important factor. Note the use of the words "often" and "rarely" in the discrimination test. This means that a few contrary examples in the historical record need not disqualify a candidate principle of war. There will always be exceptions.

Similarly, violating the concept must have rarely contributed to victory but often led to defeat *as a result of* violating the concept. Because battles are usually complex, we often encounter cases in which a commander adhered to a valid concept but still lost the battle. That is acceptable provided the defeat was not due to following the concept.

Usefulness

Commanders must have violated the concept sufficiently often in military history that it is worth paying attention to it. If the concept is something that nearly every commander follows and therefore rarely comes into play in determining victory or defeat, then it is not worth designation as a principle of war. On the other hand, a principle should not address situations that rarely occur; for example, "If there is an eclipse of the sun, then . . ."

Our definition of a principle of war is a concept that distills enduring lessons of military history to provide practical guidance. It addresses fundamental elements of warfare that often determine the difference between victory and defeat.

In subsequent chapters, we will choose candidate principles of war that best meet these criteria. We will often enumerate "tools" or "means" commanders might use to achieve the goal described by the primary principles of war. The commander must decide which tool or tools, given the particular situation, will best achieve the aim stated by the principles. Many of these tools are concepts currently included in the commonly accepted principles of war.

The following chapters present many examples from military history. Throughout this book we use such examples only to *illustrate* our points. We want to emphasize that no selection of examples can ever *prove* the validity of a candidate principle of war because one can always select examples from military history to "prove" almost any point. Indeed, it is more instructive to seek examples in which a commander followed a candidate principle of war, yet disaster ensued. By examining the reasons for failure, we can better understand the strengths and weaknesses of the concept. In the following chapters we will examine the principles of war that the United States and Great Britain currently recognize, starting with the principle of the objective.

FIVE

Objective vs. Prioritized Objectives

The main thing is to keep the main thing the main thing.

—STEPHEN COVEY (BrainyMedia)

....

Current Principle of the Objective

AS RELATED IN CHAPTER 1, WE CAN FIND THE ESSENCE OF today's commonly accepted principle of the objective in the writings of Jomini, Clausewitz, and others. Today all the English-speaking militaries of the world recognize this principle or something similar. Indeed, many of them state this is the most important of all the principles of war.

In the U.S. military, the statement of principle of the objective starts by stating the purpose of the objective:

(1) The purpose of specifying the objective is to direct every military operation toward a clearly defined, decisive, and achievable goal.

(2) The purpose of military operations is to achieve specific objectives that support achievement of the overall strategic objectives identified to resolve the conflict. This frequently involves the destruction of the enemies' capabilities and their will to fight. The objective of joint operations not involving this destruction might be more difficult to define; nonetheless, it too must be clear from the beginning. Objectives must directly, quickly, and economically contribute to the purpose of the operation. Each operation must contribute to achieving strategic objectives. [Joint Force Commanders] should avoid actions that do not contribute directly to achieving the objective(s).

(3) Additionally, changes to the military objectives may occur because national and military leaders gain a better understanding of the situation or they may occur because the situation itself changes. The [Joint Force

Commander] should anticipate these shifts in national objectives necessitating changes in the military objectives. The changes may be very subtle, but if not made, achievement of the military objectives may no longer support the national objectives, legitimacy may be undermined, and force security may be compromised.[1]

One thing missing from this explanation of the principle of the objective is that the extreme importance of the objective mandates that one should put much careful thought into choosing an objective. The corresponding British principle at least acknowledges this in the title of its principle, "selection and maintenance of the aim":

Selection and maintenance of the aim provides a focus for coordinated effort and a reference point against which to measure progress. Following this principle prevents unnecessary activity and conserves resources. The single aim must pervade subordinate operations so they contribute to achieving the end-state. To ensure that they remain valid, plans must be checked continually against the strategic objectives. Uncertainty, political reality, and insufficient understanding of a situation may prevent being able to select a single aim from the outset.[2]

As already mentioned in chapter 1, the original British version of the principle of the objective in 1920 emphasized the destruction of the enemy's forces.[3] This was in contrast to such things as capturing territory. The current American principle of the objective states that destruction of either the enemy's capacity or the enemy's will to fight is "frequently" the objective, but allows for other possibilities.

This principle gives good guidance, but it could be much better. To illustrate the shortcomings of the prevailing principle of the objective, we will look at examples of the various ways in which commanders have gone astray in using it in the past. We must then decide whether the problem comes from the principle or from its poor implementation.

Poor Choice of Objective

Poor objectives include those that are unattainable, poorly formulated, or mismatched with higher-level objectives. The problem of unattainable objectives starts at the top.

Prioritized Objectives

Clausewitz believed that the first, most momentous decision political leaders must make is whether to go to war. One does not have to be an anti-war activist to caution that wars rarely turn out the way either side expected. Examples of unattainable political objectives may be found in many cases in which a country started a war that it ultimately lost. Political leaders should carefully consider whether their objective is attainable at a price they are willing to pay. Choosing unattainable military objectives can also be a problem. For example, as Lt. Gen. James Longstreet warned Lee, the objective of Pickett's Charge was not attainable with the force available.[4]

The problem may be magnified by the poor formulation of objectives. A belief that a situation is undesirable is not sufficient reason to employ military force if a clear, attainable political objective cannot be identified. Poorly defined political objectives make it difficult to choose military objectives that will achieve them.

Lower-level objectives should contribute to attaining the higher-level object. In 410 CE Alaric was the leader of the Goths. His political objective was to have the Roman Empire grant official status to the Goths. To pressure Emperor Honorius, Alaric laid siege to Rome. Honorius, however, was in Ravenna, which had become the Empire's political center, and he did not much care what the Goths did to Rome. In fact, there is a story that when aides told Honorius that Rome had fallen, he misunderstood and started weeping because he thought that his favorite cockerel named "Rome" had died. He was much relieved when he found out "just" the city had been captured. Thus, Alaric's military objective of Rome could not attain his political objective.[5]

For a positive example of matching military with political objectives, consider Anwar Sadat's objectives in the 1973 war with Israel. His political objective was to break the deadlock and create the conditions for negotiating the return of the Sinai to Egypt. To achieve this, his military objective was limited to seizing a small amount of territory then held by Israel along the Suez Canal, not to gain territory but to create a crisis that would change attitudes.[6]

An objective should be the goal, not an activity. Effectiveness should be measured by the assessment of progress toward attaining the objective, not a gauge of effort, such as sortie rates or tons of bombs dropped. Nor should a measure be something not directly related to the objective.

Counterinsurgency forces face particularly difficult problems in choosing objectives. One of the top objectives must be, just as the cliché says, to win the hearts and minds of the people. Forming military objectives from such a general objective is extremely difficult. Nevertheless, it is of vital importance and merits careful thought. Many experts believe it is essential that counterinsurgency forces target the support of the people for the insurgents and should give higher priority to winning the support of the people than to defeating the insurgents directly.[7]

During the Vietnam War, a war without front lines, leaders faced the problem of figuring out how much progress they had made—and also convincing the public that the war was going well. Most people on the ground and throughout the chain of command understood that, in order to win, the United States had to win the hearts and minds of the Vietnamese people. However, it is hard to count the number of hearts and minds won every week. Relying on individual opinions of people on the scene for their evaluation of progress would be subjective and unreliable. In an era that overemphasized quantitative measures, leaders wanted a quantitative measure that would indicate progress. Thus, the "body count" measure was born.

Body counts were a poorly chosen measure of effectiveness. If real progress were made, then at some point the number of enemy killed would go down. Pressure to increase the body count passed down the chain of command. In effect, one objective for lower-level commanders became increasing the body count. Because the people being judged were the people doing the reporting, this objective generated unreliable information and put in place the wrong incentives. Inevitably, people exaggerated the body counts. Thus, suboptimization due to a poorly chosen measure of effectiveness damaged the overall cause of the United States.

Losing Sight of the Primary Objective

Losing sight of the objective, during either planning or execution, is a common problem. During the planning stage, one might start out with an excellent objective. Then, as the planning progresses, one might get greedy and attempt to do too much. During the execution phase, because things never work out exactly as planned, there is a strong temptation to change the plan in accordance with events. Sometimes this is appropriate. Often, however, secondary considerations drive the change. Commanders should make all decisions with the primary objective in mind.

In 207 BCE the Carthaginian Hasdrubal led fifty thousand men over the Alps to join his brother Hannibal in southern Italy. Although Hasdrubal's primary objective should have been to join forces with his brother, he stopped to lay siege to Placentia. He not only failed to capture the city but also lost many days. Due to this delay and the capture of a message from Hasdrubal to Hannibal carrying the former's plans, the Roman consul Caius Claudius Nero led part of his army on a 250-mile march in seven days to join Marcus Livius's forces facing Hasdrubal. The Romans then destroyed Hasdrubal's army in the Battle of the Metaurus. Had Hasdrubal not been distracted from his primary objective, he might have been able to join Hannibal before the Romans could stop him.[8]

Early in the siege of Petersburg during the American Civil War, some soldiers who had been Pennsylvania miners devised an idea to tunnel under the Confederate fortifications and set off explosives to blow a hole in the defensive line. Army experts told them that what they proposed was impossible. No army engineers assisted in the project, and the army provided no tools or other help. However, their corps commander, Maj. Gen. Ambrose Burnside, supported the Pennsylvanians. The miners persisted and with great ingenuity completed the tunnel, while Burnside devised a plan to exploit the explosion.

Burnside chose the freshest of his four divisions to lead the attack, even though its men had no combat experience. For a week, he trained this division for the mission. Their assignment was to fan out and seize the Confederate trenches on either side so that Burnside's other three divisions could pass through the gap and capture the high ground behind the Confederate lines. The lead division rehearsed the attack until each man knew exactly what to do. Their morale was high and they were eager to show the rest of the army what they could do. One unique thing motivating this division was that the troops were African Americans. Many officers in the Union Army thought it was impossible for black men to be good soldiers and scorned them. Burnside, to his credit, thought otherwise.

Coincidently, about this time Lt. Gen. Ulysses Grant and Maj. Gen. George Meade were looking for a way to break the siege, other attempts having failed. Finally, they latched onto Burnside's project. However, Meade, with Grant's approval, made a critical change just hours before the attack. He ordered Burnside to select a white division to lead the attack instead of the black division. As a result, a division that had done no training for the attack and whose troops did not appear to understand their assignment led the attack. On the morning of 30 July 1864, the huge explosion was set off, resulting in a crater two hundred feet long by sixty feet wide and thirty feet deep. Because the Confederates on either side of the explosion ran away, there was a huge gap in the Confederate lines. At the sound of the explosion, the lead Union division also ran away. It took at least ten minutes to round them up. When they finally made it to the crater, instead of fanning out, they went down into the crater or stared at it in amazement. This gave the Confederates time to recover and close the gap.

Meade had ordered a change in the lead division because he lacked confidence in the inexperienced black troops. Grant later said that he worried that if the attack failed and the colored division suffered heavy casualties, then abolitionist critics might say that Grant shoved them out in front to die because he did not care about them.[9]

During the Battle of Britain in 1940, Germany shifted its air attacks from target to target. From 10 July to 11 August, they attacked channel ports and shipping convoys, hoping to draw the Royal Air Force out to fight. On 12 August, the German air assault began in earnest. They attacked the vital radar stations on the coast and damaged five. After a few more attacks on the radar stations over the next week, Hermann Goering ordered the cessation of attacks on the radar because they seemed hard to knock out permanently, and he did not realize how vital they were to Britain's defenses. Also starting on 12 August, the Germans began heavy raids on British fighter airfields in southeast England, seeking to destroy British fighter aircraft either on the ground or in the air. Goering soon diverted some assets to ineffective night attacks on aircraft production facilities. The Germans finally deduced the importance of the sector stations that used radar and other data to guide fighter aircraft. For two weeks starting on 24 August, the Germans sent a daily average of one thousand aircraft against England, focusing on the airfields with sector stations. British aircraft losses exceeded their replacement rate, the sector stations were degraded, and Fighter Command came close to defeat. However, on 7 September German aircraft switched to bombing cities. They changed their objectives in part because they thought they had nearly destroyed the British fighter command. In addition, the bombing of civilians was in retaliation for a British air raid on Berlin, which in turn was in retaliation for a probably mistaken bombing of London by the Germans. Previously, Hitler had forbidden attacks on civilian areas because such attacks

Prioritized Objectives

would only harden the British resolve at a time when he hoped to get a negotiated peace with Britain.[10]

The preceding narrative indicates a lack of focus in picking targets. The initial objective was clear: eliminate British fighter aircraft as an effective force so that Germany could have air supremacy over southeast England and the English Channel, which was a necessary condition for an invasion of Britain. Nevertheless, Goering repeatedly vacillated in choosing the targets to achieve that objective. Some of the switching came from a lack of understanding of the situation. For example, the Germans did not appreciate how well the British had integrated their radars into an air defense system and greatly overestimated the damage they had done to the British Fighter Command. Finally, they allowed anger and embarrassment over the bombing of Berlin to cloud their judgment in choosing to target civilians in cities. They justified this with the hope that the terror would cause Britain to sue for peace, but that was contrary to what Hitler had said just a month previously. The bottom line is that whatever the reasons, Germany's efforts in the Battle of Britain suffered from shifting objectives.[11] Having flexibility is good; being indecisive is not.

Another way to lose sight of the objective is to allow carrying out the plan to become the objective. That is, if the planners do not stay focused on the primary objective, the effective objective can morph into carrying out the plan, rather than attaining the original objective. In 1944 the Americans planned to capture Peleliu Island, which they believed was a prerequisite for the invasion of Leyte. After carrier raids on the Philippines found little opposition, Adm. Bill Halsey recommended skipping the invasion of Peleliu and moving up the invasion of Leyte by two months. The Joint Chiefs of Staff quickly agreed to invade Leyte two months sooner than previously planned. However, they did not cancel the invasion of Peleliu, because the invasion force was already at sea. Although the capture of Peleliu was not necessary, 1,950 Americans died capturing the island.[12]

After making a plan, the planners inevitably encounter problems as they work on the details or obtain new information. As they alter one part of the plan to address the problem, they find that they also need to adjust other parts of the plan to allow execution of the problematic piece. This process is repeated several times. If they are not careful, by the time the operation begins, all the adjustments to the plan will preclude attainment of the primary objective.

Operation Anaconda in Afghanistan in 2002 was an attack on al-Qaeda in the Shahikot Valley. One part of the plan involved inserting men using helicopters. After planners introduced helicopters into the plan, they spent much time worrying about load factors, fuel, and so forth, rather than what the enemy was doing. Although the best time to attack was in bad weather, the helicopters required good weather. This caused a delay in the operation that was inconvenient for other forces.[13]

The plan assumed that the enemy was on the valley floor, not in the surrounding mountains, and would flee when confronted. Thus, the plan called for the helicopters to land in the valley to establish a blocking position. When on-site intelligence indicated the enemy occupied the high ground and appeared prepared to stand and fight, the plan was not changed. The reasons given were that it was too late and alterations could not be made because the final plan was the outcome of a long process of compromise and negotiation among various interests. As Capt. Pete Blaber put it, "The tyranny of the plan was trumping updated information from the man on the ground." The mission had become executing the plan on time. When the helicopters entered the valley, they came under heavy fire and the plan became obsolete.[14]

Multiple Objectives

At the political level, leaders sometimes argue over which alternative is more desirable and deserves to be the primary objective. In the end, they may decide to include both objectives. In ancient Athens, the public assembly set the military priorities and objectives. This was the ultimate in setting objectives by committee, compounded by demagoguery. During the Peloponnesian War, the assembly sent an expedition to Sicily in 425 BCE (ten years before the better-known expedition). The initial objective was to support Athenian allies in Sicily. When word arrived that Corcyra on the west coast of Greece was under attack by a Peloponnesian fleet, the assembly directed the expedition to assist in the defense of Corcyra as they sailed by the city. Finally Demosthenes requested the use of an Athenian fleet. The assembly voted to give him permission "to use the fleet, if he wished, upon the coast of the Peloponnesus."[15] Thus the Athenian assembly gave the expedition three commanders and three objectives.

Fortunately for the Athenians, instead of resulting in the disaster that such orders deserved, events turned out quite well in spite of the muddled instructions. Demosthenes exercised his prerogative and required the fleet to detour to Pylos. According to Thucydides, the other two generals in the expedition objected, but a storm drove the fleet there anyway. Ultimately, this resulted in the greatest Athenian victory in the war at Pylos.[16] Nevertheless, having three commanders and three objectives is not a good way to conduct military operations.

During a war against the British in 1652 Dutch Lt.-Adm. Maarten Tromp received orders that his principal objectives were to inflict all possible harm on the English by damaging the English fleet and to convoy Dutch merchantmen to the west. When Tromp asked for clarification of these conflicting objectives, he received no additional instructions. An indecisive campaign ensued. Tromp later received orders to escort a large convoy. He finally resolved his conflicting objectives by sending the convoy back to Ostend. Freed from this burden, he then attacked the British fleet with some success.[17]

Often, multiple objectives are inherent in the situation. World War II in the Pacific saw a series of amphibious assaults that were fiercely opposed by Japanese defenders. In each case, one objective of the fleet was protection of the amphibious ships as they approached the landing area and then as they unloaded equipment and supplies. Another objective of the battle fleet was to defeat the enemy's naval power. These objectives often conflicted.

After the Guadalcanal landings, Vice Adm. Frank Jack Fletcher prioritized preserving his fleet over supporting the U.S. Marines ashore. Therefore, in the face of Japanese threats, he pulled his forces out of the area, thus leaving the Marines without air support or all their supplies. The Marines bitterly resented their abandonment. This is one of the reasons why the Marine Corps today wants organic air and logistics support that they control.

During the American amphibious assault in the Mariana Islands in June 1944, Adm. Raymond Spruance commanded the Fifth Fleet. As the Japanese fleet approached, he stayed in a position where he could defend the amphibious ships if the Japanese fleet tried an end run around him. He did this even though that meant he had to forego a good chance to close on the Japanese fleet and defeat them. Spruance received much criticism for his failure to take advantage of an opportunity to wipe out the Japanese fleet.[18]

Later, during the invasion of Leyte in October 1944, Adm. Bill Halsey, commanding the Third Fleet, made a different decision. Based on some faulty information, he assumed the landing force was in no danger and took much of his combat power to attack the Japanese carriers, which were decoys in this operation.[19] Halsey's decision left the way open for a Japanese force to attack the landing area. Only heroic defense, together with Japanese lack of resolve, averted disaster.[20]

At a more mundane level, priorities must be assigned and followed in devising the loading plan for the ships in an amphibious operation. Everything might be important, but you still have to set priorities. Unloading artillery shells early in the operation is not useful if the artillery pieces are the last things to come off the ships.

A nation at war always seems to have multiple political goals leading to multiple military objectives. Even at the tactical level, one usually has the constraint of wanting to minimize one's own casualties while defeating an enemy force or capturing a position. Time is often an additional constraint. In today's world, avoiding civilian casualties is frequently a constraint. Generally, the more limited the political goals, the greater the constraints on military operations.

Constraints of the type described in the previous paragraph are in effect objectives, but are usually not absolute. Reducing one's own casualties to zero would be great, but is rarely attainable. The question then is whether the mission is so important and so time-critical that capturing the position is imperative regardless of casualties or whether one should proceed carefully.

Planning by committee or competing interests results in compromises that either blur the primary objective or result in additional objectives. Then when the plan enters the execution stage, secondary objectives can drive the problem. The following examples show how problems can arise when competing interests compromise to set objectives.

SCHLIEFFEN PLAN

Prior to World War I, Germany faced the prospect of fighting a two-front war against the Russians in the east and France in the west. Field Marsh. Alfred von Schlieffen devised a plan for dealing with this situation. Because of its vast area and backward transportation system, Russia's mobilization would be slower than that of France. Therefore, Germany planned to attack and defeat France first. Then they would switch much of their army to the east and defeat Russia. The drawback was that Russia, even considering their slower mobilization, would be able to overrun part of East Prussia. Von Schlieffen was willing to allow that outcome in order to achieve the main objective, which was to defeat France quickly.

On the western front, von Schlieffen's plan put most of Germany's forces into a right-wing sweep through Belgium and into France. The farthest part of the right wing would "brush the Channel" and then sweep south of Paris to trap the French Army. One drawback of this part of the plan was that France could attack Germany's lightly defended left wing and advance into Germany. Von Schlieffen was willing to accept that to achieve his primary objective. Supposedly, von Schlieffen's dying words admonished keeping the right wing strong. His successor, Gen. Helmuth von Moltke (the Younger) altered the plan by strengthening the left wing to avoid Germany territory being invaded and perhaps in hope of achieving a double envelopment. This weakened the right wing.

When war came, the Russians mobilized faster than expected and threatened East Prussia. Moltke transferred two corps from the west to the east. He took these two corps from the right wing, thus further weakening the main thrust. Ironically, Germany repelled the Russians at the Battle of Tannenburg before these reinforcements arrived. Thus, the two corps did not participate in the decisive action on either front.[21]

Had Moltke not weakened the right wing both in planning and in execution, it might have improved Germany's chances of a quick victory in France. However, Germany might not have succeeded even then. Her troops were exhausted, and Germany faced severe logistic constraints that probably would have precluded a sweep south of Paris.[22] Historians disagree about whether Moltke's changes to the von Schlieffen plan were good or bad, as well as about whether the plan could have succeeded. However, it is clear that by losing sight of Germany's primary objective at the time—a quick victory in France using a sweep by a strong right wing—Moltke reduced the strength of his best chance for victory.

MERCHANT SHIP ANTI-AIRCRAFT GUNS

A different type of example, this one from World War II, illustrates how one can lose sight of the real objective and instead latch onto a seemingly obvious objective that leads one astray. Early in the war, Axis aircraft attacked merchant ships in the Mediterranean Sea and damaged or sank many. As a response, the British provided some merchant ships with anti-aircraft guns. This cost a lot of money and, even more important at this stage of the war, consumed scarce resources badly needed elsewhere. The merchant ship gun crews, with only limited training, proved to be terrible shots. A review of attacks revealed that they shot down only 4 percent of all the aircraft that attacked merchant ships equipped with guns. It seemed clear that this dismal rate of success did not justify the considerable expense of equipping the merchant ships with anti-aircraft guns.

However, what was the objective in arming the merchant ships? It was not to destroy Axis aircraft. The real objective in arming merchant ships was to reduce the number of ships damaged or sunk by Axis aircraft. Subsequent analysis of the reports revealed that whereas 25 percent of all aircraft attacks against ships not firing anti-aircraft guns resulted in the ships being sunk, only 10 percent of such attacks against merchant ships firing anti-aircraft guns resulted in the ships being sunk. The reason is that anti-aircraft fire, even when ineffective, affects the accuracy of the air attacks. The correct objective was not to shoot down attacking enemy aircraft but rather to help merchant ships survive attacks. Thus, arming the merchant ships achieved the real objective of the program.[23]

JAPANESE AT MIDWAY

In the spring of 1942 commander of the Japanese Combined Fleet, Adm. Isoroku Yamamoto, the Japanese Naval General Staff, and the army disagreed about what Japan should do next to exploit its unbroken string of victories. Yamamoto argued that Japan's primary objective should be the destruction of the remaining U.S. aircraft carriers. To accomplish this, he devised the Midway operation. In the original plan, the threat to capture Midway Island was the means to achieve the aim of destroying the American carriers by luring them to defend Midway Island.

After the Doolittle Raid on Japan the Japanese perceived a need to extend their defenses out farther to prevent future raids. This increased the importance of capturing Midway. With this incentive plus an implied threat to resign if he did not get his way, Yamamoto won. Nevertheless, he had to compromise by including an Aleutian operation in the plan and by sending two aircraft carriers to support a limited incursion into the Southwest Pacific in early May.

Japan ended up with three operational objectives spread across the Pacific with none of the forces able to support another. This illustrates how multiple interest groups can cause a drift from the primary objective. Plans made by committees or competing interest groups are quite susceptible to this type of error.

The Aleutian operation siphoned off two moderately capable carriers and a considerable number of fighter aircraft, diverting forces from the primary thrust without any benefit to the primary objective. The Southwest Pacific operation ran into unexpected American opposition. In the Battle of Coral Sea, American aircraft damaged the Japanese aircraft carrier *Shokaku* and mauled *Zuikaku*'s air wing. As a result, neither of these two modern carriers was available for the Midway operation. By not subordinating everything to the primary objective, the Japanese allowed two clearly secondary objectives to drive their operations and deprive them of one-third of the Kido Butai at the Battle of Midway. Although the Japanese emphasized tactical concentration, they failed to concentrate their aircraft carriers operationally.

In the detailed planning and execution of the operation, the Japanese again lost sight of their primary objective. When the carriers had to delay sailing by a day, the Japanese did not adjust the invasion timetable, supposedly due to the tides needed for landing on Midway Island. This upset the relative position of the various forces. Due to the constraints of the secondary Aleutian operation, the Japanese did not consider delaying the operation a month to wait for the next occurrence of the optimal tides. Thus, the Japanese allowed two secondary objectives to drive the timing of the operation. Finally, overhanging everything else were the dual objectives of supporting the invasion and fighting the American carriers. The Japanese dealt with that by assuming the American carriers would not appear until after the Japanese secured Midway Island. This also compromised achieving the primary objective by favoring a secondary objective.[24]

DESERT STORM

The reader might argue that, although choosing multiple objectives may have been a problem in the past and selecting vague objectives may have been a problem as recently as the Vietnam War, nowadays the United States faithfully follows the principle of the objective. Such a reader might point to Operation Desert Storm as having a clear objective. The general perception is that the United States planned and executed the first Gulf War, Operation Desert Storm, in 1991 in a commendable manner, especially when compared to the Vietnam War. However, multiple, conflicting objectives were a serious problem even in this case.

In August 1990 Saddam Hussein invaded Kuwait and poised his Republican Guard divisions on the border of Saudi Arabia for a possible invasion of that country. This naked invasion threatened to set the tone for the post–Cold War period. Saddam Hussein controlled both Iraq's and Kuwait's large oil reserves. If he also captured Saudi Arabia, or even just the northeast part of the country that contained most of the oil reserves, then he would control nearly half of the world's oil reserves. For Saddam Hussein to have such power over the world oil market on which the entire developed world depended was contrary to the vital interests of the United States. In addition, Saddam Hussein's success upset the balance of power in the region. If no one checked him, he would not only control the oil market but would also have the money to build up his military and develop chemical, biological, and nuclear weapons. The Iraqis also held hostage Americans and other foreign nationals in Iraq and Kuwait. Finally Iraq was looting Kuwait and there were stories of atrocities against the Kuwaiti people.

In National Security Directive 54, Pres. George H. W. Bush set the nation's political objectives as:

> To effect the immediate, complete, and unconditional withdrawal of all Iraqi forces from Kuwait;
>
> To restore Kuwait's legitimate government;
>
> To protect the lives of American citizens abroad; and
>
> To promote the security and the stability of the Persian Gulf.[25]

Elimination of Saddam Hussein as leader of Iraq, though widely considered desirable, was not included as an American goal here. Later in National Security Directive 54, however, it states that if Iraq used chemical, biological, or nuclear weapons, supported terrorist acts against the United States or coalition partners, or destroyed Kuwait's oil fields, then replacing Iraq's leadership would become an explicit objective.

National Security Directive 54 assigned six missions to the military:

Defend Saudi Arabia and the other Gulf Cooperation Council states against attack;

Preclude Iraqi launch of ballistic missiles against neighboring states and friendly forces;

Destroy Iraq's chemical, biological, and nuclear capabilities;

Destroy Iraq's command, control, and communications capabilities;

Eliminate the Republican Guards as an effective fighting force; and

Conduct operations designed to drive Iraq's forces from Kuwait, break the will of Iraqi forces, discourage Iraqi use of chemical, biological or nuclear weapons, encourage defection of Iraqi forces, and weaken Iraqi popular support for the current government.[26]

National Security Directive 54 went on to specify two constraints on military action:

Minimize U.S. and coalition casualties and

Reduce collateral damage incident to military attacks, taking special precautions to minimize civilian casualties and damage to non-military economic infrastructure, energy-related facilities, and religious sites.[27]

At the next step down the chain of command, the Central Command's operations order for Desert Storm defined six theater objectives: attack Iraqi political and military leadership along with command and control assets; gain and maintain air superiority; sever Iraqi supply lines; destroy chemical, biological, and nuclear capability; destroy Republican Guard forces; and liberate Kuwait City. The Central Command assigned four or more objectives to each component command. Furthermore, the coalition identified three Iraqi "centers of gravity": Iraqi leadership, weapons of mass destruction, and the Republican Guard.[28] This is a long, long way from *"Direct every military operation toward a clearly defined, decisive, and achievable goal."*

Although there was considerable correlation among the various lists of objectives and missions, the situation was far from the ideal of a single well-defined political objective from which flowed a single military objective and a single objective assigned to each component commander within the Central Command.

In fact, contrary to the picture painted by the paper trail, the two primary objectives were understood to be expelling Iraqi forces from Kuwait and eliminating the Republican Guard as an effective fighting force. The latter objective was considered synonymous with promoting regional stability by eliminating Iraq's ability to mount an offensive threat against its neighbors.[29]

After an extensive air campaign and then four days of ground attack, coalition forces had driven the Iraqi army out of most of Kuwait. Coalition armored forces had decisively defeated part of the Iraqi Republican Guard and were about to cut off the retreat of a substantial portion of the remaining divisions of the Republican Guard. Because of televised scenes of the "Highway of Death" north of Kuwait (where there were actually not many deaths) and a misperception in Washington that the Iraqi retreat had already been cut off, the president made a political decision to call a cease-fire after one hundred hours of the ground war. As a result, several Republican Guard units were able to escape the trap, thus preserving an important core of Saddam Hussein's strength.

This came about because there was no clear statement by the U.S. national command authority of a single primary objective, nor apparently an understanding that the two primary objectives might conflict. If the national command authority had understood a clear priority—and presumably set promoting regional stability as the primary objective—then they would not have ended the war until they were *certain* that they had trapped all possible Republican Guard units.

OPERATION ENDURING FREEDOM—UNITED STATES IN AFGHANISTAN IN 2001

After the attacks of 11 September 2001, the United States attacked the Taliban in Afghanistan, who were the protectors of Osama bin Laden and much of the al-Qaeda leadership. American forces on the ground were mainly special forces and CIA agents who advised the Northern Alliance and provided ground spotting and direction for the air power. They were incredibly successful. Soon the leadership of al-Qaeda and the core of their fighters were cornered in the mountain caves of Tora Bora. The end of Osama bin Laden seemed near. However, bin Laden and much of the al-Qaeda leadership escaped the trap.[30] As a result, Osama bin Laden was on the loose for an additional ten years, having moved to a new sanctuary in the tribal areas of Pakistan and then later to Abbottabad. Thus, an otherwise brilliant military campaign resulted in a seriously flawed victory.

The primary objective was to capture or kill the leadership of al-Qaeda, especially Osama bin Laden. Nevertheless, the quarry escaped because the United States had not put greater numbers of American troops on the ground despite requests from those on the scene. In large part this was due to a desire to minimize the size of the "footprint" on the ground. For their part the Afghani forces failed to capture bin Laden, most likely because they took bribes. If there had been more American ground troops at Tora Bora, they might have blocked the escape route used by bin Laden.[31] In the end the United States failed to achieve its objective because it allowed itself to be distracted by secondary considerations.

IN SUMMARY, THE SECONDARY OBJECTIVES OF STOPPING THE Russians and preventing even temporary occupation of German territory distracted Moltke. In the merchant ship example the secondary objective of shooting down enemy aircraft distracted from the primary objective of saving merchant ships. The Japanese at Midway allowed multiple interest groups to undermine the initial primary objective. Operations Desert Storm and Enduring Freedom show how far the United States has strayed from the ideal of "direct every military operation toward a clearly defined, decisive, and achievable goal." All of these examples suggest the necessity of replacing the commonly accepted principle of the objective with a new principle that addresses situations in the real world.

The New Principle of Prioritized Objectives

While commanders rarely fail to choose any objective at all, they sometimes choose poor objectives. They may think the proper choice of objective is obvious, so that careful thought is a waste of time. However, inadequate deliberation typically results in unattainable, poorly conceived, and mismatched objectives. Because the objective drives all military activity, the determination of the objective deserves a great deal of thought—far more than it is usually given. Hence, any principle of war concerning the objective should emphasize the overwhelming importance of selecting it carefully.

Losing sight of the objective during the planning or execution of the plan is another problem. Frequently, this occurs because secondary considerations sidetrack the commander. This is partially due to commanders not following the principle and partially due to defects in the principle itself.

Finally, problems frequently arise either because a commander is assigned multiple objectives or because multiple objectives are an inherent part of the specific situation. The current principle of the objective does not address this real-world situation. Although the current principle of the objective correctly emphasizes the desirability of a single objective that is clearly defined, decisive, and achievable, it fails to address what to do when that ideal is not feasible. In the real world, the situation is usually more complex and multiple objectives are an unfortunate fact more often than not.

Therefore, we propose a new principle of *prioritized* objectives:

> Explicit prioritization of objectives in an operation facilitates the best distribution of effort and avoids the pursuit of secondary objectives interfering with achievement of the primary objective.
>
> Ideally, every operation should have a single clearly defined, decisive, and attainable objective, because that focuses all activity and encourages subordination of everything else to achievement of the objective. However, history shows commanders are frequently assigned multiple objectives, some of which may be implicit. In those cases, explicit prioritization of objectives is advantageous.
>
> Because the objective drives all military activity, careful study of the situation should precede selection of one or more objectives, as well as the prioritization of multiple objectives.
>
> At each level, all actions should contribute to the objectives of higher levels, up to and including national political goals.
>
> When there are multiple objectives, commanders assigning objectives to subordinates can convey their intent by explicitly prioritizing the objectives. At the very least, they should give subordinates general guidance on how to handle potential conflicts among the objectives. Commanders should also tell subordinates how their assigned objectives interact with those of other units, again indicating how they should handle conflicts.

As in the commonly recognized principle of the objective, military objectives are derived from the political objective, operational objectives are based on strategic objectives, and tactical objectives are derived from operational objectives. The hierarchy of objectives from the national command authority down to small units forms the organizing principle for all military plans and operations.

The new principle of prioritized objectives differs from the old principle in that although it emphasizes designating a *single* primary objective if possible, it recognizes that assignment of multiple objectives occurs frequently. The new principle of prioritized objectives suggests explicitly prioritizing all objectives and making decisions based on their ordering.

Military history is replete with cases in which commanders ran into problems because they failed to prioritize their objectives, either in their own minds or in the minds of their subordinates. The classic example of this is the Japanese at the Battle of Midway. The new principle addresses this problem.

When commanders give subordinates multiple objectives, they should also give guidance on how to resolve potential conflicts. If commanders do not give clear guidance to their subordinates and ensure they understand the higher-level objectives all the way up to the political level, as well as the objectives assigned to others, then problems result. In the previous sections of this chapter, we examined cases in which commanders failed to prioritize their objectives. Now we give some examples of the right way to do so.

From the time of the Armada, British policy dictated that when an invasion threatened, the primary target was the invading army, not the enemy's battle fleet. In 1759 during the Seven Years' War, France threatened to invade Britain and gathered a large invasion fleet. First Lord of the Admiralty George Anson made the priority of fleet commander Admiral Edward Hawke's objectives clear. His primary objective was to intercept the embarkations of the French invasion fleet at Morbihan; preventing French ships of war from coming out of Brest was a secondary consideration.[32]

When Napoleon threatened invasion in 1803, the British Admiralty gave fleet commander Lord Keith clear instructions about his priorities:

> Directing your chief attention to the destruction of the ships, vessels, or boats having men, horses, or artillery on board (in preference to that of the vessels by which they are protected), and in the strict execution of this important duty losing sight entirely of the possibility of idle censure for avoiding contact with an armed force, because the prevention of debarkation is *the object of primary importance to which every other con-*

sideration must give way. (Admiralty Secretary's In-letters, 537, 8th August 1803, emphasis added)[33]

In World War II Prime Minister Winston Churchill and Pres. Franklin D. Roosevelt faced multiple enemies in a worldwide struggle. They gave clear guidance on their priorities: Germany first, then Japan. This governed the allocation of resources to each theater.

Two examples of correctly prioritizing objectives deserve more detail.

Pres. Abraham Lincoln wanted both to free the slaves and to preserve the Union. After his inauguration, the secession movement was well underway, but the slave-holding border states of Maryland, Missouri, and Kentucky were on the fence. Keeping the Border States in the Union seemed vital to defeating the rebellion. However, many of Lincoln's supporters expected him to free the slaves *immediately* and pressed him to do so. In eastern Virginia, Maj. Gen. Benjamin Butler declared that slaves deserting their owners and entering Union lines were "contraband of war" and put them to work. Lincoln approved the policy. Maj. Gen. John Fremont, military commander in Missouri, got ahead of Lincoln by issuing a proclamation on 30 August 1861 freeing all slaves belonging to Confederates in Missouri. Such actions threatened to drive the Border States into the Confederacy. When word of Fremont's emancipation reached Kentucky troop companies, some threw down their arms. Lincoln rescinded Fremont's order and removed him from command.[34]

How did Lincoln resolve his two objectives? In a 22 August 1862 letter to Horace Greeley, which was also widely disseminated, Lincoln declared,

> My paramount object in this struggle is to save the Union, and is not either to save or to destroy slavery. If I could save the Union without freeing any slave, I would do it; if I could save it by freeing all the slaves, I would do it; and if I could save it by freeing some and leaving others alone, I would also do that.[35]

He was clear about which objective was primary and how conflicts between the objectives were to be resolved. Decisions about whether to free any slaves would depend solely on their effect on the preservation of the Union.

In fact, Lincoln had already drafted the Emancipation Proclamation, but he did not release it until the Union Army had won a victory, or at least something less than a defeat, at Antietam. Then, from a position of relative strength, Lincoln issued the proclamation, which in fact freed relatively few slaves immediately. It freed only those slaves in areas that were under Union control but declared to be in rebellion. The proclamation did not affect slaves in the Border States. He never lost sight of the fact that his primary objective was to preserve the Union. Because of Lincoln's proclamation, Britain and France were faced with the fact that if they chose to recognize the Confederacy, they would also be supporting slavery, a politically impossible choice for their own domestic reasons. Thus, in issuing the Emancipation Proclamation when he did, Lincoln employed it in support of preserving the Union.[36]

IN ANOTHER EXAMPLE, WHILE THE JAPANESE FAILED TO prioritize their objectives in the Midway operation, Adm. Chester Nimitz's instructions prior to the Battle of Midway provide an example of how to do so.

In mid-May 1942 Commander in Chief of the U.S. Fleet Adm. Ernest King sent Nimitz a message urging him to use "strong attrition tactics" and not to risk a battle that would likely result in heavy losses of American carriers and cruisers.[37] In his 26 May estimate of the situation, Nimitz defined attrition tactics as "submarine attacks, air bombing, attack on isolated units." He reasoned that if the attrition of Japanese forces were successful, then the Japanese would have to accept failure of the operation or fight the Americans at a disadvantage.[38]

Nimitz's Operation Plan 29–42 listed two objectives: hold Midway Island and damage the enemy forces. In particular, the plan ordered the carrier striking forces to "inflict maximum damage on enemy by employing strong attrition tactics." In accordance with King's urging, it also warned, "Do not accept such decisive action as would be likely to incur heavy losses in our carriers and cruisers."[39] Taken on its face value, the latter statement seemed to be an order not to risk the carriers (or even the cruisers), regardless of the prospects of damaging the Japanese forces. This is an example of the frequent contradiction between written orders and the commander's real intentions.

Nimitz would have liked to achieve three objectives:

Damage or sink the Japanese aircraft carriers.

Preserve American aircraft carriers.

Defend Midway Island.

The first two objectives were inherently in conflict. Nimitz would be sending into the battle three of the four operational aircraft carriers in the Pacific theater. One of the three, the *Yorktown*, had suffered severe damage just a month previously in the Battle of the Coral Sea and had been hastily patched together. Furthermore, Nimitz could not expect substantial reinforcements for quite a while. Meanwhile, the Japanese would send four fleet aircraft carriers into the Battle of Midway and could additionally expect the two modern aircraft carriers *Shokaku* and *Zuikaku*, casualties of the Battle of Coral Sea, to rejoin the fleet soon. Therefore, losing one or two of the three carriers that Nimitz was sending into battle, without inflicting substantial damage to the Japanese fleet, would leave the Japanese free to roam the entire Pacific Ocean with near impunity.

How could Nimitz convey to his subordinates how they were to judge whether to engage in battle? The currently recognized principle of the objective demands that Nimitz should have given his subordinates a single objective. However, giving them a single objective of damaging the Japanese fleet was not sufficient. Giving them a single objective of not risking the American carriers would not do either. That would inevitably result in avoiding a battle. If one had learned nothing else from prewar fleet exercises and the one previous carrier-carrier battle in history—Coral Sea—it was that such battles had the potential to be extremely lethal. (Five aircraft carriers participated in the Battle of Coral Sea. Two were sunk, two were badly damaged, and the fifth had its air wing devastated.[40]) Therefore, Nimitz gave his subordinates guidance on how to make the trade-off between those two conflicting objectives.

The written Operation Plan stated that the objective of the American forces was to hold Midway. However, it also directed the carriers to position themselves well off the island. It did not say to interpose themselves between Midway and the Japanese force. Verbally, Nimitz told Rear Adm. Raymond Spruance that the objective was not to hold Midway at all costs, because Midway was not worth losing the carriers.[41]

In a separate Letter of Instructions that was hand-delivered to the task force commanders, Nimitz told them:

> In carrying out the task assigned in Operation Plan 29–42 you will be governed by the principle of calculated risk, which you shall interpret to mean the avoidance of exposure of your force to attack by superior enemy forces

without good prospect of inflicting, as a result of such exposure, greater damage to the enemy. This applies to a landing phase as well as during preliminary air attacks.[42]

Thus he gave his admirals general guidance for resolving potentially contradictory objectives, but without hamstringing them with unnecessary specifics that might sap their initiative and judgment.

The third objective—defending Midway Island—was obviously paramount in the minds of the U.S. Marines defending the island. It could easily have become an issue with the American carrier forces and resulted in compromising their ability to achieve the first two objectives. With the Marines located on Midway Island and the Navy at sea, one would expect the different services to have different views on the primary objective. Nimitz's written operation plan said the objective was to hold Midway, but other directions made that task difficult. Thus, the written orders seemed to reflect muddled thinking. However, what he told his subordinates verbally demoted the objective of defending Midway Island to a subordinate position with respect to the first two objectives. In addition, although some accounts of the Battle of Midway omit the last sentence of Nimitz's Letter of Instructions, it is vitally important. Even if the Japanese landed on Midway Island, Nimitz still wanted his fleet commanders to apply the principle of calculated risk.

The American carriers positioned themselves well off Midway Island, in good position to attack the Japanese carriers but not in good position to defend Midway Island against the initial Japanese attacks on Midway. When the Americans received reports of Japanese aircraft heading toward Midway, the carriers did nothing to defend the island. Furthermore, they never considered waiting to attack the amphibious force instead of the carriers covering the operation.

The Marines on the island might have held back their woefully inadequate force of attack aircraft until the landing force appeared, but they did not do so. As soon as the Japanese fleet had been located, they sent their mostly obsolete aircraft out in a heroic but futile attempt to attack the Japanese carriers. They did so in accordance with Nimitz's instructions.[43] This resulted in occupying the Japanese fighter defenses and forcing the carriers to maneuver, thus delaying the recovery and rearming of the aircraft that had attacked Midway Island. Consequently, the Japanese carriers were in an extremely vulnerable condition (with fueled aircraft aboard and bombs strewn around) when the U.S. dive-bombers attacked them.

Although the written operation plan by itself stated that defending Midway Island was the primary objective of the operation and that the carriers were not to be risked, Nimitz's verbal instructions and separate Letter of Instructions conveyed that defending Midway was the least important of the three objectives. He also gave guidance to his carrier task force admirals on the conditions under which they should risk their carriers; namely, if they thought they could inflict more damage than they received.[44] During the battle the American carrier task force commanders took advantage of a fleeting opportunity to send all available attack aircraft against the Japanese carriers when they were most vulnerable in the midst of refueling and rearming aircraft after their initial strike on Midway Island. As a result, the Americans, with great skill and courage as well as a lot of luck, sank four Japanese aircraft carriers while the Japanese sank only one American carrier. This victory was the turning point in the Pacific War.

CHOOSING GOOD OBJECTIVES REQUIRES KNOWING YOUR enemy, as well as knowing your own forces. (Later chapters in this book consider these topics as candidate principles of war.) Having a single objective is the ideal toward which one should strive. However, multiple objectives are frequently necessary, if only in the form of constraints on operations. Although the military would prefer to be given free rein to do their job without constraints and political interference, this is not in keeping with Clausewitz's "war is a continuation of politics by other means."

A commander with multiple objectives must take care not to allow secondary objectives to drive the problem. Likewise, commanders should avoid assigning multiple objectives to a subordinate if possible. If a commander does assign multiple objectives to a subordinate, however, the commander should give guidance on how to prioritize those objectives. A commander should ensure that the subordinate understands how the assigned objective interacts with objectives assigned to the commander's other subordinates, as well as how it fits into the big picture. In this manner, when subordinates encounter unexpected conditions, they can exercise their on-the-scene judgment in a manner consistent with the commander's intent. The next chapter addresses one of the primary methods for gaining an objective.

SIX

Mass vs. Relative Advantage

To apply one's strength where the opponent is strong weakens oneself
disproportionately to the effect attained. To strike with
strong effect, one must strike at weakness.

—BASIL H. LIDDELL HART, *Strategy*

....

Current Principle of Mass or Concentration

ALL MILITARIES THAT ENUMERATE PRINCIPLES OF WAR LIST
concentration or mass as one of the principles. Most militaries use the
term concentration of force or concentration of effort, but the United
States uses the term mass:

> The purpose of mass is to concentrate the effects of combat power at the
> most advantageous place and time to produce decisive results.
>
> In order to achieve mass, appropriate joint force capabilities are integrated
> and synchronized where they will have a decisive effect in a short period
> of time. Mass must often be sustained to have the desired effect. Mass-
> ing effects of combat power, rather than concentrating forces, can enable
> even numerically inferior forces to produce decisive results and minimize
> human losses and waste of resources.[1]

The first paragraph of the official description of mass is just a definition
of the concept, not the purpose. The second paragraph of this description
suggests the decisive effects should occur in a short period—concentration
in time as well as place—yet says one must often sustain the effort. That
sounds like a committee wrote it. The description goes on to say that it is
"effects" rather than "forces" that should be concentrated, thus accounting for long-range weapons, though the point is muddled by discussing
numerically inferior forces in the same sentence.

The British define the principle of concentration of force as follows:

> Concentration of force does not necessarily require the physical massing of forces, but needs balance to deliver sufficient fighting power at critical points and times. Success depends upon subtle and constant changes of emphasis in time and space to realise effects. Commanders must accept that concentration of force on the main effort may mean economy elsewhere.[2]

This definition tells us what the principle does not require, that concentration needs to be balanced in some unspecified manner, and that the emphasis must be constantly changed in some way. This explanation does not explain very much.

Problems with Current Principle of Mass

Mass (or concentration) is the most misleading of the currently recognized principles of war. There are many problems with this seemingly obvious principle. First, the durability of the principle is questionable because it has been necessary to restate and reinterpret the principle over the years. Second, until recently, the principle referred to "*the* critical place and time," which assumed a fixed situation that was to be passively accepted. Third, the prevailing principle of mass emphasizes the wrong factor when it advises commanders to bring maximum combat power to bear at a critical point and time. Let us now examine each of these problems.

Mass Reinterpreted over the Years

Initially, the principle of mass advised commanders to keep their forces together at all times. That way, an enemy could not attack one part of your force, and you would always be prepared to attack with your entire force on short notice.

As armies grew larger, logistics and transportation problems prevented keeping the entire army together at all times. If an army gathered much of its sustenance from the land, it needed to draw on a large area, much larger than it could access if marching along a single road or staying in one camp. If an entire army of substantial size marched along a single road, it spread out over many miles, thus negating the attempt to keep all parts of the army within supporting distance at all times. Therefore a revision to the principle of mass advised commanders to bring forces together at the critical point and time. Napoleon was a master at marching parts of his army separately but bringing them together on the battlefield. Of course, if the enemy caught you dispersed, that became the critical point and time.

As combat force came to mean not just numbers of men but also accounted for longer range weapons, another update to the principle of mass recommended commanders concentrate *combat power* at the critical point and time. Yet another revision to the principle of mass advised commanders to concentrate *combat effects* at the critical point and time, with the definition of combat effects expanded to include anything that might contribute to achieving the objective (every kind of weapon; electronic, information, and psychological warfare; and other things such as humanitarian aid).[3]

Decisive Time and Place Not Fixed

Until recently, the American principle of mass described concentrating the effects of combat power at "*the* decisive place and time."[4] Assuming the critical place and time are fixed puts the cart before the horse. On occasion, especially if the enemy has the initiative, the enemy may dictate the decisive place and time. If you possess the initiative, however, you can influence the place and time that are decisive. The decision to commit forces in a major way may well make that place and time decisive. This is one more reason to seize the initiative. Thus, Robert E. Lee's decision to commit the Army of Northern Virginia to battle at Gettysburg made that the decisive place and time, both of the Gettysburg campaign and of the Civil War. Had Lee retreated to fight another day under more favorable circumstances, however, the critical time and place would have been somewhere else, and the outcome of the campaign might have been quite different.[5]

The latest American formulation of the principle does better in this regard by advocating the application of power at the *most advantageous* place and time. Commanders have several ways to influence which place and time are decisive. Often, the key decision a commander has to make is when to attack. However, the American and British principles give no guidance for how to select the critical or most advantageous place and time. If a commander waits until all his or her forces arrive on the scene, the enemy has time to prepare defenses. Therefore, it is often better to attack sooner to catch your enemy unprepared and thus achieve a more favorable force ratio. Of course, commanders must estimate whether they or the enemy will improve their position more if they wait.

When the Persians landed at Marathon in 490 BCE, the Athenians and a small force from Plataea blocked the route to Athens. They might have attacked the Persians immediately but, being badly outnumbered, they waited for the Spartans to join them after the Spartan religious observance ended. When the Athenians saw the Persians embarking on their ships to sail to Athens, however, they decided to attack. Attacking while the Persians were embarking gave the Greeks better odds because they faced only the Persian covering force, not the entire army. The result was the great victory at the Battle of Marathon.[6] Timing was extremely important.

Mass Emphasizes the Wrong Factor

The currently recognized principle of mass emphasizes the wrong factors. There is validity to the idea that when you reach a critical place and time in a battle, you should apply as much force as possible. However, the principle of mass errs in considering only one's own forces, while ignoring the enemy's capabilities. It is not the absolute level of combat power that decides the outcome but rather the relative advantage in combat power over that of the enemy at the point of contact. It is sometimes better to attack with part of your force against part of the enemy force than to attack all of the enemy force with all of your force. Generally, it is better to attack weak points, rather than strong points.

Suppose you are leader of a tribe of twenty cavemen engaged in a dispute with another tribe of thirty cavemen. Assume that all cavemen are equally good at combat. You are leading a party of fifteen men and come upon ten men from the enemy tribe. Would you prefer to (a) attack immediately with a 3:2 advantage or (b) wait for your other five cavemen to join you even though that means all thirty cavemen from the opposing tribe will have gathered, thus putting you at a 2:3 disadvantage? Of course, given the stated circumstances and everything else being equal, you attack immediately while you have an advantage. You do so even though you have not concentrated your entire force.

The logic of that example is equally applicable if you are in command of a fleet of twenty starships in a war with aliens who possess a fleet of thirty equally capable starships.

These simple hypothetical examples illustrate that the key element is not the size of your own combat force but the ratio of your effective combat force to the enemy's effective combat force. You would prefer to fight when the odds are most favorable. In other words, when entering battle, you would like the relative advantage to be as great as possible. This is the essence of the idea behind the old principle of mass. However, it is not what the principle of mass actually says.

New Principle of Relative Advantage

Sun-Tzu advised attacking the enemy's weakness, rather than his strength.[7] Jomini repeatedly stated that the primary "fundamental truth" of warfare was to attack a portion of the enemy's force with the bulk of one's own force. Furthermore, one should try to attack at a place and time where the consequences of defeat are worst for the enemy and least dangerous for oneself.[8] These ideas are missing from the current principle of mass or concentration.

We propose a new principle of relative advantage that captures the essence of the original idea behind the principle of mass:

> Increasing the relative advantage in effective combat power that is engaged over that of the enemy at a critical point and time improves the chances of victory. It is not the absolute level of combat power that determines the outcome but rather the relative advantage in effective combat power over that of the enemy.
>
> Relative advantage consists of concentrating the bulk of one's own effective combat power against a portion of the enemy's power, or attacking in an unexpected location where the enemy's defenses are not so strong, or attacking in a manner that is unfavorable for the enemy, or attacking at a time when the enemy is not as well prepared.
>
> Effective combat power takes account of everything that affects the outcome of a battle or campaign. It includes not just the numbers and effectiveness of people and weapons but also the effects of surprise, deception, initiative, environmental factors, morale, and so forth.
>
> Therefore, commanders should concentrate all means to achieve the maximum relative advantage in effective combat power at a critical time and place that will enable them to achieve their objective and exploit success. If possible, they should choose the time, place, and manner of attack

for which defeat of the enemy has the direst consequences for the enemy, but their own defeat would have minimal bad consequences.

To provide the force to achieve a large relative advantage at a critical point and time, commanders must use economy of force in non-vital areas, taking prudent risks in doing so. They should allocate sufficient force in such areas to prevent the enemy from achieving a large relative advantage in a way that might have dire consequences.

WE CHOSE TO NAME THIS PRINCIPLE "RELATIVE ADVANTAGE," but "combat dominance" or "combat ratio" might also be used. "Relative advantage in the ratio of effective combat power" would be more precise, but rather cumbersome.

The commonly accepted principle of mass emphasizes the wrong factor. It is not the absolute level of combat power that is decisive, but rather your relative advantage in combat power over that of your enemy. It is better to have a combat power ratio of four to one than to have a ratio of ten to nine.

That the combat ratio is a mathematical term does not mean we expect commanders to quantify the amount of effective combat force; however, commanders must have a feeling for the force ratio going into a battle. That is, commanders should know whether they are significantly inferior to their enemy, about equal, or superior. If inferior or superior, they should have an idea of whether the difference is large or small and whether the balance is shifting toward or away from them.

We want to emphasize that the principle of relative advantage does not mean that one should never concentrate all of one's forces. Just make sure that the time needed for concentration does not allow the enemy to take action to degrade your relative advantage.

Insofar as they can, commanders should choose a time, place, and manner of attack that maximizes their relative advantage. It may not always be possible to obtain a situation in which you have a relative advantage in effective combat ratio. You may have to settle for getting the smallest disadvantage. If you have to fight, then it is better to do so at a small disadvantage than at a large disadvantage.

Commanders might not be certain they would obtain a better relative advantage in the future, but often they should be able to judge the situation accurately. Thus, Lee should have known that he could do better than to attack a numerically superior Union Army in a good defensive position at Gettysburg. On the third day of the battle, Lee should have known that a frontal attack against a prepared enemy in a good defensive position was not his best option. Longstreet recognized the lack of relative advantage in both cases and strongly advised Lee to take different actions.

The commander must also keep in mind, however, that war is not just about winning battles, it is about reaching objectives. Winning meaningless battles or gaining unimportant territory does not do much good. Thus, a commander cannot always choose to fight only when and where odds are most favorable. His or her choices are subject to the restraint of reaching his or her objective. Nevertheless, a commander must seek to reach the objective using the approach that gives him or her the best relative odds. The idea of fighting when relative combat advantage is most in your favor applies to those situations that will achieve decisive results in a manner that makes it the critical point and time. The critical point and time is not, however, static, but depends on the situation.

Commanders might select the *manner* in which to attack so that it gives them a relative advantage. At a time when infantry and cavalry were the two primary arms of combat, if one side had more of an advantage in cavalry, then they preferred to do battle out in the open where the cavalry could have the maximum impact. Conversely, if one side were at a significant disadvantage in cavalry, then it would be better for them to fight on broken ground or in a wooded area where the enemy could not effectively employ its cavalry.

THE CURRENT PRINCIPLE OF MASS GIVES TERRIBLE GUIDANCE for guerrillas. Because the enemy is likely better-trained, better-equipped, and more numerous, the guerrillas should avoid engaging them in a major battle. If the guerrillas concentrate a large portion of their forces, they likely are playing into the enemy's hand.

On the other hand, the new principle of relative advantage provides excellent advice for guerrillas to seek situations in which they have an advantage. Guerrillas should avoid fighting on the enemy's preferred terms. They must avoid the enemy's strengths and only fight when they have a significant relative advantage. In any "fair fight" the counterinsurgency forces will have a great relative advantage. Guerrillas try to avoid such "fair fights," just as the Americans, after Lexington and Concord in the American Revolution, shot at the British redcoats from behind trees rather facing them in an open field.

One standard guerrilla tactic is to attack an enemy outpost, then ambush the relief force under favorable conditions. The Chinese People's Liberation Army, and later the Viet Cong, used this tactic repeatedly.[9] Once again this is a way of seeking a relative advantage.

Mao Tse-tung summarized the guerrilla's mode of operation as follows:

> When guerrillas engage a stronger enemy, they withdraw when he advances; harass him when he stops; strike him when he is weary; pursue him when he withdraws.[10]

We encapsulate Mao's picturesque guidance as telling guerrillas to fight only when they have a relative advantage.

The converse of the principle of relative advantage suggests that commanders avoid having part of their force attacked by superior enemy forces or being attacked when they are not as well prepared. Similarly, commanders should avoid battle if their relative advantage is not as good as they can hope to obtain, or if they might be attacked in a manner that gives the enemy a relative advantage.

Liddell Hart's theory of using indirect attacks is a subset of the concept of relative advantage. Rather than attacking at the obvious point, where the enemy's defenses are usually strongest, attack at an unexpected point where the enemy's defenses are much weaker. Better yet, attack in a psychologically unexpected manner. In this way, you often obtain a much better effective combat ratio. However, the concept of relative advantage is more broadly applicable than Liddell Hart's indirect attack.

We will now clarify some aspects of the principle of relative advantage with historical examples from ground, naval, and air warfare. These examples illustrate how the principle of relative advantage provides guidance for choosing the best time to attack, the best location to attack, and the type of attack to use.

Timing of an Attack

Military history provides many examples in which commanders attacked before all their forces were in place and won as a result. History also gives us numerous examples in which an attack failed because it was premature. We can also find instances in which commanders delayed initiating action until they had all their forces gathered and prepared, but then lost because their enemy had gained more from delay than they had. On the other hand, commanders have delayed action until all their forces were fully ready and then attacked successfully. In the following sections, we present a sampling of these examples throughout history and examine whether the current principle of mass gives useful guidance. We also consider whether the concept of seeking relative advantage is better.

ATTACKING IMMEDIATELY SUCCEEDS

In 334 BCE Alexander the Great and his army crossed the Hellespont and arrived at the steeply banked Granicus River. On the other side, part of the more numerous Persian army faced them. The Persian cavalry guarded the riverbank, but the Persian infantry were some distance behind them on higher ground. Alexander saw this and attacked immediately with his cavalry and infantry. It was not just that his relative disadvantage in numbers would get worse if the Persian infantry came up to defend the riverbank. Alexander's relative advantage was much greater at this moment because the Persian infantry were a key part of their defensive capabilities. In their absence, Alexander's combined arms force crossed the Granicus and routed the unsupported Persian cavalry, the Macedonian lance proving superior to the lighter Persian spear. After scattering the Persian cavalry, the Macedonian cavalry soon caught up with the Persian infantry, who were then unsupported by their own cavalry. The Macedonian cavalry held them in place until the Macedonian infantry came up. Surrounded, the Persian infantry was defeated.[11]

In 56 BCE one of Caesar's generals, Crassus, led a small force into Aquitania. The local tribes prepared their defenses and began gathering a large force. "Finding that . . . the enemy's numbers were increasing daily, [Crassus] thought he had better lose no time in bringing them to a decisive action," and defeated them.[12] The force ratio was moving in a direction unfavorable to the Romans, therefore Crassus attacked immediately in accordance with the principle of relative advantage.

Early in World War II both American and Japanese naval doctrine strongly emphasized mass at the tactical level. When the Americans located the Japanese fleet at Midway, Rear Adm. Raymond Spruance faced a critical decision. Should he wait for the *Yorktown* to recover her aircraft and move forward to join Spruance's two aircraft carriers in a mass attack from all three American carriers? Alternatively, should he attack immediately with the aircraft from the two carriers that he had available? Despite the principle of mass, Spruance launched an immediate attack with aircraft from only two of the three American carriers because he hoped to achieve an especially favorable relative advantage. The first reason was the highly perishable nature of their offensive assets. Planes destroyed on the decks of Japanese carriers could not subsequently attack the American carriers. In carrier warfare early in World War II, due to relatively weak defenses and the inability of carriers to sustain serious battle damage, there was an extremely high premium to making the first effective attack.[13] The second reason an immediate attack gave the Americans a relative advantage was their attack would arrive while the Japanese were in an extremely vulnerable state refueling and rearming aircraft. The quest for relative advantage overrode the flawed principle of mass.

ATTACKING PREMATURELY FAILS

At the start of 378 CE Valens, emperor of the eastern Roman Empire, asked Gratian, the western emperor, for help in subduing the Goths in Thrace. Gratian agreed to bring his army there personally. When the Alamanni started making trouble, however, Gratian felt he had to subdue them first before assisting Valens. After waiting two months for Gratian's army to arrive, Valens instead received a letter enumerating Gratian's victories over the Alamanni. About the time Gratian wrote to Valens that he was on his way to Thrace, Valens learned a Gothic force was near Hadrianople. Valens was jealous of the praise Gratian had received for his victories and did not want to share with Gratian the glory of his upcoming victory. Therefore, Valens attacked the Goths before Gratian arrived on the scene. When the enemy force turned out to be the entire Goth army, it destroyed Valens's army and killed him. Hadrianople was the worst defeat for the Romans since Cannae in 216 BCE.[14]

During the Hundred Years' War the French fought the English invaders, led by King Edward III, on 26 August 1346 in the Battle of Crecy. The French had unity of command under King Philip VI and had a three-to-one advantage in numbers. Coming upon the English position late in the day, Philip ordered a halt. Because his army was spread out along a long line of march and would continue to arrive throughout the day, Philip ordered the attack to occur on the next morning. The French knights, however, reputed to be the finest cavalry in Europe, were eager for glory and uncontrollable. The undisciplined and disorganized army pushed forward. Seeing that he was losing control, Philip ordered his crossbowmen to attack, but they were repulsed. The impatient knights charged through the retreating crossbowmen, ignoring the fact that the English had chosen the battlefield and prepared it by digging ditches to protect their flanks and "horse trap" holes in the fields in front of their position. The knights' charge through the crossbowmen left both groups in confusion while the English arrows rained down upon them. Each part of the French column rushed into battle as it arrived on the scene. The English repulsed fifteen attack waves and destroyed the French army while suffering few losses themselves. By their piecemeal, premature attack, the French threw away their three-to-one advantage in men. Instead each of the fifteen attack waves was probably outnumbered by the English fighting on a battlefield prepared for defense.[15]

The Battle of Crecy is a lesson in the value of relative advantage: If they had waited until the next day to attack, they would have enjoyed a much greater relative advantage. The Battle of Crecy also illustrates the limits of unity of command. By losing control of the situation King Philip showed that unity of command does not necessarily result in unity of effort. At the very least, we must conclude that unity of command under an ineffective commander is not the route to victory.

DELAYING ATTACK FAILS

At the Battle of Midway, in contrast to the Americans, when the Japanese discovered American carriers nearby, rather than launching a partial attack unescorted by fighters as soon as possible, they followed the principle of mass and waited until they could launch a strong attack. Disaster ensued. Experts still debate whether a partial attack would have been better.[16]

In 415 BCE the Athenian expedition arrived off Sicily. Unexpectedly, the allies they had counted on would not provide any help. With leadership of the expedition divided three ways, there were three plans for how to proceed. The one of interest here came from Lamachus. He proposed immediately attacking Syracuse, the largest city in Sicily and an ally of Sparta, before Syracuse could prepare its defenses and while it was panic-stricken at the appearance of the Athenian fleet. Although Lamachus's plan had problems, it might have worked. As it turned out, although the Syracusans dithered, they eventually prepared their defenses. After trying other things, the Athenian expedition finally attacked Syracuse. They landed unopposed and defeated the Syracusans in battle, but did not capture the city. Then they left the area to resupply. Now thoroughly aroused, Syracuse took extreme measures to improve their defenses: they gave arms to poorer men, reduced the number of generals from fifteen to three, and curtailed their democracy to allow more effective leadership and secrecy. They also appealed for help from others. When the Athenians finally returned and attacked Syracuse again, they were not successful against the improved defenses and help from outsiders. In the end, the entire expedition was killed or captured. Athens never fully recovered from this disaster.[17]

In April 1862 Maj. Gen. George McClellan achieved operational surprise by moving a large force to the peninsula east of Richmond. When he approached the Confederate forces at Yorktown, even though his forces were much stronger, McClellan hesitated. Fooled by a theatrical deception, he decided not to attack immediately when he had a great advantage. Instead, he waited a month to bring up his siege train. By that time Gen. Joe Johnston had shifted most of his Confederate army to the peninsula and McClellan faced defenses that were far more formidable.[18]

A few months later, on 15 September 1862, with the aid of Lee's famous lost order (used to wrap cigars and left behind) that revealed the dispersion of his forces to the Union, McClellan faced Lee with a better than two-to-one numerical advantage. Again misled by exaggerated estimates of Lee's strength, McClellan dithered for two days while Lee gathered most of his forces. When the Union forces finally attacked, they still had a substantial advantage, but it was no longer overwhelming. Late in the day, just as Union forces were about to overrun Lee's right wing, A. P. Hill's division arrived from Harper's Ferry after a forced march and beat back the Union attack at nearly the very last minute.[19]

When you have an advantage there is often a temptation to make sure all your forces are in place, prepared, and organized so that your attack is not premature. Sometimes this is the correct decision; sometimes it is not. The current principle of mass advises always waiting until you concentrate all your forces. However, many times this is not the correct course of action. To make the correct decision you must judge whether your enemy is likely to improve its position faster than you can improve yours. That is, is your relative advantage increasing or decreasing?

DELAYING ATTACK SUCCEEDS

During World War II, fighting in the North African desert shifted back and forth, as each side successively gained the upper hand and pushed the other side back, but then had their offensive stall because of overextended supply lines. In 1942 Field Marshal Erwin Rommel drove the British back into Egypt. In July of that year, however, his attack stalled at El Alamein, fifty miles short of Alexandria.

Lt. Gen. Bernard Montgomery took over the British Eighth Army in August and methodically made his preparations. He integrated massive reinforcements into his army and trained them specifically for the battle to come. He also gathered tanks and artillery and stockpiled supplies. Meanwhile, the Germans were not sitting idle. They laid barbed wire and half a million mines in front of their positions and generally erected formidable defenses. Finally, on 23 October 1942, Montgomery opened the attack. It was slow going against the German defenses and took until 5 November to achieve a breakthrough. Once the British drove the Germans out of their prepared defensive positions and had them on the run, things went faster. Even with the methodical Montgomery in command, the British captured Tripoli in January 1943.[20]

Was Montgomery correct is delaying his attack until late October? Probably. First, while Montgomery increased men, equipment, and supplies, the Germans received little of the same. German-held ports in North Africa had limited capacity, and the main port of Tripoli was thirteen hundred miles behind the front line. For much of the distance, there was just a single road for moving supplies. As a result, by the time of the battle Rommel was down to one-third of his desired ammunition supply and just one-tenth of the fuel he wanted.[21] Montgomery had four times as many troops, three times as many tanks, and four times as many aircraft as Rommel.[22]

Thus, despite the German mines and barbed wire, the force ratio shifted toward the British. This was mitigated by the strengthened German defenses but not enough to compensate for the increased combat power of the British. By building up his supplies and gathering additional combat power,

Montgomery was able to exploit his success at El Alamein. Unlike previous offenses, this one did not peter out halfway across the desert.

We conclude that following the principle of mass worked in this case, but Montgomery's decision was also consistent with the principle of relative advantage. The improved relative advantage at the time battle began was the most important factor, not simply the amount of force that Montgomery had gathered. What made his delay work was the fact that the Germans were not able to improve their position comparably. Another important lesson from this example is to emphasize that in considering whether you have a relative advantage, you must consider your ability to exploit any success you achieve.

In World War II, Britain and the Soviet Union became allies after Germany invaded the Soviet Union in June 1941. Stalin immediately started urging Churchill to open a second front in the west to relieve pressure on the Soviet Union. His demands became more and more strident. When the Americans entered the war in December 1941, they were inclined to open a second front more quickly than Britain, which had been chastened by its recent experience fighting the German army. The Americans wanted a second front in 1942 but settled for landings in North Africa. They also urged a second front in 1943 but agreed instead to use the landing craft to invade Sicily. By 1944 Britain and the United States had built up enough force to invade France. The delay also gave the Germans time to improve the defenses in France. An invasion in 1942 almost certainly would have failed, as evidenced by the disastrous Dieppe raid on 19 August 1942. The outcome of an invasion in 1943 was less clear, but probably would have been at best a limited success that the Allies could not have exploited. Although the delay gave the Germans a chance to improve their defenses, the Americans and British built up their forces more than the Germans did, and the Americans gained invaluable experience during the delay until 1944. Thus, the relative advantage shifted significantly toward the Allies between 1942 and 1944.

GUIDANCE FOR TIMING OF ATTACK

In deciding when to attack, you cannot focus solely on your own forces, nor can you consider only the state of the enemy. You must consider both sides and judge whether your relative advantage is improving or wasting away. This might mean attacking quickly if the relative advantage is shifting toward the enemy, or it might mean delaying the attack if the relative advantage is becoming more favorable. Important factors include judging which side will receive more reinforcements and supplies, how much the enemy will improve its defenses, how much delay will improve your offensive potential, and whether you will be able to exploit any success.

In August 1941 the Japanese Navy's general staff estimated that if Japan had a naval force ratio with the United States no worse than five to ten, this would give them a chance for victory provided they used surprise. They estimated that this force ratio would improve to seven to ten at the end of 1941, but then decline to six and a half to ten in 1942, five to ten in 1943 and three to ten in 1944. In part because of their increasing relative disadvantage after the end of 1941, Japan decided to go to war in 1941.[23]

Attack at a Favorable Location

A "slog it out" attack with types of forces similar to those of the enemy is unlikely to yield decisive results, even with a numerical advantage. World War I trench warfare in France illustrates this point. You must seek some way to gain a relative advantage. One way is to pick the location that gives you the best chance.

ATTACK WEAKNESS

It is often better to attack with part of your force in a weakly guarded area or time than to attack with all of your force against the enemy's full strength. If your enemy is defending a front, he has to decide how to allocate his forces along that line. Ideally he would allocate his forces along the front so that, when combined with the natural defensive properties of the terrain, the defense is equally strong everywhere. However, the enemy commander will probably consider where he thinks you are most likely to attack and reinforce that sector at the expense of weakening other sectors. The solution is obvious: attack where he does not expect you to attack.

For example, in planning for the Battle of France in 1940, the French *correctly* believed that the Ardennes Forest provided natural obstacles to a German assault there. Therefore they made the defense in that area weaker so they could strengthen their forces elsewhere. However, they overdid it. Had they provided a modest amount of additional defense in the Ardennes sector, those forces could have used the terrain and limited road system to delay the German attack. This would have allowed the French to move other forces to that sector before the Germans got to the Meuse River.

After the Battle of Guadalcanal ended with the evacuation of the Japanese on 7 February 1943, the Americans had to decide how to fight their way across the central Pacific. They might have identified the heavily defended Japanese bases on Rabaul and Truk as the critical places at which the Americans should mass their forces. That would have been correct if the objective were to kill as many Japanese troops as possible. Fortunately, the American strategists realized that the operational objective was to capture islands that they could use as naval and air bases from which they could launch attacks against the next target and eventually the Japanese mainland. Because there were a larger number of suitable islands than the Japanese could occupy and defend, for the most part, the Americans could pick weakly defended islands to capture. As for the Japanese bases, the Americans just needed to neutralize their ability to harass the American supply lines. Air power could do that. There was no need to destroy Japanese ground forces, because with American sea control, the Japanese could not go anywhere. Thus was born the leapfrog (island-skipping) campaign, instead of a methodical island-by-island campaign. The lesson is that by keeping the objective in view, the Americans could choose the critical places to attack. Thus, they avoided following the flawed advice of the principle of mass and instead fought the campaign in accordance with the new principle of relative advantage.

At a tactical level in ground warfare, if the enemy leaves their flank exposed, this is a weakness. Attacking them there gives good chances for success because their defenses are not oriented to receive an attack from the side and the attacker can achieve a local superiority in force that can continue for some time. Lee and Jackson's flank attack at Chancellorsville is a prime example, but similar attacks have played a role in many battles.

In air-to-air gunnery dogfights during World War II pilots tried to get on their enemy's tail. That way the enemy could not shoot back, while they could fire at the enemy and have a good chance of continuing in a favorable position where they had the initiative. This was a great relative advantage.

Early in World War II U.S. Navy pilot Jimmy Thach devised the Thach Weave to counter the superior Japanese Zero fighters. This tactic prevented an enemy plane from getting an unopposed position on someone's tail. Two two-plane sections would fly parallel courses separated by at least their turning diameter. Each section could scan the other section's blind spot behind them. If a Japanese fighter got on one section's tail, then the two sections would turn toward each other, giving the second section a shot at the Japanese fighter on the tail of the first section. This tactic negated much of the relative performance advantage enjoyed by the Japanese Zeros over the American fighters. The Thach Weave enabled fighter aircraft to defend each other's weak spot. This illustrates that the principle of relative advantage can be applicable in defense as well as offense.[24]

If two opponents are roughly equal but have different capabilities, each force will have an advantage under certain circumstances. If you can arrange to fight a battle in an environment that favors you, then that improves your chances of prevailing. The more asymmetric the forces are, the more advantage that can potentially be gained by fighting in an environment that favors your forces.

During the Saratoga Campaign of the American Revolutionary War, on 19 September 1777, the British force under Gen. John Burgoyne advanced toward the American position. The American commander, Gen. Horatio Gates, wanted to wait passively inside his fortifications for the British to attack him. Maj. Gen. Benedict Arnold recognized that the British soldiers were best in open fields that allowed them to keep in formation, fire by volleys, and use bayonet charges. On the other hand, the American soldiers were best at frontier-style fighting in wooded areas. Therefore, Arnold persuaded Gates to allow him to advance with some of the American troops and fight the British at Freeman's Farm. The hard-fought battle stopped Burgoyne's attack and stalled his march south.[25]

Insofar as there are asymmetries in the forces, seek to exploit those asymmetries that favor you. For example, currently the United States has unmatched capabilities at night. This ranges from ground troops having night-vision goggles to stealth bombers. Any advantage American forces have during the day is magnified at night. Hence, they often prefer to fight at night because that usually maximizes their relative advantage.

Create Asymmetry

If your forces are similar to those of your enemy, you should seek to gain a relative advantage by creating asymmetries. One way is to create an asymmetric position where you can apply part or all of your force against just part of the enemy force.

In the age of twentieth-century big-gun battleships, fleet commanders sought to "cross the tee" of the enemy. If both fleets steamed in a column, then each commander sought to cross in front of the enemy on a perpendicular course. In that way, every gun on one side of every ship could fire at the leading enemy ship, but only the forward guns of the leading enemy ships could effectively fire back. The Japanese did this to the Russians at the Battle of Tsushima in 1905 and won a great victory. At the Battle of Jutland in 1916 the British crossed the tee of the German fleet twice. At the Battle of Leyte Gulf in October 1944, American battleships, several salvaged from the mud of Pearl Harbor, won an anachronistic battle in the age of the aircraft carrier. They crossed the tee of a Japanese fleet emerging from Surigao Strait by steaming across the mouth of the strait as the Japanese emerged from it.

In the age of sail Rear Adm. Horatio Nelson, commanding thirteen British ships of the line, attacked thirteen more heavily armed French ships of the line at Aboukir Bay on the night of 1–2 August 1798, in the Battle of the Nile. Because the French ships were at anchor, Nelson's ships could pass on both sides of the anchored French line. In several cases two British ships attacked a single French ship. Because of the direction of the wind, the unengaged French ships would not be able to come to their aid. For those French ships the British attacked, it was as though the British had a two-to-one advantage in ships. Compounding Nelson's advantage, part of the crews of the French ships were ashore. By attacking as soon as he sighted the French force, Nelson did not give them time to get their crews to their ships. This increased his relative advantage.[26]

At the Battle of Trafalgar on 21 October 1805, Nelson repeated the feat at sea. To "break" the enemy line, he endured the temporary disadvantage of sailing his twenty-seven ships toward the French-Spanish line of thirty-three ships. As he closed the enemy, Nelson in effect did what in the twentieth century would be called "crossing his own tee." This succeeded because at that time weaponry was shorter ranged and not sufficiently lethal to inflict fatal damage during his approach to the enemy's battle line. The punishment he endured in this phase of the battle paid off because when his ships broke through the enemy line, they then had a chance to rake the bow of the enemy ship on one side and the stern of the ship on the other side. The advantage was that the fire was unopposed and the bows and sterns of the ships of the line were more vulnerable than the sides. After half his ships passed through the enemy line, they then ganged up on some of the enemy ships, again typically having two British ships engaging each French ship. Meanwhile, the unengaged ships in the van of the French-Spanish line did not turn around and join the fight in time to affect the outcome.[27]

A common weapon asymmetry is greater range. The advantage is obvious: you can hurt your opponent while he cannot harm you. The trick is to have tactics that allow you to exploit the advantage. You must be able to maintain the situation long enough to inflict serious damage.

In 53 BCE Marcus Crassus, a member of the First Triumvirate and the richest man in Rome, led a Roman army in the area now called Turkey. A former friend now in the pay of the Parthians persuaded Crassus to lead his army into a waterless desert, where he encountered the Parthians near Carrhae. His subordinates advised resting before seeking battle, but Crassus, eager to gain glory, decided to march on.

The Parthian army comprised nine thousand light cavalry and one thousand heavy cavalry. The light cavalry were horse archers, whose tactic was to gallop toward the enemy and shoot arrows when they got within one hundred yards, then close to fifty yards and parallel the enemy while firing more arrows. This kept them out of range of infantry javelins. In addition, they were adept at firing arrows over their shoulder as they rode away, hence the term "Parthian Shot."

Crassus formed his troops, consisting of at least twenty-eight thousand heavy infantry, four thousand light infantry and four thousand cavalry, into a battle square. The Parthian horse archers attacked from a distance. It soon became apparent that if the Romans stayed, they would suffer many casualties from the rain of arrows. However, when they advanced, the Parthians simply moved out of range. Because a train of many camels resupplied the Parthians with arrows, they could continue this tactic for a long time. Needing to break this situation, Crassus ordered his son Publius to leave the infantry square and attack the Parthians with a mixed force of infantry and cavalry. The Parthians induced Publius to chase them a substantial distance. Then the Parthians turned and surrounded the Romans. They annihilated Publius and his force. When the Parthians finally withdrew that night, Crassus abandoned his wounded men and retreated. The Parthians slaughtered the wounded. In the dark, four cohorts of the Romans lost their way. The Parthians cut them down. Two days later, the Parthian commander invited Crassus to a parley and then killed him. Carrhae was the worst Roman loss since Cannae, with twenty thousand killed and ten thousand men captured.[28]

The primary reason for Romans' devastating loss was their poorly balanced combined arms. The longer range of the Parthian weapons, combined with tactics to take full advantage of the situation, was a decisive advantage. Their light cavalry's mobility and tactics took advantage of the terrain favoring horses. This gave them an initiative and a relative advantage that they fully exploited.

In 9 CE Publius Quintilius Varus commanded three Roman legions with which he was to keep the peace in Germania. Arminius, one of Varus's advisors, secretly fomented insurrection to induce Varus to march through difficult terrain that was supposedly friendly. Varus disregarded tips about a plot. Two days into the march, Varus's German allies disappeared. Thick woods, marshes, and gullies confined the legions to the road. Rain made the roads muddy and impeded their progress.

Suddenly, Germans attacked the Roman rearguard from the woods using javelins. Because the Germans that deserted had comprised Varus's light infantry, Varus now had an unbalanced force of heavy-infantry legions. When the Romans got ready to counterattack, the Germans simply disappeared into woods. When the Roman column reached open ground, the Germans continued to harass them with javelins but did not approach within range of the Roman weapons. The column again entered woods and Romans made camp. Arminius made good use of this time to cut down many trees that fell across the road to slow the Roman column to a crawl. Later, Germans on the hillside threw missiles at the Romans trying to get through Doren Pass in a heavy rain. A controlled retreat turned into a rout. Over several days of fighting, the Germans destroyed three legions and killed ten thousand camp followers.

The Germans defeated the Romans because the German light infantry used terrain effectively and utilized their longer ranged weapons to attack from a distance at which the Romans were unable to retaliate. Other factors that contributed to the Roman defeat included deception, surprise, and the initiative.[29]

A modern example is Operation Desert Storm. In the poor visibility of a sandstorm, the American tanks were able to detect and kill the Iraqi tanks before the Iraqi tanks were able to detect the American tanks. That made for a one-sided battle.[30]

At sea, there is also an obvious advantage to having longer-ranged guns, which give you the option of fighting at a range at which your guns are effective but the enemy's guns are not. In the Battle of the River Plate on 13 December 1939, one heavy and two light British cruisers confronted the German pocket battleship *Admiral Graf Spee*. The heavy cruiser *Exeter* had 8-inch guns, the light cruisers, *Achilles* and *Ajax*, mounted 6-inch guns. In contrast, the *Graf Spee*'s 11-inch guns gave her much greater range and killing power. Once Capt. Hans Langsdorff on the *Graf Spee* recognized the nature of his opponents and saw that they were intent on joining battle, he should have turned away. The British ships would have been exposed to the *Graf Spee*'s 11-inch gunfire for a long time before they could get within effective range of their smaller guns. Instead, Langsdorff headed directly for the British ships, quickly closing the range. The lesson is that possession of an advantage does no good if you do not employ tactics that exploit your advantage.[31]

Relative Advantage Applies to Defense

The converse of the principle of relative advantage suggests that a commander avoid getting part of his force attacked by superior enemy forces, getting attacked when he is not as well prepared, being forced into battle if his relative advantage is not as good as he can hope to get, or getting attacked in a manner that gives the enemy a relative advantage.

In the Battle of Thermopylae, the Persians were more numerous and more mobile than the Greeks, but fighting in the narrow pass negated both their numbers and mobility. The Greeks' advantages of better protection and longer lances were more important factors in the pass than in other settings. The advantage of fighting at Thermopylae was more than just a matter of the numbers of men who could participate in active fighting at any one time.[32]

A Caveat

An unwritten premise of the principles of war is that they apply to situations where the two sides are somewhat evenly matched, or at least there is a possibility either side might win. If one side has an overwhelming force advantage, then concentrating forces to achieve the maximum relative advantage might no longer make sense. During the Battle of Leyte Gulf in 1944, Adm. Bill Halsey should have known that a portion of his force of aircraft carriers and battleships would be more than sufficient to deal with the depleted Japanese force. Nevertheless, he took his entire force of carriers and battleships to attack a Japanese force of carriers that was a shadow of its former self. In doing so, Halsey left the San Bernardino Strait unguarded. As Milan Vego points out, it is acceptable to divide one's force when each part of one's force is sufficient to defeat any possible combination of the enemy force with a large margin of safety. It is also acceptable if the two parts of one's divided force stay within a mutually supporting distance of each other.[33]

Means to Achieve and Enhance Relative Advantage

Tools a commander can use to achieve a relative advantage include surprise, deception, maneuver, knowing and using the environment, knowing your enemy, economy of force, threats, and, especially, the initiative. Because examples earlier in this chapter covered most of these factors, here we will summarize just a few points briefly.

The element of surprise helps achieve relative advantage if the enemy is, as a result, unable to bring as much combat power to the situation as he would otherwise be able to do, or if his power is less effective. If surprise has a psychological impact of the enemy, then that can be a major plus. (In chapter 9 we will discuss surprise further and explore whether it deserves designation as a principle of war.)

Deception often leads to surprise, which in turn leads to a relative advantage. Alternatively, deception can lead directly to a relative advantage, even without the benefit of surprise. (Chapter 10 explores the role of deception in more detail.)

Economy of force in secondary sectors helps achieve relative advantage at the critical point and time. Economy of force is a useful tool, but does not rate status as a primary principle because it is just a tool for achieving dominance. How else would one concentrate combat power at the critical point if not by reducing forces elsewhere?

Combined arms warfare is one of the best ways to gain a relative advantage. The essence of combined arms is to use threats from one type of force to improve the relative advantage enjoyed by another type:

In modern warfare, any single system usually is rather easy to overcome; combinations of systems, with each protecting weak points in others and exposing enemy weak points to be exploited by other systems, make for an effective fighting force.[34]

Robert Leonhard considers combined arms doctrine to be one of the few "immutables" in the history of warfare, claiming that the most important effect of a weapon system is not how much it destroys but rather the enemy reaction it causes. When the enemy reacts to one weapon system, it often leaves him vulnerable to another system.[35] We agree with Leonhard about the effectiveness of combined arms warfare but believe combined arms warfare is a means to achieve relative advantage, which is the true "immutable" from military history.

SEVEN

Offensive vs. the Sustained Initiative

> The initiative in a chess position belongs to the player who can make threats
> that cannot be ignored. The opponent is required to meet the threats before
> considering his own threatening moves. The initiative usually arises from an
> advantage in time = better development, or space = better mobility.
> The player with the initiative can often parlay the threats into a
> material advantage or an attack on the opponent's King.
>
> —MARK WEEKS, *Chess "Initiative"*

Current Principles of the Offensive

EXCEPT FOR FRANCE, THE MILITARIES OF ALL WESTERN COUN-
tries that list principles of war include as such the offensive (United States),
offensive action (most countries), offensive spirit (NATO), or initiative
and offensive (Israel). China uses the principle of initiative. Various lists
of Soviet principles included combat aggressiveness, activeness, and defeat
of the enemy to full depth, all of which stress intense high-tempo offense
and gaining the initiative. The Soviet principle of tactics closest to the
principle of the offensive is that of aggressiveness and resolve. This principle advises seizing and retaining the initiative using a high level of aggressiveness and resolve to keep constant pressure on the enemy, depriving the
enemy of the option of choosing whether, when, where, and how to attack.[1]

American joint doctrine's explanation of its version of the principle of
the offensive starts by explaining the purpose of the offensive:

> The purpose of an offensive action is to seize, retain, and exploit the
> initiative.
>
> Offensive action is the most effective and decisive way to achieve a clearly
> defined objective. Offensive operations are the means by which a military

force seizes and holds the initiative while maintaining freedom of action and achieving decisive results. The importance of offensive action is fundamentally true across all levels of warfare.

Commanders adopt the defensive only as a temporary expedient and must seek every opportunity to seize or regain the initiative. An offensive spirit must be inherent in the conduct of all defensive operations.[2]

The British definition of their equivalent principle of offensive action is similar:

Offensive action delivers the benefits inferred by action rather than reaction and the freedom to force a decision. Offensive action is often decisive. However, its application should not preclude defensive action when required. Underpinned by an offensive spirit, offensive action implies a vigorous, incisive approach to exploiting opportunities and seizing the initiative.[3]

A RELENTLESS OFFENSIVE KEEPS THE ENEMY ON ITS HEELS, always reacting to your attack and unable to implement its own schemes. During the German Blitzkrieg in 1940, the rapid advance of armored forces assisted by close air support never gave the Allies time to organize a counterattack even though the German flanks seemed quite vulnerable at times.

In 1864 Pres. Abraham Lincoln appointed Lt. Gen. Ulysses S. Grant as commander in chief of the Union forces. Grant joined the Army of the Potomac amid talk of how he had achieved success in the western theater only because he had never had to face Robert E. Lee. Grant was not concerned with what Lee would do; he wanted Lee to be concerned about what Grant would be doing. Once Grant went on the offensive in the Battle of the Wilderness, he kept on the attack continuously so that Lee never had a chance for a significant counterattack.

Sometimes, the offensive is effective in raising morale. Bruce Catton tells a marvelous story about the aftermath of the Battle of the Wilderness in 1864. The exhausted veterans of the Army of the Potomac gloomily expected they once again would be retreating and regrouping after a bloody battle. As part of Sedgwick's Corps glumly trudged in the night, they came to a fork in the road at Chancellorsville. The left fork led north to rest and safety; the right fork led south toward hardship, more fighting, and death for many. When the column turned right, the veterans suddenly relaxed and their gloom lifted. Another group of veterans from Warren's Corps also expected to retreat. As they marched along the Brock Road, they eventually realized they were heading south, not retreating. When Grant rode by, the men cheered him wildly because they now knew Grant was a different kind of general, one who would stay on the offensive until the war ended.[4]

BECAUSE "THE PURPOSE OF AN OFFENSIVE ACTION IS TO seize, retain, and exploit the initiative," we need to clarify what we mean by the initiative, a word with multiple meanings. When this book uses the term "the initiative," it does *not* mean the resourcefulness, enterprise, cleverness, and ingenuity that subordinates exhibit when, upon encountering an unexpected situation that requires quick action, they make decisions and take action without the help or advice of higher authority. Instilling such individual initiative accompanied by good judgment is often critical to the success of any military organization, but it is not the concept referred to in this chapter.

In this book, we use possession of the initiative to mean being able to make threats to which the enemy must respond. One way to distinguish the two militarily relevant definitions is to distinguish between an individual showing "initiative" and a force seizing "the initiative." Possession of the initiative means that you can keep the enemy occupied countering threats that the enemy cannot ignore. As a result, the enemy does not have time to pose its own threats, perhaps does not even have time to think about such things. While the enemy attempts to counter the first threat, you devise additional serious threats that the enemy must counter. This favorable situation allows you to control events (as much as this is ever possible in warfare) and precludes many of the enemy's options for action. Having the initiative means that, to a large degree, you get to decide when, where, and how to engage the enemy.

Problems with Current Principle of Offensive

History has often demonstrated the value of being on the offensive, but exceptions exist. The currently recognized principle of war advises one to stay on the offensive as much as possible in order to seize and retain the initiative, but it misses two crucial points. First, the offensive is not always *sufficient* to gain the initiative. Second, the offensive is not always *necessary* to gain the initiative. Other methods for gaining the initiative are sometimes preferable.

In addition, although the currently accepted principle of the offensive does not explicitly encourage constant offensive, people often interpret the principle that way. While the principle of the offensive should not mean, "Attack at all times," as Marshall Fallwell observed, many have incorrectly interpreted the principle of the offensive as a "rule of war" that means one must always be on the offensive. This susceptibility to misinterpretation is one of the inevitable drawbacks of using single words or terse phrases to denote principles.[5]

Furthermore, because offense and defense enjoy alternating ascendancy in warfare as new weapons and methods are introduced, we should expect the value of the offensive to vary with time. The desirability of the offensive was greatest in the minds of the commanders at the start of World War I, yet the introduction of the machine gun and other factors greatly favored the defensive at this time.

The Offensive Is Not Always Sufficient to Gain the Initiative

One reason the offensive is not always good is because offensives often fail. Alternatively, an offensive might succeed but fail to gain, or even lose, the initiative.

FOOLISH OFFENSE FAILS

In many doomed offensives, the men involved behave with great courage—sometimes so great that poets commemorate the event. In the Battle of Balaclava on 25 October 1854 during the Crimean War, because of an ambiguous order, Lt. Gen. Lucan ordered the Light Brigade to charge guns that were a mile away at the end of the "Valley of Death." Russian infantry and artillery lining both sides and the end of the valley exacted a terrible toll on the Light Brigade. After hearing an account of the action, Alfred Lord Tennyson wrote "*The Charge of the Light Brigade.*" His lines made the charge a symbol of warfare at its most courageous and at its most tragic.[6]

The American Civil War provides many examples of foolish offensives. In December 1862, Maj. Gen. Ambrose Burnside ordered the Union army to charge Confederate forces holding a strong defensive position behind a stone wall on Marye's Heights near Fredericksburg, Virginia. The Confederates repulsed fourteen Union charges, with brutal losses.[7] In June 1864, Lt. Gen. Ulysses S. Grant pressed his troops to charge the Confederates in a frontal attack on a strong defensive position well supported by artillery at Cold Harbor. In less than an hour, the Army of the Potomac suffered nearly seven thousand casualties. In both cases, many lives were lost in attacks that had no hope of success.[8] As Liddell Hart has declared, "In the face of the overwhelming evidence of history, no general is justified in launching his troops to a direct attack upon an enemy firmly in position."[9]

The attack from the Civil War that has become most famous for being "magnificent" was Pickett's Charge at Gettysburg. As related in the prologue of this book, on the third day of fighting, Gen. Robert E. Lee sent three divisions toward the Union center in Pickett's Charge. With incredible courage, they marched half a mile or more to the Union lines under heavy fire most of the way. A few men actually reached the Union lines before the defenders drove the Confederates back. Many have described this as the "high water mark of the Confederacy." However, the carnage was great. When the remnants of the attack straggled back to the Confederate lines, Lee rode out to meet them, saying, "All this has been my fault."[10] He was correct.

Lee's entire conduct of the Battle of Gettysburg was uncharacteristically un-Lee-like. The master of the flank attack and mobile warfare just slugged it out with the superior Union forces for three days. He never had a relative advantage. The Army of the Potomac had occupied excellent defensive ground, yet Lee attacked on all three days, suffering casualties the smaller Confederate army could ill afford. Lt. Gen. James Longstreet had advised Lee to fight on the tactical defensive during the Gettysburg campaign, but Lee did not heed his advice.

The current principle calls for using the offensive to seize and hold the initiative. Lee's invasion of the North was strategically offensive; his engagement of the Union Army at Gettysburg was operationally offensive; and his tactics, especially Pickett's Charge, were offensive. Yet at the end of the third day Lee had lost the initiative and never regained any significant initiative for the rest of the Civil War.

Perhaps the most horrific illustrations of the foolish offensive come from World War I. The battles were gory with so little gained that no one calls the battles magnificent. Repeatedly, men charged across No Man's Land toward the enemy trenches in the face of artillery fire, barbed wire, and miles of echeloned, interlocking machine-gun fire. A gain of a few yards constituted "victory." With multiple lines of defense to fall back on, a meaningful breakthrough was unlikely. Verdun was the most foolish offensive in a war of foolish offensives. The battle went on from February to December 1916. First, the Germans attacked and made minor gains. After the German attack petered out, the French attacked to regain the lost ground. During the ten-month battle, the Allies suffered about 542,000 casualties, the Germans 434,000. Just think about the human loss and agony represented in those incredible numbers. At the end of the battle, nothing much had changed in the military situation.[11]

Surely, none of the offensives described here are something we want a principle of war to encourage and hold forth as an ideal way to conduct war. Yet the current principle of the offensive does not say anything about only conducting successful offensives.

In each foolish offensive described here, the attacker briefly gained the initiative in the sense that the attacker forced the enemy to react. However, in none of these cases did the attacker impose his will on the enemy. Any gained initiative was short-lived. At the end of the battle, the attacker had fewer options than before the battle. Thus, these foolish offensives lost the initiative. Part of the calculus in deciding whether to attack is that one must consider the possibility that the offensive might fail, leaving one weakened and more vulnerable than before. A principle of war must in some way warn against such offensives and provide guidance for how to recognize these situations.

OFFENSIVE SUCCEEDS BUT LOSES THE INITIATIVE

In view of the previous section, a defender of the current principle of war might concede that unsuccessful offensives usually are not helpful and suggest revising the current principle to call only for successful offensives. Even that concession, however, is not enough to salvage the principle. Even if an offensive is successful, your force might not be able to exploit your success because success might come at too high a cost, preventing exploitation of victory; alternatively, victory might leave you overextended and thus vulnerable to counterattack.

In 281 BCE the Greek city of Tarentum at the heel of Italy was at war with Rome and facing certain defeat. They brought in Pyrrhus, the king of Epirus and a well-regarded general. Pyrrhus brought with him a substantial, largely mercenary, army that included strong cavalry and twenty war elephants. At the Battle of Heraclea in 280 BCE Pyrrhus's army deployed the Macedonian Phalanx against the Roman Legion—the first time these two forces had met in battle. Each charged and failed to break the line of the other. Finally, when Pyrrhus employed his war elephants, these strange creatures terrorized the Romans and created havoc. The Roman cavalry horses became frightened and ran away, throwing the legions into disorder. Pyrrhus then deployed his cavalry to complete the rout. Of the roughly thirty thousand men on each side, Pyrrhus lost four thousand according to one source, thirteen thousand according to another source, and Roman losses were even heavier.[12]

About a year later, the two armies met again at the Battle of Asculum. Again, the two forces were roughly equal in numbers. Despite Roman anti-elephant weapons that included chariots with long spikes, Pyrrhus defeated the Romans once again, though again with heavy casualties. When someone congratulated him after one of his victories, however, Pyrrhus reportedly said, "One more such victory and I am lost."[13]

Although the Romans lost more men than Pyrrhus in each battle, they could draw on a much larger pool of recruits. Pyrrhus had few sources of additional manpower. Thus, he could ill afford his losses. Indeed, the Romans gradually gained the upper hand and defeated Tarentum. The term "Pyrrhic Victory" has come to mean a victory gained at such a high cost as to be worthless.[14]

On 11 September 1709, during the War of the Spanish succession, the duke of Marlborough and Prince Eugene of Savoy met the French under Marshal Villars at the Battle of Malplaquet, near the Belgium border. It was the bloodiest battle of the eighteenth century in Europe. Marlborough and his allies won the battle but suffered far more casualties then the French. The defeated French commander Marshal Villars wrote to King Louis XIV, "If God gives us another defeat like this, your Majesty's enemies will be destroyed."[15]

The real tragedy of Pickett's Charge is that even if it had been successful and broken the Union line, the chances were quite remote that it would have led to a meaningful victory. The Confederates would inevitably have suffered great casualties and therefore would have had difficulty

Offensive vs. Initiative

exploiting any gap in the Union lines. The main body of the Confederate Army of Northern Virginia was too far away to have provided any significant help in time. Meanwhile, Union forces on the right and left would have hurried to seal the gap. Unused divisions held in reserve could have contained any breakthrough. Nothing much would have changed from what actually happened. Even if Pickett's Charge had briefly broken the Union line, the Confederates would not have held the initiative at the end of the day.

Even if you do not suffer crippling losses like Pyrrhus or Lee, victory can leave you overextended and thus vulnerable to counterattack. A few examples from World War II come to mind. The initial Japanese string of victories left the empire in control of a vast area, but with too few forces to defend such a large area, the Japanese had little hope of hanging onto their gains unless the United States gave up the struggle. The war in the North African desert was a series of alternating offensive drives by each side. Every time (except the last), however, the attacker was left with long supply lines and vulnerable to counterattack. In 1942 the Germans drove deep into the southern Soviet Union, all the way to Stalingrad. Their drive left them with a long, exposed flank defended by weak divisions supplied by Germany's allies. When the German drive stalled at Stalingrad, the Soviet army had time to regroup and then attack and cut off the German Sixth Army.

The Offensive Is Not Always Necessary to Gain the Initiative

The offensive frequently gains the initiative, but it is not always the only way to gain the latter. In certain cases, maneuver, defense, deception, or surprise can gain the initiative. In addition, acting defensively at one level might gain the initiative at a higher level. Finally, commanders sometimes can sacrifice a short-term initiative to gain a longer-term initiative.

MANEUVER

Maneuvering can force the enemy to react to your moves and thus help gain the initiative. It is one way to obtain a favorable position from which to launch an offensive. Commanders can also use maneuver to achieve mass or relative advantage. However, here we wish to consider whether maneuver can substitute for the offensive to gain the initiative.

On occasion, maneuver may create a threat that by itself may be sufficient to gain your objective. For example, when Gen. George Washington's Colonials fortified Dorchester Heights overlooking Boston in March 1776 and placed cannons there overnight, the threat itself was sufficient to persuade the British to evacuate Boston.[16]

In 1588 as the Spanish Armada approached, the English fleet sailed out of Plymouth. The Spanish, with large numbers of troops on board, preferred to engage in a close-range battle where they could grapple the English ships and fight them hand-to-hand. Conversely, the English preferred to fight at a distance using their long-range guns. On 31 July the more weatherly English ships—meaning they could sail closer to the wind—managed to get on the windward side of the Armada, thus gaining the "weather gauge." Consequently, the English could decide whether to close the Spanish ships or not. Because they held the weather gauge and their ships were more maneuverable, the English could choose the time to fight and the range at which to fight, or they could choose not to fight. The Spanish did not have those options because the English ships were upwind from them and were more maneuverable. Thus, before they took any offensive action, the English had gained the initiative without firing a shot. Although initially the English greatly overestimated the damage they could inflict with long-range gunnery, because this was the first naval gun battle, that does not change the fact that they had the initiative.[17]

In the spring of 217 BCE Hannibal marched his army through a swamp past the Roman army under Caius Flaminius. As a result, Hannibal gained the initiative. Flaminius then pursued Hannibal. This might seem like Flaminius had the initiative. The Romans were offensive while pursuing, but Hannibal had the initiative because he chose the route and could decide when and where to fight. He stayed ahead until he wanted the Romans to catch up. This initiative gave Hannibal the opportunity to choose Lake Trasimene for the site of one of the all-time great ambushes.[18]

In May 1864 Maj. Gen. William Sherman and a Union Army of about a hundred thousand men faced a Confederate Army of sixty thousand men led by Gen. Joe Johnston in northern Georgia. Because the Confederates held a good defensive position, a frontal attack would have been bloody, with no success assured. Sherman moved some of his forces around Johnston's left flank and threatened an attack from a more favorable position. This forced Johnston to retreat. Sherman repeated this process two more times. On 27 June Sherman made a frontal assault on Johnston's position at Kennesaw Mountain. Johnston repelled the attack with a loss of eight hundred men to three thousand for Sherman. Evidently learning his lesson, Sherman once again turned Johnston's left flank, forcing him to retreat to a position just north of the Chattahoochee River. Yet again, Sherman turned his position and Johnston retreated to Peachtree Creek, just north of Atlanta. At this point, the Confederate administration replaced Johnston with Gen. John Bell Hood, who was extremely offensive-minded. Hood made a series of attacks, each of which failed.[19]

Sherman's series of flank moves, each threatening an attack likely to be successful, forced the Confederates to retreat to where they could no longer avoid the Union Army. As if to reinforce this point, Sherman's one use of the direct offensive—the frontal assault at Kennesaw Mountain—was a failure.

Throughout the campaign, Johnston was reacting to Sherman's moves. Without using the direct offensive, Sherman gained and held the initiative. The Confederates could disrupt Sherman and gain some temporary initiative only by fighting on unfavorable terms. As a telling counterpoint, Hood's impetuous offensives cost the Confederates dearly. Because of those failed Confederate offensives, Sherman had an even stronger initiative that enabled him to capture Atlanta and make his march to the sea.

DEFENSE

The current principle of the offensive allows one to go on the defensive, but only temporarily, and advises one to look for opportunities to resume the offensive to regain the initiative. While one certainly should not wait passively to see what the enemy will do and hope to parry the blow, defense can play a role in gaining the initiative. In part, this is the inverse of the foolish offensive losing the initiative.

Hidden in the explanation of the principle of the offensive is a good point, but it has it backwards. Instead of looking for opportunities to resume the offensive in order to gain the initiative, what the principle should have said was that, when on the defensive, look for opportunities to seize the initiative and then exploit your initiative by attacking on favorable terms.

During the reign of Byzantine Emperor Justinian in the sixth century, Belisarius reconquered much of Italy. Though badly outnumbered most of the time, he won many battles by acting operationally offensive and tactically defensive. Belisarius would goad the enemy into unsound attacks and then, when the enemy was off-balance, he would counterattack with great success.[20]

We previously considered the Battle of Gettysburg from the Confederate point of view; now we look at it from the Union's perspective. Except for a few tactical counterattacks to parry Confederate attacks, the Army of the Potomac was on the defensive for the three days of the battle. At the end of the battle the Army of Northern Virginia was badly hurt and limping back to Virginia. Continuing the battle or taking any offensive action was not an option for General Lee. Although the Army of the Potomac had also suffered heavy casualties, many units were still in good condition. Although Maj. Gen. George Meade chose not to pursue the Confederates as vigorously as he might have done, he had the option to do so. Thus after three days of being almost exclusively on the defensive, Meade had gained the initiative.

Meade did so because Lee chose to use offensive methods against a strong enemy in good defensive position. If the enemy forces are foolish enough to wear themselves out by attacking your strong position, let them.

Contrary to the current principle of the offensive, it was *not* necessary for Meade to resume the offensive in order to gain the initiative—he *already* had the initiative at the end of the battle. Possession of the initiative gave *Meade* the option of attacking Lee or choosing not to do so. Understanding this difference is crucial.

Basil Liddell Hart expressed a similar idea in another way: "Instead of seeking to upset the enemy's equilibrium by one's attack, it must be upset before a real attack is, or can be successfully launched."[21] In this context the terrible losses Lee's army had suffered in three days of offensive action had upset its equilibrium.

In President Lincoln's opinion Meade should have pursued Lee more vigorously. Had he done so he might have thoroughly smashed the Army of Northern Virginia and conceivably could have ended the Civil War in 1863. We will let others debate whether this outcome was likely or not. The point here is that, if your enemy is willing to oblige you by attacking a strong position and if a successful defense will gain the initiative, then defense may be the correct course of action. However, usually you need to exploit the initiative by going on the offensive at some point. Generally the initiative is a wasting asset: if you do not exploit it, then your initiative fades. After Gettysburg, Meade had the option of following Lee and attacking him, perhaps cutting off his retreat. If this attack were successful, Meade's initiative would have expanded. By not attacking, Meade allowed Lee to get what was left of the Army of Northern Virginia across the Potomac River to a place where he could rest and regroup. Because of Lee's heavy losses at Gettysburg, the Union still had a long-term initiative—the Army of Northern Virginia was no longer capable of large-scale invasions of the North, and its capability for other bold moves was limited. Nevertheless, they survived to resist for another twenty-one months.

Naval battles are rarely won without being offensive. Yet it can happen. In 1944 the American Fifth Fleet under Adm. Raymond Spruance had the task of defending the amphibious landings in the Marianas. In part because of his perceived need to stay close to the landing area and in part due to the prevailing winds being from the east, Spruance was unable to launch an effective attack on the Japanese fleet. The Japanese on the other hand were able to launch numerous attacks against the American fleet. Due to American radar and good direction of their fighters, the American fleet withstood all the Japanese attacks and downed many aircraft. Aviators referred to the Battle of the Philippine Sea as the "Great Marianas Turkey Shoot." Nevertheless, many criticized Spruance for not having closed with and attacked the Japanese carriers.[22] However, it turns out that Japan's center of gravity was not its aircraft carriers, but its cadre of skilled carrier aviators. Because of the losses of the last of Japan's experienced naval aviators, the Japanese carriers never again posed a dangerous threat to the American carriers. The Americans, despite being on the defensive, came out of the battle with the option of using their fleet of aircraft carriers without fear of serious opposition from their Japanese counterparts. That is, the Americans emerged from the Battle of the Philippine Sea with a greatly increased initiative.

New Principle of the Sustained Initiative

If the purpose of the offensive is to gain and hold the initiative, but the offensive is neither necessary nor sufficient to gain the initiative, then the initiative is a more fundamental concept than the offensive. Our focus should be on the initiative, rather than the offensive, which is just a tool for gaining and holding the initiative.

The currently accepted principle of the offensive gives bad guidance in some cases and confuses the means and the goal. During the Vietnam War, Ambassador Henry Cabot Lodge suggested redefining "offensive operations" from "seek out and destroy" to "split up the Viet Cong and keep them off balance."[23] Keeping someone off balance is a good working definition of maintaining the initiative. However, rather than redefining "offensive operations," we believe we should deemphasize the offensive and instead emphasize seizing and expanding the initiative. Because a sustained initiative is the essential idea underlying the current principle of the offensive, we propose a new principle of the sustained initiative:

> Seizing and maintaining the sustained initiative greatly increases the chances of success.
>
> Constant pressure from attacks and threats that the enemy cannot ignore keeps the enemy occupied so that they do not have time to prepare and execute threats of their own. The side holding the sustained initiative often can decide whether to fight and, if so, when, where, and how.
>
> The initiative must be expanded and exploited or the advantage will dissipate. Pressure must be constant. Usually, at some point one must use the initiative to attack on favorable terms and gain a long-lasting advantage.
>
> The offensive is neither necessary nor sufficient to gain the sustained initiative. The direct offensive frequently is the best way to gain the initiative, but it is not always the only way to gain the initiative. Maneuver, defense, threats, deception, and surprise can also gain the initiative. Offensives can lose the initiative if they fail, if they succeed at too high a cost, or if they cannot be exploited. However, the offensive is usually necessary to exploit the initiative and gain a long-lasting advantage.
>
> Having a sustained initiative can be good even without the offensive. This does not mean that the offensive is always bad, but the offensive is good only if it gains or holds the initiative.
>
> An attack will usually gain at least a short-term initiative, but if that initiative cannot be sustained and expanded, then that attack is probably not wise.

ONE CHARACTERISTIC OF THE INITIATIVE IS HAVING MORE good choices than the enemy has. Of course, one excellent choice trumps

Offensive vs. Initiative

a large number of moderately good choices. The ideal situation is when the enemy has no good choices because they have to respond to threats that they cannot ignore.

We might have chosen a different title, such as "constant pressure," for this principle. Alternatively we might have used the title of the Soviet principle of "combat aggressiveness" or that of "activity." We selected the title "the sustained initiative" because it most precisely captures the concept. However, the title is far less important than the concept behind the principle.

Gaining the Initiative

We gain the initiative by forcing the enemy to respond to our actions. We must pose a threat that the enemy cannot ignore without peril. Because the enemy is occupied responding to our threat, it is less likely to be able to threaten us in a similar manner.

One can gain the initiative using maneuver, deception, surprise, security, threats, and the offensive. Surprise can gain the initiative by posing a threat to which the enemy must respond. Security helps gain the initiative by allowing you freedom to take various actions without having to worry about enemy attacks on your own potentially vulnerable points. Security also deprives the enemy of good options for attacking you.

If deception deceives the enemy into responding to a nonexistent threat, this gains the initiative and allows you time to take other actions to expand your initiative. An example of this is Magruder's theatric deception at Yorktown during the Peninsular Campaign. Another use of deception is to induce the enemy to expose a weakness that you can exploit.

MULTIPLE THREATS

In general, we seize and increase the initiative by making a series of threats to which the enemy must respond. An even better way is to make simultaneous multiple threats. By confronting your enemy with a series of dilemmas, you gain the initiative. To do this you should be flexible about intermediate aims and your plan should have multiple branches that threaten alternative intermediate objectives. The idea is to expand your good options and reduce your enemy's good options. Multiple threats can place the enemy on the horns of a dilemma about how to defend. Suppose you are in a position where three roads lead to potential intermediate objectives A, B, and C, each of which leads to your primary objective. Each of these three positions is important to your enemy, but the latter does not have sufficient resources to defend all three positions. Only bad choices remain. The enemy can divide its force into thirds even though one-third of its force is insufficient to stop your attack. Alternatively, the enemy can guess which position you will attack and hope you are not able to detect its decision. At this point you clearly have the initiative.

A threat is sometimes more powerful than the execution of that threat. During the Napoleonic wars Napoleon complained that thirty thousand men in transports in England, with the potential to land anywhere along the French coast, tied down three hundred thousand Frenchmen to defend the entire coast. This improved the allied force ratios everywhere else. However, when the British concentrated on the Iberian Peninsula, this threat dissipated, freeing the three hundred thousand Frenchmen for other duties.[24]

Going on the offensive against one objective may remove the threat to other potential objectives, thus paradoxically relieving pressure on the enemy. After you commit to attacking, say, objective A, then much of the pressure is relieved. The enemy can shift forces from B and C, either to reinforce objective A or to backup position A if you break through; alternatively, the enemy can attack some position of yours. Unless you achieve a breakthrough at A that gives you several options, your initiative has diminished. Therefore, when exploiting the initiative, one should prefer to do so in a way that increases the number or severity of your threats to sustain and increase the initiative.

Let us alter our hypothetical example a bit to illustrate the relationship between the offensive and the initiative. Suppose that when you attack and seize position A, it gives you several options. If you still have the option of attacking B and C, and in addition now threaten D and E, then your threats have expanded and increased your initiative. You have used the initiative to go on the offensive at a relative advantage and in turn used that offensive in such a way as to increase your initiative. If you can continue in this way, the pressure on your enemy grows ever greater. Thus, our point is not that you should not be offensive but that you should use the offensive in a manner that you can reasonably expect to increase your initiative. Moreover, often you must gain the initiative before effective offense with a great relative advantage is feasible.

Offensive vs. Initiative

For a real-world example of a commander placing his enemies in a series of dilemmas, consider Sherman's actions in 1864 after he captured Atlanta. Sherman's subsequent march through the South utilized a series of alternative objectives. Therefore the few Confederates left to oppose him could never be sure what Sherman intended to do next. First, Sherman threatened both Macon and Augusta but bypassed both and went through Milledgeville instead. Then he threatened both Augusta and Savannah. After capturing Savannah Sherman headed north and threatened both Augusta and Charleston but bypassed both and instead went between them to Columbia. He next threatened both Charlotte and Fayetteville, then both Raleigh and Goldsborough.[25] Even had the Confederacy had more troops available to oppose Sherman, they would have had a hard time getting in front of him.

Of course, if you can destroy a large part of your enemy's force with relatively light losses to your forces, then most likely you have increased your initiative. Furthermore, if the enemy cannot replace its losses, then your initiative may be long lasting. If, however, the enemy can replace its losses, even though large, but you cannot replace your light losses, then the initiative is swinging toward your enemy. That was Pyrrhus's predicament.

Another way in which a threat is powerful is that people often greatly fear a threat with unknown consequences, whereas they can deal with the reality of an attack, even though repulsing it might be difficult. Often, we fear the unknown more than that which we know.

DEFENSE AT ONE LEVEL MAY GAIN THE INITIATIVE

AT A HIGHER LEVEL

The hierarchy of objectives from the National Command Authority down to small units forms the organizing principle for all plans and operations. Obviously an objective at a high level takes precedence over a lower-level objective. One should not gain a tactical objective on terms that preclude reaching an operational or strategic objective. Similarly, having the initiative at a higher level is more important than possessing the initiative at a lower level. Therefore, being on the defensive at one level might be a better way to gain the initiative at a higher level than going on the offensive. For example, being on the tactical defensive is desirable if it gains the operational initiative. Being on the operational defensive is acceptable if it leads to seizing or holding the strategic initiative.

Cases abound in which the tactical defensive leads to an operational initiative. A common tactic is occupying a key position that threatens to cut the enemy's supply line, and then going on the defensive at that point. This leads to an operational initiative because it forces the enemy to spend effort to remove your tactical force from the threatening position. This costs the enemy time and weakens the enemy in other areas. Your other forces then have more good options than they would otherwise have.

During the Peloponnesian War the Athenians made a strategically offensive move by establishing a base at Pylos in the western Peloponnesus. This gave them the initiative because they could have used this as a base from which to raid territory near Sparta. This forced the Spartans to counter the threat and send a force to attack Pylos. Part of the force landed on the island of Sphacteria. This was a major tactical miscalculation because the Spartans could not maintain control of the surrounding sea. When the Athenians gained control of the sea, they trapped the Spartans on the island. At this point, the Athenians went into a defensive containment mode, merely patrolling the surrounding sea to prevent escape or resupply of the Spartans. In spite of being tactically defensive and without the tactical initiative—the Spartans could try to escape at any time of their choosing—the Athenians held the operational initiative and, because hundreds of the trapped Spartans were from the highest ranks of the Spartan nobility, the strategic initiative. Sparta sued for peace. To be sure, peace negotiations broke down because Athens demanded too many concessions, and the Athenians eventually had to storm the island and defeat the Spartans.[26]

OFFENSIVE

Although we emphasize that the offensive without the initiative is generally not good, we do not wish to discourage the offensive in all cases. When the offensive gains, holds, or increases your initiative, it is good. In addition, the offensive is usually the only way to exploit the initiative. In many cases, however, rather than using the offensive to gain the initiative, you first need to gain the initiative in order to launch a successful offensive under conditions of relative advantage.

After the Army of the Potomac invested Richmond during the American Civil War, Lee made no flanking attacks or quick marches. It was not that Lee suddenly lost his preference for the initiative and the offensive.

Lee could not take the offensive effectively because he had lost the initiative. He did not lose the initiative because he failed to attack. Lee could take the offensive; he just could not do so effectively. In March 1865, just before the end, he tried one attack on the Union lines at Fort Stedman in an attempt to gain a modest initiative so he could get his army out of Richmond. After some initial success, it ultimately failed.[27] Grant held the initiative not because he was on the offensive—for the most part he was not—but because he maintained the threat to break Lee's lines. The reason Lee could no longer make a flashy flank attack is that the Army of Northern Virginia was pinned to its trenches. If Lee thinned his lines, the Union Army would then be able to break his line. To keep increasing the pressure—and thus increase his initiative—Grant continuously extended his lines to the south of Richmond, threatening to cut the railroad to the south.

Short-Term vs. Sustained Initiative

In war, as in life, often it is good to suffer short-term pain for long-term gain. Relinquishing a short-term initiative is acceptable provided it leads to a long-term initiative. Of course, one of the best long-term initiatives is to have an overwhelming force advantage over the enemy, and the offensive is one way to reduce the enemy's forces. If the enemy is willing to attack a strong position and if a successful defense will gain a sustained initiative, then defense may be the correct course of action. When on the defensive, look for opportunities to seize the initiative and then exploit the initiative by attacking on favorable terms.

Trading space for time is a well-known strategy when facing superior forces. After the Germans invaded the Soviet Union in 1941, they proved vastly superior in battle. They drove the Soviets back and captured huge numbers of men. Not that they had much choice, but the Soviet Union traded space for time. As the front lines moved deeper into Russia, the German supply lines grew longer while the Soviet supply lines grew shorter. With the time gained, the Soviets moved troops from Siberia, ramped up munitions production, and received supplies from Britain. Moreover, with each day winter came closer. During this retreat the Soviet Union did not have the initiative. They could have tried to go on the offensive, but any such effort would certainly have failed and made the situation worse. All they could do was to try to slow the German advance and hope they could use the time gained to change the situation. Finally, with the onset of winter, the German advance stopped just before reaching Moscow. This relinquished the initiative. The Soviets then gathered sufficient forces to give them the option to counterattack effectively. This gave them the initiative, which they expanded and exploited by counterattacking.

Another variation of this theme is the Fabian strategy, which seeks to avoid decisive pitched battles and is content to merely harass the enemy in order to gain the initiative eventually. If time is truly on your side and no near-term path to victory is apparent, then you should first avoid losing in the near-term. During the Second Punic War, after Hannibal soundly defeated Roman armies in the Battles of Trebia and Lake Trasimene, Quintus Fabius Maximus Verrucosus, dictator of the Roman Republic, had to deal with Hannibal. He believed that Carthage could not defeat Rome as long as Rome's allies remained loyal. Fabius thought that Rome's allies would desert her only if Hannibal defeated the Romans in another decisive battle. Therefore, Fabius harassed Hannibal but refused to fight him in a major battle. Hannibal could not force the Romans to fight a major battle. Because Hannibal did not have the siege engines necessary to take cities, he could only ravage the countryside. He did not have an achievable military objective that would achieve his political objective. Time was on Rome's side. Without plunder, Hannibal's largely mercenary army would lose interest. In addition, Roman control of the sea restricted Hannibal's ability to get reinforcements from Carthage. Thus Fabius avoided a decisive battle and let time work against the Carthaginians. Hannibal then had no good choices.

This was a good strategy for Rome but not for Fabius. The Romans called Fabius the "Cunctator," that is, "the Delayer." Because of his failure to win big victories, the Roman people turned against him and the Senate removed him from command. That led to the disaster at Cannae. Thereafter, Rome reverted to the Fabian strategy. That strategy, plus Roman successes against the Carthaginians in places such as Iberia and political problems for Hannibal in Carthage, eventually forced Carthage to recall Hannibal.[28]

After the disaster and near annihilation of the Continental Army at the Battle of Long Island, George Washington had the flexibility of mind to adopt a Fabian strategy—and took much criticism for doing so. He made a few attacks when he saw a good opening with a relative advantage, for example at Trenton, but would not allow the British to draw him into a possibly fatal battle. The British eventually wore down.

Regaining the Initiative

If the enemy has a strong initiative, how do we reverse the situation? We could hope the enemy does not fully exploit his initiative and allows us the opportunity to do something. We must be alert for any opening. When the German army turned short of Paris in World War I, they exposed their flank. The French responded quickly to this opportunity in the Battle of the Marne.

Of course, we prefer not to wait passively for the enemy to make a mistake. We should explore generating serious threats by the use of deception, surprise, maneuver, or counterattack to get the enemy off-balance. A counterattack can gain time for the rest of one's forces to escape the enemy's pressure and thus give the opportunity to pursue one's own schemes to gain the initiative. For example, during the Battle of Jutland, the second time the Germans had their tee crossed by the British Grand Fleet, the bulk of the German High Seas Fleet escaped when four badly damaged German battle cruisers made their "Death Ride" directly at thirty-three British capital ships.[29]

Sustaining the Initiative

A short-term initiative is generally not useful. You can nearly always attain a short-term initiative by attacking, but this can leave you less capable after the defeat—and less in possession of the initiative.

We want a sustained and increasing initiative. We want to keep the enemy responding to our actions, rather than pursuing their own schemes. If at any point we pause, the enemy may have a chance to respond in a way that regains the initiative. We must use the enemy's occupation with our initial threat to threaten the enemy in additional ways to which they must respond. We want to use our initiative to prevent the enemy from adjusting and to present them with increasingly serious situations that force them to scramble to defend. This gives us an increasing initiative. We call this the sustained initiative.

You can gain the initiative by, for example, a maneuver to get into a favorable position. However, if you just sit there, the enemy adjusts so that your position is no longer favorable and you no longer have the initiative. This is a general concept. If you do something to gain the initiative, the enemy will, if allowed, adjust to relieve the pressure and allow him to seize the initiative from you.

At the Battle of Chancellorsville during the American Civil War, Maj. Gen. Joe Hooker surprised Gen. Robert E. Lee by moving a large Union force across the Rappahannock River near the town of Chancellorsville in May 1863. This maneuver gained Hooker the initiative. Hooker's forces quickly expanded their beachhead south of the Rappahannock while Lee scrambled to move Confederate forces to meet the Union troops. Having gained the initiative, Hooker needed to exploit the initiative. If Hooker had continued to push forward, he might have gained an important positional advantage astride Lee's supply routes between Lee and the Confederate capital of Richmond. Instead, inexplicably, Hooker halted his troops. Perhaps he simply lost confidence, worrying what Lee, with his great and growing reputation, might do. Whatever the reason, the halt gave Lee's army time to reach defensive positions. By failing to exploit the initiative, Hooker had lost it. As soon as the Confederates stabilized the situation, Lee and Stonewall Jackson started planning to regain the initiative. They did so by maneuver—Jackson's long march around the Union's right flank. The Confederates then exploited that initiative by taking offensive action.[30] The lesson is that slowing an offense can give up the initiative and allow the enemy to pursue its schemes.

The Gallipoli landings during World War I illustrate the high cost of failing to exploit the initiative. The primary Allied assault at Cape Helles on 25 April 1915 involved five landing beaches. The two largest were well-defended and resulted in heavy casualties to the invading force. On the other hand, the Turks left Y Beach, narrow and dominated by cliffs, undefended due to the difficult terrain. The British force landing on Y Beach was able to reach high ground quickly. Had they exploited this opportunity and attacked the Turks in the rear, the entire Gallipoli campaign might have been different. This failure to exploit the opportunity gave the Turkish forces time to reach the area and drive the invaders from the high ground. As a result, the Allies evacuated the force from Y Beach the next day, and the entire campaign bogged down.[31]

Conversion of Initiative into a Durable Advantage

We need to convert the initiative into a non-wasting asset, that is, a long-term advantage. The durable advantage might be gaining an important objective, such as a geographic point, or the infliction of many enemy casualties to improve the force ratio. In most cases, we convert our initiative into something long-lasting by reducing the enemy's combat power. Possession of the initiative allows us to attack with a relative advantage. For example, we might use our initiative to engage a portion of the enemy's force with the bulk of our force. You must exploit the initiative, but you should be sure to get fair value in return for giving up the multiple threats.

Military history has many examples of commanders who gained an initiative but then saw their advantage dissipate when they failed to exploit it. The Japanese Navy in World War II provides three examples. First, Vice Adm. Chuichi Nagumo might have made a third attack at Pearl Harbor that targeted repair facilities and oil storage tanks. This had the potential to set the United States farther back than sinking the battleships. Second, in the Battle of Savo Island on the night of 7–8 August 1942, Vice Adm. Gunichi Mikawa's Japanese force sank four Allied heavy cruisers. However, fearing air attack after daybreak, they withdrew without attacking the supply ships unloading on the beaches of Guadalcanal. Destroying those ships would have put the U.S. Marines on Guadalcanal in extreme difficulty.[32] Third, on 25 October 1944, at the Battle of Leyte Gulf, when the Americans left San Bernardino Strait undefended, Vice Adm. Takeo Kurita's strong force of battleships and heavy cruisers broke through unopposed and encountered just escort carriers and destroyers. They were in position to maul the supply ships unloading in the landing area. However, they failed to exploit their initiative and retreated after courageous attacks by the overmatched American forces.[33]

COMMANDERS MUST FOCUS ON GAINING, KEEPING, EXPANDING, and exploiting the sustained initiative. The offensive has an important role to play in this, but commanders should focus on the sustained initiative, not on the offensive. Gaining and exploiting the sustained initiative requires using all one's forces in a coordinated manner. This is the subject of the next chapter.

EIGHT

Unity of Command vs. Unity of Effort

> Coordination, therefore, is the orderly arrangement of group efforts, to
> provide unity of action in the pursuit of a common purpose. As coordination
> is the all inclusive principle of organization it must have its own principle and
> foundation in authority, or the supreme coordination power. Always, in every
> form of organization, this supreme authority must rest somewhere,
> else there would be no directive for any coordinated effort.
>
> —JAMES D. MOONEY, James D. Mooney Quotes

TO MAXIMIZE OUR CHANCE OF OBTAINING A RELATIVE ADVAN-
tage, seizing the sustained initiative, and attaining our primary objective,
we must use and coordinate all our potential power in a cohesive manner toward that end. Two possible methods include having a single commander in charge of everything or having various commands cooperate.

Current Principles of Command

As described in chapter 1, when the U.S. military first listed principles of
war in 1921, one of the principles was cooperation. When they reinstated
the principles of war in 1949, they replaced cooperation with unity of
command. Of the world's major military powers, only the United States
lists unity of command as one of the principles of war. Other English-
speaking militaries use cooperation as the corresponding principle. Many
of these militaries also state the importance of using unity of command
where possible. The Australian Defense Force lists unity of command as
one of the principles of command.[1] NATO employs unity of effort as the
principle, while noting that the usual means to achieve this is unity of
command. France and Israel do not have a similar principle.

Soviet principles used different terms to address how best to combine
forces. Three lists of Soviet principles contain the corresponding prin-

ciples of cooperation, interworking, and coordination. Cooperation is a Soviet principle of military art, with the relevant explanation stating that unity of command is necessary to achieve cooperation.[2] The principles of operational art and tactics include the principle of interworking, understood implicitly as combined arms warfare and unity of effort.[3] Coordination is one of the Soviet principles of combined arms combat tactics, which is seen as requiring unity of command, unity of purpose, and unity of action.[4]

The explanation of the American principle of unity of command begins by stating the purpose of the principle:

> The purpose of unity of command is to ensure unity of effort under one responsible commander for every objective.
>
> Unity of command means all forces operate under a single commander with the requisite authority to direct all forces employed in pursuit of a common purpose. Unity of command may not be possible during coordination and operations with multinational and interagency partners, but the requirement for unity of effort is paramount. Unity of effort—the coordination and cooperation toward common objectives, even if the participants are not necessarily part of the same command or organization—is the product of successful unified action.[5]

In contrast, the British title their corresponding principle "cooperation":

> Cooperation is based on team spirit and training. Cooperation relies on three interrelated elements: mutual trust and goodwill; a common aim (or at least unity of purpose); and clearly divided responsibilities (including understanding the capabilities and limitations of others). Within alliances or coalitions, differences must be harmonized and political/military cohesion promoted and protected.[6]

Unlike the principles of war that we previously considered, the British and American principles differ from each other significantly. The American definition declares that the purpose of unity of command is to achieve unity of effort. It even states that the paramount requirement is unity of effort, which it defines as cooperation toward common objectives. On the other hand, the British definition of their principle includes clearly divided responsibilities as one of three essential elements necessary to achieve cooperation.

The NATO principle of unity of effort bridges the gap somewhat:

> Unity of effort emphasizes the requirement to ensure all means are directed to a common goal. Military forces achieve this principally through unity of command.[7]

Thus, we have three candidate principles of war: unity of command, cooperation, and unity of effort.

THE ARGUMENT FOR UNITY OF COMMAND STARTS WITH THE observation that having multiple commanders with confused lines of responsibility and authority usually does not work. Ancient warfare gives us many examples of command relationships that, to the modern observer, seem bizarre. In the early fifth century BCE, command of the Athenian army rotated among the ten generals, each being commander for a single day.[8]

At the Battle of Aegospotami in 405 BCE, six generals (we would call them admirals) rotated command of the Athenian fleet daily. They were supposed to make decisions by consensus. The details of what happened are not clear, but the command arrangement probably contributed to the disaster that lost the war for Athens.[9]

During the Republic the Romans tried to balance political power by electing two consuls. When the two consuls combined their armies, they alternated command each day. Citizens of the Roman Republic so distrusted giving too much power to one person that no unit of the Roman army had a single commander other than the maniple commanded by a single centurion. For example, each legion had six elected tribunes. Command rotated among pairs of the tribunes.[10] This often worked, but not always.

Without a single person in charge of reaching an objective there is no continuity of plans or actions. It is easy to lose sight of the objective. If one person is responsible for meeting an objective and possesses the requisite authority, then that minimizes confusion. The commander does not wait to see if someone else will do necessary tasks. If things go wrong, the one commander is held responsible.

When there is overlapping responsibility there is potential either for some things to be left undone or for interference as two groups try to achieve the same objective. Two commanders faced with the same problem will often come up with different approaches. After all, commanders, being human, have differences of experience and style. They bring their own biases and agendas, not to mention big egos in some cases, to the problem.

Having a single commander eliminates the need to consult, negotiate, and compromise with cooperating commanders. Such consultations might result in better decisions in some cases. However, under the time constraints of combat, this can cause fatal delays in decisions. A good decision made quickly is usually better than a perfect decision made later. In addition, the need to negotiate and compromise can result in muddled thinking. Chapter 5 gave several examples of multiple, counterproductive objectives arising from the need to compromise among competing interests.

Some of the reasons for using cooperation rather than unity of command involve elements of human nature. People of one service (or branch of the same service) often do not fully trust members of other services. Julian Corbett described the reasoning behind the British practice of having coequal army and navy commanders for amphibious operations. They were willing to accept possible friction between the two commanders to avoid the more dangerous possibility of a single commander making mistakes about the employment of the other service through lack of familiarity with their limitations.[11]

What Is Unity of Command?

All services of the American military agree that unity of command is important. However, they interpret the concept differently. As indicated in chapter 3, the U.S. Air Force believes that, because airpower is uniquely flexible, a single commander should control all airpower in a theater. Furthermore, this commander should be an airman who reports directly to the joint force commander. This air commander is then in the best position to allocate air resources in accordance with the joint force commander's overall direction. The air commander can allocate airpower between support to ground units and attacks far behind the front lines and can also shift ground support to those areas most in need of such support.[12]

The U.S. Marine Corps believes that unity of command means that all forces, including airpower, in a geographic area should be under the control of a single commander. This makes airpower more responsive and allows for more training of close air support procedures. Because the Marines are a "light" force that emphasizes mobility, air power often substitutes for artillery. They want air assets readily available to support the ground forces. As stated in the Marine Corps principles, it is part of the cultural identity of the Corps that a Marine Air-Ground Task Force integrates organic air assets under a single commander. The Marines fear they might not have air support when they most need it unless they have organic air support. Marine Corps ground troops have a high degree of trust in Corps aircrews because they are fellow Marines. Recall that the U.S. Marine Corps principles state that every Marine, regardless of specialty, is first and foremost a rifleman. This is another part of the cultural identity of the U.S. Marines.

The U.S. Air Force, on the other hand, holds that although a sector commander might think he or she has great need for air support at a particular time, in the big picture it might be better to concentrate all air power in a more critical sector at that time.[13] However, the Marines do not want a distant Air Force officer, who may not have a deep understanding of ground combat, making decisions on whether to provide air support to a Marine task force. The Air Force counters that the air component commander makes decisions in accordance with the guidance of the joint force commander.

The U.S. Navy also believes in unity of command of air assets. Its position is that Navy air assets are an integral part of fleet operations and should be under the command of the naval commander. If a naval task force supplies air assets to others, it must control the timing and level of support. Furthermore, because a fleet's primary defense is its mobility, tying it to a specific location is anathema to the U.S. Navy.[14]

The different views about unity of command largely concern the level at which to unify command. Technically, there is always unity of command at the level of the National Command Authority, but that is not what most people would consider to be unity of command. In ground combat, command of artillery might reside at various levels. Below that level, command of ground troops and artillery is not unified. The U.S. Air Force wants to unify command of air power at the Joint Force Command level, but the U.S. Marine Corps and the U.S. Navy believe it should occur at the component commander level. Often, commanders resolve this conflict with ambiguous agreements that allow both sides to claim they prevailed in the dispute. (For a detailed history of this subject, see the book *Joint Air Operations*.)[15]

Thus, the disagreement is not about the principle of unity of command itself but about the implementation of the principle. As accepted by nearly all militaries of the world, commanders must consider the unique circumstances of each situation and use their judgment in implementing the principles of war.

What Is Unity of Effort?

The American definition of the principle says that unity of effort is "cooperation toward common objectives." The British definition only describes requirements for cooperation. NATO says that unity of effort consists of directing all means toward a common goal. We will define unity of effort as the efficient coordination of all assets to achieve the primary overall objective. One way to elaborate on unity of effort is to consider the factors that contribute to achieving it. We will do so later in this chapter.

Another way to explain our conception of unity of effort is to identify various levels of unity of effort. In *Desert Storm at Sea* the author presents a model hierarchy for joint operations.[16] We will adapt and modify that hierarchy to apply it to unity of effort. It is applicable to joint (multi-service) and combined (multi-country) operations, as well as the employment of multiple types of units from the same service.

We define five levels of unity of effort. In order from least effective to most effective, these are disarray, unity of purpose, deconfliction, harmony of action, and combined arms.

At the lowest level, disarray denotes a state of affairs in which there is little agreement on goals and people work at cross-purposes. This results in no unity of effort.

The next lowest level is a situation in which people share a unity of purpose. Everyone works toward the same goal, but there is no effective use of all forces in a coherent manner. Disagreement about how best to achieve the common goal sometimes can be just as bad as having different goals. Unity of purpose without coherent action gives some unity of effort, but not much. To get more unity of effort, cooperation is necessary.

Deconfliction is the next higher level of unity of effort in our model. Deconfliction separates efforts from various units geographically or temporally to avoid interference or even fratricide. The route packages of Vietnam employed geographic deconfliction by assigning separate regions to U.S. Navy and Air Force aircraft. Desert Storm used some temporal deconfliction. Attack aircraft from the various services attacked some of the same targets, but the Air Tasking Order separated the attacks by time over target. On land, different units will have separate zones of operation. Deconfliction avoids some problems but otherwise does not advance unity of effort very much.

Harmony of action, the second-highest level of unity of effort, means that each service or unit takes one part of the problem but works toward a common goal. Each unit offers a complementary capability that provides one piece of the solution. The harmony indicates that sometimes the "instruments" of various units play in unison, sometimes independently, but always in harmony with the overall objective.[17] Operating independently, different types of aircraft from various services might attack different types of targets, each attacking the target for which it is best suited. Operating together, one platform with better sensors might provide targeting information to another platform more suited to attack. Harmony of action provides true unity of effort by utilizing the most-needed features of each type of asset to reach an objective that any one type of asset could not achieve efficiently.

Finally, combined arms is the inspired use of harmony of action to achieve the highest level of unity of effort. On offense, each capability forces the enemy to react in a way that exposes vulnerabilities to other capabilities. On defense, each element protects vulnerabilities in other elements. As explained at the end of chapter 6, combined arms is a good way to gain a relative advantage over the enemy. Combined arms is the only level in which we achieve synergistic effects, where the whole is greater than the sum of the parts, whether we employ the parts separately or sequentially.

Criteria

The concept of unity of command meets most of the criteria that we set out as the requirements for an idea to qualify as a principle of war. For instance, it meets our four-part discrimination test. Following the concept has often contributed to victory and rarely caused defeat. (Incompetent commanders are a distinct problem.) Conversely, not following the concept has often led to defeat, but few victories have been won *because* there was no unity of command. The concept is applicable to all levels of warfare and all types of warfare. In addition, it has endured for centuries and most likely will be important for centuries to come. Unity of command is also relevant because military history provides numerous cases in which the lack of it resulted in a disjointed effort that led to defeat. Similarly, the concepts of cooperation and unity of effort both pass all these tests. All three concepts describe desirable features. But which of these three concepts is most fundamental?

Because the reason for having unity of command (or cooperation) is to achieve unity of effort, perhaps unity of effort is a more fundamental concept. To determine which concept is more fundamental, let us ask whether unity of command is always *sufficient* to achieve unity of effort. Conversely, is unity of command *necessary* to achieve unity of effort? If unity of command is both necessary and sufficient to achieve unity of effort, then we might regard either concept as equally fundamental. In that case, unity of command might be the preferred concept because it is more concrete than unity of effort, which can easily degenerate into platitudes. Similarly, let us ask whether cooperation is both necessary and sufficient to achieve unity of effort.

Unity of Command vs. Cooperation

To determine whether unity of command or cooperation is more important, let us examine possible combinations of having or not having unity of command and cooperation. First, having neither unity of command nor a requirement for cooperation seems to be a recipe for chaos rather than unity of effort. Second, it seems obvious that having both unity of command and full cooperation among subordinates is best, provided it is practical. What about instances in which one has either unity of command or cooperation, but not both?

Cooperation without Unity of Command Fails

In the lead up to the Battle of Trasimene, Caius Flaminius should have waited for his co-consul to join him with an additional consular army before battling Hannibal. Although the two consuls should have cooperated, they did not because Flaminius was impetuous.[18] In 378 CE emperors of the eastern and western Roman Empires were expected to cooperate. The eastern emperor Valens attacked the Goths instead of waiting for the western emperor Gratian to arrive with help. Valens did not wait because he did not want to share the glory of the expected victory. The result was the disaster at Hadrianople.[19]

In chapter 5 we described how Athens dispatched an expedition with three commanders in 425 BCE during the Peloponnesian War. The three commanders soon disagreed about the course of action to follow. Evidently, not having learned from the 425 BCE expedition, the Athenian public assembly sent a large expedition to Sicily in 415 BCE and again appointed three commanders: Nicias, Alcibiades, and Lamachus. Moreover, Nicias and Alcibiades were political rivals who apparently detested each other. When the fleet arrived in Sicily the three commanders proposed three different courses of action. The divided command might not have been the primary reason for the ensuing disaster, but it was a contributing cause.[20]

It was not only the ancients who suffered the consequences of a divided command. In 1941 the U.S. Army and Navy divided command in Hawaii between Adm. Husband Kimmel and Lt. Gen. Walter Short. They were expected to cooperate. The two men played golf together and had quite cordial relations, but apparently they did not share vital information about what each command was or was not doing. The Army had responsibility for the defense of Hawaii but assumed the Navy was doing reconnaissance. The Navy assumed the Army would take care of air defense. Each commander assumed that information sent to him was also sent to the other. When informed that the Japanese were destroying codes and code machines—a sure sign of imminent war—Kimmel failed to pass this information to Short. Thus divided American command contributed to the success of the Japanese attack.[21]

During World War II the Japanese Imperial General Headquarters had little authority compared to the U.S. Joint Chiefs of Staff. The Japanese did not have joint commands. They expected the Japanese Army and Navy to cooperate, but such cooperation was poor. As described in chapter 5, Japan's army and various parts of the navy disagreed about the proper course of action after their initial successes. They finally agreed to the Midway operation but only after Admiral Yamamoto compromised by providing two aircraft carriers to the Coral Sea operation and two more for an Aleutian operation. The four carriers siphoned off the

Midway operation might well have made a decisive difference in the outcome of the battle.[22]

During World War II in the Pacific, American command was unified only at the level of the U.S. Joint Chiefs of Staff. The United States divided the Pacific Ocean into the Pacific Ocean Area and the Southwest Pacific Area. Adm. Chester Nimitz led mostly naval forces in the central Pacific, and Gen. Douglas MacArthur led mostly army forces in the southwest Pacific. They were expected to cooperate. This arrangement worked reasonably well until the October 1944 invasion of Leyte in the Philippines. For the first time MacArthur would be operating beyond the range of land-based air power. Therefore he needed the airpower from Adm. Bill Halsey's Third Fleet carriers. However, Halsey did not report to MacArthur, even though MacArthur's forces included substantial naval forces. Instead Halsey reported to Nimitz, who ordered Halsey to cooperate with MacArthur.[23]

This divided command caused many problems. One of those problems was that each command devised its own search plans and no one surveilled some critical areas. This allowed Japanese forces to remain undetected.[24]

Unity of purpose was lacking. In MacArthur's plan, Halsey's primary mission was to provide distant cover and support for the landing force by maintaining air superiority and protecting the landing. However, Nimitz, evidently without consulting MacArthur, gave Halsey the additional objective of destroying the Japanese fleet. Nimitz's order to Halsey said that if he could *create* an opportunity to do so, then that would become Halsey's "primary task." Nimitz did not explain how Halsey was to prioritize this objective if it conflicted with Halsey's obligations under MacArthur's plan.[25] Because of difficult communications, an ambiguous message, and different views of Halsey's primary mission, no one guarded San Bernardino Strait. This nearly resulted in disaster when a powerful Japanese force went through the strait unopposed.[26]

These examples show how cooperation is not always *sufficient* to achieve unity of effort when there is no unity of command.

Cooperation without Unity of Command Succeeds

Cooperation without unity of command can work provided the people involved are determined to cooperate for the overall good of the country. The American Civil War provides several notable examples. In the east, competent army and navy officers frequently overcame interservice rivalries to produce victory.

In early 1862 the expedition to capture Roanoke Island employed two commands that were supposed to cooperate. Brig. Gen. Ambrose E. Burnside led the army part of the expedition, and Flag Officer L. M. Goldsborough led the naval part. Gunboats bombarded the two forts guarding the channel connecting Pamlico and Albemarle Sounds. The Union gunboats also drove off several Confederate gunboats that tried to oppose them. After naval gunfire silenced the guns in the Confederate forts, Union troops went ashore. As the troops were about to reach the shore, the gunboat *Delaware* provided close-in gunfire support by shelling a patch of trees behind the landing point.

The next morning the ships suspended naval gunfire when troops moved into the bombardment zone and did not resume firing until requested to do so from shore. This gunfire support helped the troops capture the fort. Because the gunboats had control of the waterways, the army was soon able to move to the mainland at New Bern. Gunboats shelled roads and strong points ahead of the troops. In the Roanoke campaign, close cooperation between army and navy forces was necessary and led to success. They achieved unity of effort without unity of command.[27]

The western theater of the American Civil War provides additional examples. First, in February 1862 Brig. Gen. Ulysses S. Grant and Flag Officer Andrew H. Foote cooperated well in capturing Forts Henry and Donelson. Technically there was a form of unity of command here because the riverine naval forces in the West were under the direction of the War Department—an early example of a unified command. However, Grant did not have command of Foote; they both reported to Maj. Gen. Henry W. Halleck, who was far from the action back in St. Louis.[28] Therefore there was no unity of command on the scene and the situation between Grant and Foote was dependent on their cooperation. The result was the first two important successes of the Union in the Civil War.

During the Vicksburg campaign the Union forces did not have unity of command because the riverine naval forces had been transferred to the Navy Department in October 1862. Grant, now a major general, led the army forces and Acting Rear Adm. David D. Porter commanded the Mississippi Squadron. They cooperated closely during the campaign, in part because Porter was on cordial terms with Grant and Maj. Gen. William Sherman. Grant effusively praised Porter and his subordinates for their wholehearted cooperation.[29] They achieved unity of effort without unity of command.

The U.S. Marines invaded Guadalcanal in August 1942. Two days after the landing, the aircraft carriers withdrew from the area, leaving the Marines with no effective air support for nearly two weeks. (This is one of the reasons that today's U.S. Marines want to control their own, organic airpower.) After that, a Marine squadron of fighters and another of dive bombers, five U.S. Army Air Force fighters, and a squadron of U.S. Navy dive bombers from the aircraft carrier *Enterprise* arrived on the island. This formed the Cactus Air Force on Guadalcanal. Aviation fuel was constantly in short supply. Japanese warships bombarded Henderson Airfield several times. Japanese ground forces repeatedly tried to capture the airfield. In short, the situation was dire and the outcome uncertain for several months. In the face of danger and urgent operational needs, everyone set aside service concerns about proper roles and command relationships. There was extensive improvisation. Aircraft were assigned to the most urgent needs without regard for the mission for which they were designed. For example, the U.S. Army's P-400 was designed as an interceptor, but it was poor at that role. It could not climb to the altitudes at which the Japanese aircraft flew. Therefore, the P-400 was used for reconnaissance and close air support. Although it was not designed for close air support, it proved to be excellent in that role.

The command situation was complex but it worked because commanders off the island provided assets and left the tactical decisions to the people on the island. Members of all services on the island cooperated wholeheartedly because the danger and the grim situation produced a strong motivation to set aside interservice disagreements.[30]

The cooperation necessary to achieve unity of effort without unity of command seems to require strong motivation to make the situation work, agreement on general ideas of how to proceed, a willingness to compromise, excellent and frequent communications, and, above all, trust. Those requirements are just as desirable—and just as necessary—for achieving unity of effort even when there is unity of command.

The ultimate example of cooperation without unity of command might be the Battle of Novara on 6 June 1513. The French expedition in northern Italy was attempting to wrest control of the Duchy of Milan from Swiss occupiers. Mercenary captains led each company of Swiss pikemen. The Swiss captains acted independently, controlling only their own company. No superior headquarters coordinated their actions, and they cooperated by consent. Even though the French outnumbered the Swiss, the Swiss captains collectively decided to attack. With the aid of surprise, the early morning attack defeated the French.[31]

International coalitions are common and often unity of command is impractical. At the Battle of Waterloo the Allies did not have unity of command. The Duke of Wellington led a multinational army of ninety-three thousand British, Belgians, Dutch, and Germans. Field Marshal Gebhard von Blücher commanded the Prussian army of 116,000 men. Blücher was not under Wellington's command but they communicated well before and during the battle and coordinated their efforts to defeat Napoleon Bonaparte.

Napoleon, with roughly 120,000 men, was a genius at placing his army between two opposing armies to prevent their juncture and then defeating each in turn. On 16 June 1815 Napoleon inflicted heavy casualties on Blücher at Ligny and forced him to retreat. He then detached Marshal Emmanuel de Grouchy with thirty-three thousand men to shadow Blücher and prevent him from supporting Wellington. Napoleon believed that he had ensured that the Prussians would not be able to join Wellington for two or three days. That interval would give Napoleon time to destroy Wellington's army.

Blücher had two choices of directions in which to retreat. Napoleon expected him to follow the safest course of action and retreat eastward toward his base at Liege. However, Blücher knew that if Wellington retired westward along his line of communications, the two armies would diverge, and Napoleon would have the situation he wanted. Therefore, Blücher retired northward to Wavre, which kept him close to Wellington. He made this fateful decision even though it increased the risk to his army.

Wellington might have retreated anyway and put the Prussians at risk of facing Napoleon alone. Nevertheless, Blücher informed Wellington he would start one Prussian corps on the march toward Wellington at dawn on 18 June and other corps would follow soon after. Relying on Blücher to keep his word, Wellington stayed and defended his position at Waterloo, in doing so running the risk that the Prussian general would be slow to support him. Meanwhile, Blücher ignored the objections of his chief of staff and sent three of his four corps toward Wellington, which entailed

Unity of Effort

the risk of having his fourth corps destroyed by Grouchy at Wavre. In the end, the three Prussian corps made a decisive difference in the battle. By trusting each other, Blücher and Wellington won the Battle of Waterloo and the campaign.[32]

During the Falklands War in 1982 the British did not have unity of command in the South Atlantic. The component commanders all reported directly to the same overall commander, who was thousands of miles away in London. The component commanders were expected to cooperate.[33] Inevitably, friction arose, but good communications with their common superior in London prevented major problems.[34]

During Operations Desert Shield and Desert Storm, national pride prevented Saudi Arabian forces from serving in their own country under Gen. H. Norman Schwarzkopf. The Americans and the other Western coalition forces were not about to place all their forces under the practical command of a Saudi Arabian inexperienced in leading such large forces. HRH Gen. Khaled bin Sultan, a Saudi Arabian, was officially coequal with Schwarzkopf. As Joint Forces and Theater of Operations commander, Khaled had command over forces from Saudi Arabia, Syria, Egypt, Kuwait, Bangladesh, Senegal, Qatar, the United Arab Emirates, Oman, Morocco, Czechoslovakia, and Bahrain. The forces of the United States, Britain, and France were each under their own national command.[35] Thus unity of command did not exist. In practice Schwarzkopf had effective command of the Western coalition forces and had to rely on cooperation from the Arab forces.

Vice Adm. Henry H. Mauz Jr. commanded the U.S. forces at sea during Operation Desert Shield. He also led the multinational force that conducted maritime interception operations to enforce the sanctions against Iraq. He had operational control over some parts of the multinational force, much less authority over others. The navies of most countries involved worked together both operationally and tactically. Operationally they coordinated schedules and patrol areas so that everything was covered. They also hammered out procedures for conducting interceptions of merchant ships. Tactically, commanding officers of ships from different navies talked over the situation and agreed on who would act as the scene-of-action commander and the part that would be played by each ship. Scene-of-action commanders could not order ships of other countries to do things such as firing warning shots, but had to "invite" them to do so.

Because every country imposed their own rules of engagement on their ships, this might have created difficulties. However, it produced opportunities for innovative solutions to problems because some nations allowed their ships to do things that other nations forbade their ships to do. For instance, one nation might allow its ships to fire warning shots, but not to enter territorial water. Meanwhile, another nation might allow its ships to enter territorial waters, but not to fire warning shots. Thus despite the lack of a formal command structure, the maritime interception operations proceeded with few glitches.[36]

When war began in Operation Desert Storm in January 1991, multinational command and control problems intensified. Most governments involved in the coalition imposed restrictions on the use of their naval forces. Some foreign navies were willing to serve under the command of Vice Adm. Stanley R. Arthur, who was then in charge of all the U.S. naval forces, but were not willing to serve under the command of certain other countries' naval officers. The British and the Kuwaitis were full partners. The forces of many other countries were not under Arthur's command so he had to use persuasion and rely on their good will to obtain their cooperation. What occurred in practice was often not that which written documents indicated. Personal relationships were often more important than pieces of paper.[37]

Throughout Desert Shield and Desert Storm, on land, at sea, and in the air, even though there was no formal unity of command due to the multinational sensitivities, there generally was cooperation, which resulted in unity of effort, which was of course more important.

During Operation Enduring Freedom in Afghanistan in 2002, personnel from the CIA, Special Forces, and the Advanced Force Operations (small teams to do high-risk reconnaissance missions) were under different chains of command. The Advanced Force Operations personnel were commanded by an organization separate from the commander of the rest of Americans in the country. Of course, command of the CIA personnel was also separate.[38] Nevertheless, people from all three groups worked together closely and effectively in advanced locations by disregarding chain-of-command issues.[39] For example, normal procedure called for CIA personnel to send raw intelligence back to CIA headquarters. They would process the intelligence and disseminate it several days later. In this instance, however, the CIA personnel in theater cut days out of the process by simply giving the raw intelligence to the person a few feet away, who needed the information immediately.[40]

Thus, we have many examples in which cooperation without unity of command achieved unity of effort. We conclude that effective cooperation does not *require* unity of command.

Unity of Command with Poor Cooperation

Having unity of command does not guarantee cooperation among subordinates. Lack of cooperation might occur in several ways: disobedience of orders, subordinates with their own agenda, ineffective commanders, rivalries, and bureaucratic inertia. Unity of command does not lead to unity of effort if subordinates, with perhaps the best of intentions, undermine the commander.

Maj. Gen. J. E. B. Stuart led the Confederate cavalry. During the Peninsular Campaign in 1862, he had gained glory by riding around the Union Army. In 1863, during the Gettysburg campaign, one of Stuart's tasks was to be Gen. Robert E. Lee's eyes and tell him where the Union forces were located. During the days prior to the Battle of Gettysburg, Stuart again rode around the entire Army of the Potomac and captured many wagonloads of supplies. This was a glorious thing to do. As a result, however, Lee did not know the location and disposition of the Union Army and stumbled into contact at Gettysburg. Had Lee been aware of the disposition of the Army of the Potomac, he might have arranged to do battle under circumstances that were more favorable. Because Lee was in charge of all Confederate forces, unity of command was not an issue. Nevertheless, Stuart's quest for glory undermined the Confederate effort.[41]

During the 1983 operation to rescue students in Grenada, some U.S. Army helicopters landed on the amphibious helicopter carrier *Guam*. Vice Adm. Joseph Metcalf, in charge of the task force, received an urgent message from the U.S. Navy comptroller in Washington, telling him not to refuel the Army helicopters until the U.S. Army arranged payment for the fuel. Metcalf ignored the message.[42]

Later during the same 1983 operation the Atlantic Command ordered the task force to rescue students located at Grand Anse. Approaching land forces were moving slowly. From the deck of the *Guam*, Brig. Gen. H. Norman Schwarzkopf, Army advisor to the task force commander, could see the building where students were being held. He wanted the

U.S. Marine Corps helicopters sitting on the deck to ferry Army personnel to the building. When he explained this plan to the Marine colonel in charge, the latter replied that he was not going to do that because "We don't fly Army soldiers in Marine helicopters." It took an ugly scene and threats of court-martial to make it happen.[43]

During the 2003 invasion of Iraq the Americans had unity of command in the theater and at the level of the Land Component Commander. The U.S. Army's V Corps and the I Marine Expeditionary Force were each assigned zones of action. The two forces were expected to cooperate, though the Land Component Commander could resolve serious disagreements. The gap between the forces grew as large as fifty miles. Because the Marines were not allowed to attack across the zone boundary, this effectively created a sanctuary for the Iraqi Fedayeen. Michael Gordon and Lt. Gen. Bernard Trainor, authors of *Cobra II*, judged that in some ways they were fighting separate wars.[44]

Unity of command alone is not *sufficient* to achieve unity of effort because it is possible to have unity of command without achieving unity of effort if subordinates do not cooperate fully.

New Principle of Unity of Effort

The preceding examples suggest that cooperation is *necessary* to achieve unity of effort, but it is not *sufficient* to achieve it. They also suggest that unity of command is *neither necessary nor sufficient* to achieve unity of effort. Nevertheless, unity of command is desirable because cooperation is enhanced, though not guaranteed, by direct orders.

Because unity of effort is the goal and neither unity of command nor cooperation is sufficient to attain unity of effort, we conclude that unity of effort is the more fundamental concept and therefore should be the principle of war. As Russell Glenn puts it, "Unity of effort is the function we require for success . . . unity of command is the form we should seek to attain it."[45] It is peculiar that the U.S. military, which universally proclaims the value of mission-type orders—tell subordinates the desired end state but do not dictate *how* to achieve it—endorses a principle of war that tells *how* to achieve unity of effort.

We propose the following definition for the new principle of unity of effort:

Unity of Effort

Unity of effort consists of the integration of forces and activities to make efficient use of all resources to achieve the primary overall objective. The highest level of unity of effort is effective combined arms operations.

Unity of command is an important factor in achieving unity of effort. Even with unity of command, however, commanders must emphasize cooperation and unity of effort to subordinates.

If unity of command is not feasible, there must be cooperation.

Good cooperation requires mutual trust, goodwill, and unity of purpose. Good cooperation also requires integrated training and interoperable systems, especially communications.

We wish to emphasize that commanders, even those having unity of command, must make a major commitment to achieving unity of effort by emphasizing the need for cooperation to their subordinates and not tolerating interservice or international rivalries. Factors discussed in other sections of this book also contribute to unity of effort. For example, commanders should set clear objectives, make orders lucid, and ensure subordinates understand how their objectives interact with higher-level objectives and the objectives of others in the organization.

The remainder of this chapter discusses various aspects of command—with or without unity of command—that contribute to unity of effort. We first discuss foundational elements and then operational factors that enable and promote unity of effort.

Foundational Elements

If each of the following foundational elements is in place prior to an operation, the chances of achieving unity of effort will be enhanced.

Every potential participant should have systems, especially communications, that are interoperable both technically and procedurally. Interoperable weapons and equipment improve the efficiency of operations. During Desert Storm, different air-to-air refueling systems and jet fuel hindered U.S. Navy and Air Force cooperation. The Air Force used a more efficient fuel, while the Navy used a less volatile fuel due to the dire consequences of fire on ships. Also during Desert Storm, the U.S. Air Force and Navy used electronic data link systems that were technically interoperable, but incompatible procedures meant they could not combine the data links reliably.[46]

All participating units should share doctrine that has been tested. people should train together, share a common approach, and use the same vocabulary. Doctrine codifies lessons from the past. Shared doctrine ensures everyone has a common approach. Testing doctrine is essential to detect flaws. To implement doctrine effectively, people need to train in using it. To ensure systems are truly interoperable, people must train together in their use.

Encourage low-level enterprise. All subordinates should be trained to show enterprise when encountering unexpected circumstances. They should be taught to act to carry out the commander's intent without orders when the situation requires quick action. No set of orders or objectives can ever anticipate every contingency. More so in warfare than any other human endeavor, one should "expect the unexpected." The complexity of warfare is one factor. Another factor is that the enemy is constantly scheming to thwart your plans, frequently by doing something you do not expect. However, if everyone in the chain of command understands the overall objective, the prioritization of secondary objectives, and the plan, then they can exercise ingenuity and enterprise when they encounter unexpected situations, as they almost invariably do.

In northern Gaul in 57 BCE the Nervii tribe attacked the Romans just as they started to make camp and before they had it fortified. Caesar was fully occupied gathering his forces. The Romans, however, prevailed, due to two factors that in Caesar's view saved the day. First, because of his soldiers' training and experience, they knew what to do without waiting for orders. Second, Caesar had ordered each of his generals to stay with his legion. The generals also did not wait for orders but acted as they deemed necessary on their own responsibility. In any event, because of the confused situation with several separate actions and thick hedges obstructing the view, Caesar could not see what his forces needed to do in the various parts of the battlefield. Therefore real-time unity of command was impossible. In the end the Romans nearly annihilated the Nervii tribe.[47]

As the three divisions in Pickett's Charge approached his position in the Union lines, Brig. Gen. Alexander Hays saw that his northern flank overlapped the Confederate left. Therefore, he swung his northern regiment "like a gate" and attacked the flank of the enemy. Thus, a regiment that otherwise would not have had a role in the fight wreaked havoc on the left flank of the Confederate attack. Maj. Gen. Winfield Scott Hancock saw this maneuver and liked it so much he immediately rode to the other end of the line to order the same maneuver at that end. When he arrived there, he found that Brig. Gen. G. J. Stannard had already pivoted two of his regiments. Soon his Vermonters were firing into the right flank of the Confederate attack. Again, a force that otherwise would not have played a role became an important part of the defense. In both cases, this was not due to orders from the top but resulted from enterprise on the part of subordinates who understood the big picture.[48]

In extreme cases, a subordinate might not just have to improvise: the situation might indicate that he should disobey his orders. Obviously, such an action is not to be taken lightly, and its widespread use would result in chaos rather than unity of effort. However, judiciously used, disobedience of orders might be just the thing needed to get true unity of effort. As Charles Cook writes,

> The best discipline does not require blind obedience. At those rare times when the subordinate can be sure, it demands the action that his commander would desire were he in possession of the same information.
>
> This kind of discipline places the burden of decision upon the subordinate. The burden lies in the need to be certain that one has all the facts and that they fully justify the decision to take action contrary to the immediate intentions of the commander.[49]

At the Battle of Copenhagen on 2 April 1801, Adm. Sir Hyde Parker was in overall command of the British forces. He was outside the harbor, but thought the battle inside the harbor was going badly. Therefore, he signaled his second-in-command, Vice Adm. Horatio Nelson, who was in the thick of the fight inside the harbor, to break off the battle. When told of the signal, Nelson raised his telescope to his blind eye and said he could not see any such signal. The battle continued and Nelson won a great victory.[50] This was a case of the superior officer not having as good a picture of the situation as the man on the scene.

During the Saratoga campaign in the American Revolutionary War, the Americans and British fought a pivotal battle at Bemis Heights on 7 October 1777. At the start of the battle, Gen. Horatio Gates took exception to Maj. Gen. Benedict Arnold's criticism of his orders and relieved

Arnold of his command: "I have nothing for you to do. You have no business here."[51] Arnold was humiliated. While in his quarters, he heard the sound of battle and could not stand it. He disobeyed his orders, grabbed a horse, and rushed to the front. Gates sent a man to order him back to camp. Arnold outran him. Arnold usurped the command of another officer and led three regiments in a wild charge. While Gates remained in his tent, two miles from the battle, Arnold and Col. Daniel Morgan led the Americans in the battle, with Arnold bravely leading the final charge that resulted in victory. Ten days later, the British commander, Gen. John Burgoyne, and his fifty-nine hundred men surrendered at Saratoga.[52]

Operational Factors

Commanders can take actions to enable and promote unity of effort during an operation.

Ensure unity of purpose. Everyone should understand and support the overall objective of the operation. They must be willing to set aside personal biases, service parochialism, and national rivalries to work for the common good.

Commanders should ensure that everyone in the chain of command knows his or her role in the big picture. Of course, everyone needs to understand the specific objectives assigned to him or her and the commander's prioritization of those objectives. They also should be familiar with the overall objectives of the operation so they know the context of their assignment. In addition, they should be aware of parallel assignments that might support or interfere with their own assignments. Clear orders facilitate this. Sharing information is vital. A subordinate does not need to know everything about everything, but should know everything he or she might need to know. This is more than just the information that they will certainly need to know. They must also be aware of information that they might need if something unanticipated occurs.

Unify commands. Although unity of command is neither a *necessary* condition nor a *sufficient* condition for achieving unity of effort, it is highly desirable. Even when circumstances—international coalitions for example—preclude unity of command overall, commanders should strive to achieve as much unity as possible throughout the command structure. In addition, commanders should not give up the benefits of unity of command when circumstances beyond their control do not conclusively preclude its use. Although unity of command is not a principle of war, it is an important and effective tool to achieve unity of effort.

Demand subordinates cooperate. It seems obvious that having both unity of command and cooperation is the best situation, provided it is practical. However, unity of command does not automatically result in effective cooperation, and cooperation is necessary to attain unity of effort.

To achieve cooperation, commanders should demand subordinates work together. They should try to develop trust—think of Horatio Nelson's band of brothers. Commanders should set an example of trust and respect for other participants. As Michael Clapp, British commander of the Amphibious Task Group during the Falklands War, observed, "It is more difficult for staff officers to squabble if they see their bosses working well together and this will filter down to the lowest ranks, inevitably with enormous benefits."[53]

A commander's conference before Desert Storm brought together the four component commanders, each from a different service, each having seen interservice rivalry impede overall operations in Vietnam and elsewhere. These four men—Lt. Gen. John Yeosock of the U.S. Army, Lt. Gen. Chuck Horner of the U.S. Air Force, Lt. Gen. Walt Boomer of the U.S. Marine Corps, and Vice Adm. Stan Arthur of the U.S. Navy—agreed they would work together. Arthur recalled, "What carried the day was that we, the component commanders, shook hands and said, 'We're not going to screw this up, we're going to make it work.' And it did."[54]

During World War II an American, Gen. Dwight D. "Ike" Eisenhower, was the Supreme Allied Commander, but his air, naval, and ground force component commanders were British. The staff was multinational. Inevitably the British and American staff officers often disagreed, sometimes heatedly. Eisenhower absolutely would not tolerate inter-ally rivalries. Reportedly Eisenhower said that it was okay to call a British officer a son of a bitch but it was not okay for an American to call an officer "a *British* son of a bitch." When an American officer did so during a dispute in a pub, Eisenhower allegedly demoted the man and sent him home on a slow boat. The story might not be true, but Eisenhower approved its dissemination.[55]

Avoid suboptimization. Any complex organization is susceptible to a situation in which a subordinate single-mindedly carries out his mission successfully but, through lack of consideration of the overall objective, impedes, or even precludes, the organization as a whole from reaching its primary objective. We call this "suboptimization."

To avoid suboptimization, commanders must consider the impact on unity of effort in everything they do. One way is to set and clearly state the primary objective, along with prioritized secondary objectives, in accordance with the new principle of prioritized objectives. Subordinates who are assigned subtasks must keep in mind the overall objective. Orders should clearly explain the interactions between the primary objective and the subtask, as well as parallel subtasks.

If subordinates, with the best of intentions, execute their assigned mission in exemplary fashion, but in the process undermine the commander's efforts, we call that suboptimization. Suboptimization is a failure to achieve unity of effort.

Suboptimization can occur when a unit hoards resources intended for other units. Lt. Gen. George Patton's Third Army in France was infamous for its use of extraordinary methods to obtain supplies. They diverted or hijacked trains and convoys intended for other units. They seized fuel that supply vehicles needed for the return trip. They even impersonated other units to obtain supplies.[56] This enhanced the Third Army's ability to advance, but it disrupted the intended distribution of supplies over all Allied forces.

Suboptimization can arise when commanders give subordinates inappropriate objectives, for example, emphasizing body counts during the Vietnam War. Due to the perverse incentives introduced by using body counts as a measure of effectiveness, not everyone was contributing to the overall goal. There was no unity of effort even though there was unity of command.

Suboptimization can also occur because commanders fail to explain to subordinates how their assigned objectives relate to other objectives, as well as the overall objective. Those cases also fall under the category of a failure to adhere to the new principle of prioritized objectives.

In Operation Desert Storm Gen. H. Norman Schwarzkopf's plan was to have the U.S. Marines ashore in Saudi Arabia invade Kuwait from the south. The expectation was that this would draw the Iraqi Republican Guard units into central Kuwait to meet the coming threat. Meanwhile, U.S. and British Army ground forces would make a wide swing through the desert west of Kuwait, enter southern Iraq, and then move east to cut off the retreat of the Iraqi forces in Kuwait. This plan required the U.S. Marines to attack but not to advance too rapidly.

Because the U.S. Marines were greatly outnumbered, the safest way for them to advance was to move rapidly and keep the Iraqi forces off-balance. To stop or slow their advance would be risky. The result was that the rapid Marine advance pushed—"like a piston," to use Lt. Gen. Bernard Trainor's phrase—some of the Iraqi forces out of the trap before the U.S. and British Army forces could close the trap. Thus the Marines accomplished their perceived goal of pushing the Iraqis out of Kuwait, but at the cost of reducing the number of Republican Guard units that the U.S. and British Army forces' left hook trapped.[57]

Choose the best force. Commanders should select the best force for each task, neither favoring their own service nor including units from every service for the sake of appearances. This requires that commanders be knowledgeable about the capabilities and limitations of all forces available.

Avoid micromanagement. A sure way to discourage individual enterprise in subordinates is to constantly micromanage them. This is also frequently self-defeating. Despite modern advances, the person on the scene almost always has a better understanding of the conditions in his immediate vicinity than does the commander. The person on the ground is immersed in the local situation and perhaps has been for days. This person gets information from many different "channels." Meanwhile, a far-off commander must continuously deal with many other problems. Therefore the person on the ground almost always will have better information about the local situation than a far-off commander.

In December 1983 the aircraft carriers *John F. Kennedy* and *Independence* received orders to prepare a retaliatory strike against Syrian anti-aircraft positions in Lebanon. Personnel on the carriers planned a coordinated strike. Then Washington sent a preemptory and very specific order that moved up the time of the attack by several hours. Furthermore, the order specified many details of the attack. With almost no time to change the plan, brief the pilots, or rearm the aircraft, the raid was a disaster and two aircraft were lost. The dangers of micromanagement and political interference were so obvious that this became a watershed event. The U.S. Navy made major changes in the way it plans and conducts air strikes. It established the Naval Strike and Air Warfare Center (the strike counterpart to Top Gun), changes that led to more thorough mission planning processes.[58]

In the twenty-first century, slow communications are not often a problem. In fact, the ability of modern-day commanders to know where their forces are, their condition, and what they see would seem like fantasy to commanders of just a few decades ago. However, the view at the end of an electronic link is often the equivalent of looking at the world through a soda straw. Commanders never have as complete a view of the situation as they think they have. Modern commanders also have an unprecedented ability to confer with subordinates many layers down in the chain of command and to convey orders directly to them. This is a double-edged sword because it makes for a nearly irresistible temptation to micromanage. It also has the potential to leave intermediate levels of command in the dark. While this situation might qualify as unity of command, it is not a good idea in general. Indeed, excessive interference by high-level commanders frequently makes things worse and hinders unity of effort.

During Operation Enduring Freedom in Afghanistan, Operation Anaconda was an assault into the al-Qaeda stronghold in the Shahikot Valley. Commanding generals far from the battlefield observed the situation via satellite links. This gave them a detailed, real-time view of some parts of the battlefield, but not all of it. It did not give them the situational awareness of all the other relevant factors that a person on the ground would know. The far-off commanders thought they knew a lot more than they actually did know. With a badly incorrect understanding of the situation, they issued orders without listening to the people on the ground, who had much better situational awareness. A general not in the country ordered a significant change to the role of the Advanced Force Operations. He replaced critical people with others who were not familiar with or acclimated to the operational environment. At one point a U.S. Air Force general who had no ground combat experience and was far from the scene ignored the advice of those closest to the situation and made ill-advised tactical decisions that violated hard-earned lessons. On three separate occasions, a few hours apart, commanders sent helicopters to the top of a mountain that the enemy occupied. People close to the scene told the commanders that the enemy was likely there, but the commanders ignored the information. The enemy shot down two helicopters and damaged a third.[59]

One way to ensure that the decision-maker has the best information is to make the person with the best information the decision-maker. Commanders should allow people with the best picture, often those closest to the situation, to make decisions in accordance with guidance about objectives from higher command.

Reduce the fog of war. Throughout history two factors that have hindered attempts to achieve unity of effort are the fog of war and friction. Commanders can take various measures to reduce both factors but should never delude themselves into thinking they can eliminate either factor. Unity of command is often useful at reducing the fog of war and may be the best way to minimize the harmful effects of friction, but unity of command by itself is not sufficient to eliminate the deleterious effects of the fog of war and friction on unity of effort.

History repeatedly records the importance of the fog of war in the outcome of battles and wars. A major cause of the fog of war is the lack of information by a decision-maker. Hence the first step is to ensure good communications both up and down the chain of command.

It is incumbent on lower-level subordinates to push information up the chain of command. If they understand the big picture, then they should understand what information is vital. Reports must be clear and unambiguous. History tells of many battles lost due to ambiguous reports up the chain of command. A variation of Murphy's Law applies here: any ambiguity that can be misinterpreted usually is. Simplicity and clarity are the vaccines to reduce misinterpretation.

It is also incumbent on commanders to seek information from the person on the scene. Nothing can substitute for getting a "reality check" from a person who can see what is in front of him or her and is intimately involved in all aspects of the situation there. Subordinates with such local knowledge often know what to do, what is possible, and what is not possible.

Reduce friction with realistic expectations and robust plans. Friction in warfare arises in myriad ways. One aspect of friction is the failure of a unit to perform a task on time because of unrealistic expectations. Commanders should know the environment and the constraints it entails. This book's chapter 16 on the environment covers this topic.

To reduce friction, commanders should devise plans and orders that are practical and realistically account for potential difficulties. One way to help ensure that plans are realistic is to solicit input from the people on the ground, who see the situation in front of them, and from those who will carry out the plan.

During Operation Anaconda in Afghanistan in 2002 a few Americans led Task Force Hammer with four hundred Afghan troops loaded onto three dozen "Jinga" trucks (brightly painted flatbed trucks). Even though this force was the main effort, requirements of a supporting force imposed a tight timeline on Task Force Hammer. Inexperienced Afghans drove the top-heavy trucks at night without lights and without night-vision goggles, along a primitive track softened by two days of rain. One truck overturned and several others became stuck.[60] One could blame this on friction, but the problem really arose because of unrealistic expectations. Inexperienced drivers, top-heavy trucks, darkness, and a muddy track are problematic by themselves. Combined with a tight timeline, things predictably went wrong. This was not bad luck.

UNITY OF EFFORT SHOULD BE THE GOAL OF ANY MILITARY organization. It is the integration of forces and activities to employ all resources in an effective manner to attain the objective. Although cooperation is necessary to achieve unity of effort, cooperation is not a principle of war because it is not sufficient to achieve unity of effort. To improve the chances of achieving a truly coordinated effort, an ultimate coordinating authority should be vested in someone, preferably someone reasonably close to the action. Thus, unity of command is very desirable. However, unity of command is not a principle of war because it is neither sufficient, nor necessary, to achieve unity of effort. Unity of effort should be the principle of war for organizing activity because it better embodies the essence of the concept and is a more fundamental and comprehensive concept than unity of command or cooperation.

NINE

Surprise

[For the French counterattack plan] to be carried out, it would have been necessary for the High Command still to have hope and the will to win. The crumbling of the whole system of doctrines and organization, to which our leaders had attached themselves, deprived them of their motive force. A sort of moral inhibition made them suddenly doubtful of everything, and especially of themselves. From then on, the centrifugal forces were to show themselves rapidly.

—CHARLES DE GAULLE, describing the situation on 20 May 1940, ten days after the German attack, from *Call to Honour*

....

SURPRISE HAS PLAYED A LEADING ROLE IN WARFARE THROUGH-out history: the Trojan Horse, Washington crossing the Delaware, Stonewall Jackson's flank attack at Chancellorsville, Pearl Harbor, the Tet attack, Egypt's attack on Yom Kippur, and the 9/11 attacks are only some of the better known examples. Just a few words of description can evoke vivid pictures of famous (or infamous) attacks.

Current Principle of Surprise

The American definition of the principle of surprise once again starts by giving the purported purpose of the principle:

The purpose of surprise is to strike at a time or place or in a manner for which the enemy is unprepared.

Surprise can help the commander shift the balance of combat power and thus achieve success well out of proportion to the effort expended. Factors contributing to surprise include speed in decision-making, information sharing, and force movement; effective intelligence; deception; applica-

> tion of unexpected combat power; [operational security]; and variations
> in tactics and methods of operation.[1]

The first paragraph of the above explanation actually does not give the purpose of surprise; it defines surprise.

The British definition of their principle of surprise is much better. Each of its five sentences addresses an aspect of the principle that we think is vital:

> Surprise limits our opponents' reaction time by affecting their ability to make decisions. Surprise may also undermine our opponents' cohesion and morale. Surprising an opponent is a significant way of seizing the initiative and may be a critical precondition for success. Surprise is transient and must be exploited rapidly. Commanders should anticipate the effects of being surprised themselves and make appropriate contingency plans to safeguard their freedom of action.[2]

WITH TWO EXCEPTIONS, EVERY MAJOR MILITARY THAT LISTS principles of war includes surprise. France does not. Israel uses stratagem as their corresponding principle, which includes both surprise and the exploitation of surprise. It has the connotation of being part of a plan to achieve a particular end.

Problems with Current Principle of Surprise

To decide whether we should consider surprise to be a principle of war, let us apply the criteria for a principle of war listed in chapter 4. Surprise qualifies as an enduring concept because it has been an essential component of many victories over the ages. The advantage inherent in the basic idea of doing the unexpected has not changed since the time of the Trojan War. It is applicable to the strategic, operational, and tactical levels of warfare. Surprise is relevant to land, sea, and air warfare, as well as space warfare, terrorism, guerrilla warfare, cyberwarfare, and nuclear warfare.

Surprise qualifies as a concept that makes a difference. Certainly, surprise has frequently contributed to victory. Admittedly, combatants have often been defeated in spite of their use of surprise. The Tet attack was a tactical defeat for the Viet Cong even though it was a major cause of the ultimate defeat of the United States. Egypt lost the Yom Kippur War, though it gained its political objective.

Nevertheless, it is difficult to find cases in which the defeat has been *because* of the use of surprise. One can make a case that the Japanese use of surprise in attacking Pearl Harbor in 1941 contributed to their ultimate defeat because it aroused the "sleeping giant." Although war between the two nations was probably inevitable and Japan's defeat was quite likely, nothing else could have united the previously pacifist American populace as successfully as this sneak attack without a declaration of war. Surprise backfired strategically because the Japanese could not exploit the successful surprise sufficiently to win the war. Similarly, the 9/11 attacks backfired on al-Qaeda. In both instances, the problem was that the tactical and operational objectives did not support attainable strategic objectives.

Violation of the concept of surprise by doing what the enemy expects has contributed to many defeats throughout history. Every frontal attack against a strong position qualifies to be on that list. Thus, commanders have violated the concept of surprise sufficiently often in military history that it is worth paying attention to it. Finally, failure to use surprise has rarely led to victory as a result, though commanders have won many victories without dramatic surprise. It is, however, difficult to find cases in which a combatant was victorious without the use of surprise unless he possessed superior force, superior will, or some other distinct advantage.

The final criterion asks whether the principle of surprise is a fundamental element of warfare. Doing the unexpected by attacking in a poorly defended location gains a relative advantage, which in turn can lead to victory. Alternatively, an unexpected action can give the attacker the initiative. If this were the only use of surprise, it might not qualify as a fundamental element. The use of surprise in this manner is important, but it only touches the surface of what surprise can achieve. In many cases, surprise has psychological effects. The essence of the principle of surprise has two parts: doing the unexpected and the psychological effect that directly attacks the mind of your enemy. The latter effect can be extraordinarily powerful. If the principle of surprise includes the psychological effect on the enemy, that allows us to call surprise a fundamental element of warfare and therefore conclude that surprise qualifies as a principle of war.

However, the American definition of the principle of surprise does not even mention any psychological effects. The British definition mentions only that surprise might affect the enemy's cohesion and morale. Thus, both the American and British principles of surprise fail to meet our criteria as a principle of war.

New Principle of Surprise

We propose a new principle of surprise that incorporates the five vital factors in the British definition and also emphasizes the psychological aspects of surprise:

> Surprise consists of acting at a time or place or in a manner that the enemy does not expect and hence for which the enemy is relatively unprepared. Even if it is not feasible to surprise the enemy about all of these factors, a surprise in just one aspect can be useful.
>
> Surprise is most effective when the enemy is psychologically unprepared. A psychological surprise can seriously disorient the enemy and damage the adversary's morale. It may force the enemy to make decisions rapidly and degrade the enemy's ability to make good decisions.
>
> More generally, surprise leads to a transient advantage that commanders must quickly exploit by using the initiative gained by surprise to achieve a relative advantage, and then converting that into a long-lasting advantage.
>
> Surprise applies also to defensive operations.
>
> The unexpected introduction of weapons, sensors, or tactics can be extremely effective, but the usage must be massive to maximize the effect, and one must be ready to exploit success.
>
> Factors leading to surprise include deception, knowledge of the enemy and the environment, speed of maneuver, and, above all, imagination and innovation.
>
> Commanders should anticipate the potential for being surprised themselves and make appropriate contingency plans.

USUALLY, DECEPTION IS INCLUDED UNDER THE PRINCIPLE of surprise, either explicitly or at least implicitly. We wish to distinguish the two concepts. Deception is a tool to achieve surprise. It often aids surprise, but surprise can be achieved without deception. In the next chapter we will discuss whether we should consider deception to be a principle in its own right.

Surprise is an important tool that is useful for gaining the initiative or relative advantage, but it also has unique aspects that cause us to consider it as a principle on its own. The psychological effect on the enemy can be greater than the physical effect. We will first discuss doing the unexpected and then examine the psychological aspects of surprise.

Doing the Unexpected

Lord Thomas Cochran was the real-life inspiration for Patrick O'Brian's Jack Aubrey and other fictional heroes. In January 1820 Cochran was the commander in chief of the Chilean navy, as Chile was fighting for its independence from Spain. Cochran intended to attack the strong fortress of Valdivia, defended by nearly 2,000 men.[3] Regarding the wisdom of attacking Valdivia with only 250 men, Cochran made the following observation before the attack:

> Cool calculation would make it appear that the attempt to take Valdivia is madness. This is one reason why the Spaniards will hardly believe us in earnest, even when we commence; and you will see that a bold onset, and a little perseverance afterwards, will give a complete triumph; for operations unexpected by the enemy are, when well executed, almost certain to succeed, whatever may be the odds; and success will preserve the enterprise from the imputation of rashness.[4]

Because the fortress's guns guarded against an attack from the sea, Cochrane attacked from the landward side and succeeded.[5]

If something is obvious to you, then the enemy probably will see it also. If you do what the enemy expects you to do, the enemy will do its best to counter you. If the enemy expects you to attack point A, it will strongly defend point A. In that case you can only win if you have greater combat power that you can bring to bear or if your army has greater skill or courage or some similar factor pertains. By attacking in an unexpected manner or at a time or place where the enemy does not expect you to attack, you will likely face something less than the enemy's best defense. Similarly, if an attacking enemy expects to encounter a certain type of defense, it is likely to be prepared to counter it, but if your defenses are not what it expects, it may be unprepared. Doing the unexpected so that the enemy is not as well prepared as he would otherwise be is one part of the essence of the principle of surprise. Sun Tzu, Vegetius, and Machiavelli all advocated doing the unexpected. Frederick the Great said, "Everything which the enemy least expects will succeed the best."[6]

Liddell Hart, in his explanation of the theory of indirect attack, claims that nearly all success is due to doing the unexpected. He gives dozens of examples throughout history in which a commander tried the direct attack and failed, perhaps multiple times. Then, out of desperation, the commander tried an indirect attack that was successful.[7] The concept is quite simple: gain a relative advantage by doing the unexpected. Even if you cannot bring all of your combat force to bear at the point of attack, if you face a significantly less formidable defense, then you have a better relative advantage. Note that the principle of mass argues against making attacks where you cannot bring your full force to bear. The new principle of relative advantage clarifies why that proposition is incorrect.

Because a wise enemy commander will also be seeking a relative advantage, the best way to achieve a relative advantage is to do something that the enemy does not expect. Better still, do something the enemy considers impossible. Best of all, do something the enemy has not even considered. One bias of the human mind is that we tend to think that things of low probability are in fact impossible.

However, surprise normally gives an advantage that is only short-term, so you must exploit that temporarily favorable position in order to make the gain permanent. If the use of surprise gains the initiative, then you must expand and use it to achieve a relative advantage. If you simply "sit" on the initiative gained by surprise, your advantage dissipates as the enemy recovers. We will discuss a few examples of surprise at the strategic, operational, and tactical levels of warfare. We will then consider technological surprise, such as the use of new weapons or sensors, which deserves to be in a separate category all to itself.

STRATEGIC SURPRISE

One way to achieve strategic surprise is to move forces in ways that the enemy thinks are impossible; for example, Hannibal's crossing of the Alps. Another way to achieve surprise is to do something that no reasonable person would do. Then, even if the enemy sees indications of the coming attack, it will put a more benign interpretation on the evidence.

Iraq's invasion of Kuwait in August 1990 achieved surprise in spite of satellite surveillance and numerous clues that became clear only in retrospect. Intelligence agencies dismissed Iraqi troop movements to the border near Kuwait as just blustering to pressure Kuwait and Saudi Arabia into forgiving some of Iraq's debts. Saddam Hussein might have exploited his advantage by invading Saudi Arabia and at least occupying the northeast part of the country where many of the oil fields are located. Most likely, Iraq's limited logistics capability prevented this. If it had been possible, that would have given Iraq a much better position than it obtained by simply sitting on its gains in Kuwait and allowing the coalition to build up its forces unhindered for more than five months.

OPERATIONAL SURPRISE

By 210 BCE Roman fortunes in the Second Punic War were at a low point. Hannibal had been ravaging Italy for seven years and defeating every Roman army sent to confront him. In Spain the Carthaginians had also inflicted a series of defeats on Roman armies. Scipio Africanus recognized that Spain was Hannibal's real base because it was the source of his reinforcements. The Romans elected Scipio proconsul for Spain, in part because no one else wanted such a seemingly hopeless job. When he arrived to take command, he found that three widely separated Carthaginian armies were in Spain, each about the same size as Scipio's own army. Because of personal animosity among the Carthaginian commanders, Scipio knew they would be slow to come to each other's aid. If Scipio attacked one of the enemy armies, he might be able to defeat it before one of the other two came to its assistance. However, if Scipio could not induce the opposing army to engage in a decisive battle quickly and then defeat it decisively, he would run the risk of two or three enemy armies trapping him.

Which of the three armies would Scipio attack first? None. Instead, he chose to attack New Carthage (modern Cartagena). As one of the few Spanish cities with a harbor fit for a fleet, it was the Carthaginians' main base. They kept their treasury there and a large stash of supplies. Because Carthage controlled most of Spain, they did not think the Romans would dare attack the strong fortress at New Carthage. Hence only a thousand trained soldiers defended it. The rest of the large population was comprised of artisans, tradesmen, and sailors—none of whom were trained defenders.

In spring 209 BCE Scipio's army made a forced march of five hundred miles to reach New Carthage and achieved surprise. He had arranged for his fleet to arrive offshore at the same time. Scipio's operational surprise was supported by tactical surprise. Before setting out, Scipio had gathered information about New Carthage and its defenses. The city was on a narrow isthmus guarded on two sides by the sea and on a third side by a lagoon. High walls guarded all sides of the city. From local fishermen, Scipio learned that the "protective" lagoon was in fact fordable under certain conditions. Scipio first made simultaneous attacks from the seaward and landward sides to distract the enemy's attention and draw the Carthaginian reserves to those sectors. Guides led five hundred Romans through a passage across the lagoon. Because the other attacks had drawn the defenders away from that sector, this group of Romans scaled the wall without opposition, worked their way along the wall, and attacked the Carthaginians near the gate. The entire city soon fell.

Scipio now controlled the best port for communications with Carthage. He had captured the Carthaginian treasury and supplies. In addition, by being generous to the captives, Scipio soon had the artisans making weapons for the Romans. Psychologically, he had shocked the Carthaginians by capturing their main base.[8]

The capture of New Carthage was a triumph of imagination—the ability of Scipio to see that he could capture it, in contrast to the Carthaginians who thought the city was so impregnable and their control of Spain so complete that no one would dare attempt to capture the city. It also illustrates the value of knowing the environment (the passage through the lagoon) and relative advantage (drawing defenders away from the lagoon side of the wall).

Similarly, the surprise of the Pearl Harbor attack was not the fact that the Japanese would attack American interests, but the location they chose. The Americans expected Japan to attack the Philippines and other places in the Far East. They did not imagine that Japan could project power halfway across the Pacific. However, the Japanese did not fully exploit their advantage. Tactically, they did not bomb the repair shops or the oil storage tanks, which would have hindered the Americans' ability to fight for the next six months or more. Instead, the primary result of the

Japanese attack was the destruction or damage to eight American battle-ships. Ironically, the Pearl Harbor attack demonstrated that battleships were obsolete as the core of a navy. While the Japanese periodically caused themselves unnecessary problems by giving their battleships a larger role than they should have, the Americans initially had no choice but to build their naval forces around aircraft carriers. At the operational level, for five months the Japanese navy ranged from the Indian Ocean to the middle of the Pacific Ocean with limited opposition. Yet the Japanese did not fully exploit their advantage by seeking battle with the American aircraft carriers while the Japanese had the advantage in numbers and in the quality of their aircraft and pilots.

If you can find a way to do something that seems physically impossible or stupid, that is an excellent way to gain surprise. By the time the Americans were about to invade Tinian in 1944, the Japanese had deduced the Americans' requirements for a landing beach. They knew Tinian Town and Asiga Bay were the only two suitable beaches. Accordingly they defended those two areas. The Americans astutely chose a *third* option. The third area, near the northwest tip of Tinian, consisted of two tiny beaches with little more than a trail leading inland. This seemed an impossible landing site because it was too small. Nevertheless, the Americans chose that beach. They conducted a feint at Tinian Town that involved battle-ship fire support and landing craft getting within four hundred yards of the beach. When the landing craft turned back, the Japanese jubilantly reported they had repelled the invasion. When the Americans landed in the northwest part of the island, they encountered just one machine-gun nest guarding the beach, while six thousand Japanese defenders guarded the other two beaches. Fifteen thousand Marines landed on the first day, with only fifteen Marines killed.[9]

The amphibious landing at Inchon in 1950 also involved doing the unexpected. There were many reasons not to land there. One reason was a thirty-two-foot tidal range that would leave the first landing units iso-lated until the tide came back in. Mud flats extended two miles out to sea at low tide. Currents were ferocious. An amphibious assault at Inchon would be nearly impossible. Nevertheless, the landing was successful.[10] The Americans exploited their success at Inchon by threatening to cut off the North Koreans in the southern part of the Korean peninsula. Before long, they crossed the thirty-eighth parallel and drove deep into North Korea.

In 371 BCE Sparta attacked Thebes at Leuctra in a battle of the best two Greek armies of the time. In Greek warfare of that time, the place of honor for the best hoplites was on the right wing. Typically, victory went to the side whose right wing first broke the weaker left wing of the enemy. This time, however, Epaminondas, commander of the out-numbered Thebans, put his best hoplites in an over-weighted left wing with more than fifty rows of hoplites, as opposed to the usual twelve rows. The powerful Theban left wing drove into the Spartan right wing and defeated them.[11]

In the same vein of choosing to do battle in a manner that maximizes one's relative advantage, one might choose to precede an infantry attack with an artillery barrage to soften up the enemy. Alternatively one might forgo a lengthy artillery barrage to achieve some measure of surprise. During World War I stupendous artillery barrages up to nineteen days long preceded most infantry attacks. When the artillery started firing, the enemy hunkered down in their bunkers until the barrage moved away. Then they came out of their bunkers and manned the trenches to meet the infantry attack they knew would be coming. Meanwhile, higher headquarters, alerted by the artillery, would get reserves ready to move if needed. In March 1915 at Neuve Chapelle the British tried an intense thirty-five-minute artillery barrage. Four divisions of infantry immediately followed and achieved tactical surprise and a breakthrough. Unfortunately, the British high command had not planned for any exploitation of the attack. The Germans counterattacked and limited the British gains to one square mile. The British found that the element of surprise gave them a better relative advantage than the concentration of combat effects with a massive artillery barrage.[12] Later the Germans also discovered that surprise was better than lengthy artillery barrages.[13]

The Battle of Ilipa during the Second Punic War illustrates a more elaborate surprise. It also touches on deception and the use of surprise to gain a relative advantage. In 206 BCE Scipio Africanus encountered the more numerous Carthaginians under Hasdrubal at Ilipa in Spain. The Carthaginian cavalry attacked the Romans while they were constructing their camp, but Scipio, anticipating the move and utilizing surprise in defense, had hidden some cavalry behind a hill. They attacked the flank of the Carthaginian attackers and disrupted the attack.

Over the next several days Scipio "conditioned" Hasdrubal. According to the custom of the day, the Carthaginians advanced to the valley between the two camps and offered to give battle. After the Carthaginians moved out, the Romans formed up on the opposite side of the valley but did not advance to give battle. Scipio remained there until the Carthaginians returned to camp late in the day. Scipio repeated this every day for several days. The Carthaginian formation had the Carthaginian and African regulars in the center and their less-dependable Spanish allies on the wings. The Romans mirrored this formation each day. After several days of responding to his opponent in the same way, Scipio had conditioned his opponent's expectations.

One day Scipio fed his army early and marched out at dawn. His cavalry attacked some Carthaginian outposts. Caught by surprise, the Carthaginians had to march out hurriedly without eating. There, Hasdrubal found a second surprise. Scipio had altered his formation so that now his less-steadfast Spanish allies occupied the center and his more-reliable Roman legions were on the wings. The Carthaginians could not alter their formation because of the danger of the Romans attacking them in the midst of repositioning. Scipio waited several hours while Hasdrubal's army lamented their missed breakfast.

When Scipio moved, he had a third surprise. His center advanced very slowly, just enough to fix the Carthaginian center in place. Meanwhile, on each of his wings, the well-trained Romans marched obliquely and then attacked the end of the Carthaginian line. Simultaneously, Roman cavalry attacked the rear of the end of the Carthaginian line. This rolled up both ends of the line back toward the center and caused first an orderly retreat and then panic. The surviving Carthaginians fled the battlefield.

Scipio exploited his advantage. With foresight, Scipio had approached the battlefield to cut off the Carthaginians' natural line of retreat. Harassment by Roman cavalry and light infantry slowed the Carthaginians flight and allowed the Roman legions to catch up and complete the destruction.[14]

Ilipa illustrates the use of deception in conditioning his opponent to expect the usual timing and formation. Scipio achieved surprise in timing, formation, and tactics. This gave him a relative advantage where he could attack the enemy's weakest units with his best units, and could do so from an unexpected direction. The attack on the Carthaginian's rear was psychologically devastating and caused panic. Finally, Scipio exploited his resulting initiative by pursuing and destroying his enemy.

TECHNOLOGICAL SURPRISE

New weapons can be quite lethal. Furthermore, sometimes the first use of a weapon can be decisively effective because the enemy has not yet devised countermeasures. In addition, the fear of the unknown can affect the enemy. The next two examples illustrate the failure to maximize the initial effect and to exploit success.

During World War I the Germans used poison gas for the first time on the eastern front at the Battle of Bolimov on 31 January 1915. The gas did not work very well in the freezing temperatures, but the Russians did not report the use of gas to their Allies.[15] Thus the Germans still had the advantage of surprise when they used gas on the western front for the first time, at the Second Battle of Ypres in April 1915. In a surprise attack they used five thousand cylinders to form a cloud of chlorine gas. Two French divisions were understandably terrorized. Two German corps broke through them to create a wide gap in the Allied lines. However, the Germans had not expected such great success and had few reserves available to exploit the penetration. Thus their initiative rapidly faded and the Allies closed the gap in their line.[16] On subsequent occasions when the Germans employed poison gas, the Allies were better prepared and the effects were not so dramatic.[17]

Also during World War I both Col. Ernest Swinton and, more famously, First Lord of the Admiralty Winston Churchill played roles in Britain's development of the tank. The First Battle of the Somme raged from 24 June to 13 November 1916. It was the usual succession of costly assaults for minor gains. The British secretly shipped forty-seven tanks to the front in France. The British first used their tanks in an attack on 15 September. The tanks, however, were underpowered, slow, unreliable, and small in number. Few of the forty-seven tanks actually made it into the battle. The appearance of the tanks caught the Germans by surprise and the British made gains. However, the number of tanks was insufficient and the design was too immature to enable a decisive breakthrough. Swinton had urged his superiors to wait until they had enough tanks to gain a decisive advantage, but they ignored his advice.[18] Again, the advantage gained from surprise quickly dissipated.

Late in World War II Germany deployed jet aircraft, V-1 buzz bombs (cruise missiles), and V-2 rockets (ballistic missiles). All were effective weapons, far ahead of anything the Allies possessed, but too few in numbers and too late in the war to reverse Germany's fortunes.

We can learn valuable lessons by examining the successful technological surprises that did not lead to greater success. First, commanders should hold back the new weapon until they can deploy it in sufficiently large numbers and at a sufficient stage of development to make a *decisive* difference. If you introduce it piecemeal, then the enemy has time to devise countermeasures before you have sufficient numbers of the new weapon to deploy. Second, you should plan for success and be prepared to exploit your advantage quickly before the enemy can recover. You should quickly follow the initial use of the new weapon with additional attacks before the enemy has time to devise countermeasures. You must also have a good concept of how to employ the new weapon. In the case of the British introduction of tanks, they tied the tanks too closely to the infantry, thus partially negating their advantage. In the case of the German development of jet aircraft, Hitler caused a delay of many months by trying to develop it as a jet bomber. Had Germany instead developed it from the start as a jet fighter, it might have been available soon enough and in sufficient numbers to make a significant difference in stopping the Allied air raids over Germany.

In a counterexample, the Long Lance torpedo developed by the Japanese Navy prior to World War II was hugely successful. The Long Lance was a high-speed long-range torpedo that packed a large explosive warhead. Unlike the American torpedoes, the Long Lances were reliable, especially the triggering device. They had a decisive impact on several surface battles in the Solomons because the Japanese deployed large numbers of weapons, crews had good training in use of the torpedoes, and the Japanese had developed good tactical doctrine to take advantage of the capabilities of the Long Lance. The Japanese surprised the Americans by launching the torpedoes at much longer ranges than the Americans thought was possible. When the Long Lances hit American ships, the large explosive warhead often proved to be a ship-killer. After a few losing battles the Americans adjusted their tactics to avoid giving the Japanese good opportunities to employ the Long Lance, but those tactics carried their own detriments in other areas.[19]

In all these examples of technological surprise, the new technology afforded not only a substantial physical advantage but also a psychological impact. The latter was due in part to being put at a disadvantage and in several cases to the feeling of being helpless or trapped. For instance, poison gas has a terrorizing effect because the very air is poisonous and you cannot breathe.

TECHNOLOGICAL SURPRISE: BAY OF BISCAY

Submarine and anti-submarine operations in the Bay of Biscay during World War II illustrate the complex interplay between technological and tactical innovations, countermeasures, counter-countermeasures, and so forth. In this ongoing battle, codebreaking greatly helped the Allies by showing them what the Germans knew. Operations research analysis by some of the best scientists of the day helped the Allies quickly detect trends and inefficiencies.[20] Central direction of the U-boats by Adm. Karl Dönitz both helped and harmed the Germans. The complexity of the interactions merits a detailed examination of the events.

In June 1940 the Germans captured French ports on the Bay of Biscay. This greatly helped them in the Battle of the Atlantic because U-boats sailing from occupied France were much closer to the shipping lanes than U-boats based in Germany. This enabled the U-boats to spend more time on station in the North Atlantic shipping lanes.

In September 1941 British aircraft started patrolling the Bay of Biscay. They wanted to make the search area sufficiently wide that the U-boats could not transit the entire search zone underwater. Because this search was only effective in daylight, the Germans countered by surfacing only at night, starting in December 1941. The British countered that by installing a Mark I radar on their aircraft and flying patrols twenty-four hours a day beginning in January 1942. However, because the radar could not guide the aircraft all the way into the final attack, this did not help much. From January to May 1942 the British did not sink a single U-boat in the Bay of Biscay. However, the slower transit time of the U-boats due to daytime submergence saved merchant ships by reducing the U-boats' time in the shipping lanes.

The aircraft needed a searchlight that the aircrew could turn on at the radar's minimum range to guide the pilot on the aircraft's attack run. The military establishment produced a light for the aircraft, but it was not effective because its brightness and location in the nose of the aircraft dazzled the aircrew. Without any official sanction or help, Squadron Leader Humphrey Leigh devised a steerable carbon arc searchlight mounted in the belly of the aircraft. This proved to be effective. Aircraft equipped with the Leigh Light began flying patrols in June 1942.

Also in June 1942 the British introduced the Mark II radar, which transmitted in the same one-meter radar band as the Mark I radar but had an improved range. The Germans recovered a Mark II radar unit from a British bomber that crashed in Tunisia. They immediately deduced that the British would probably use this radar in their antisubmarine patrols. Therefore the Germans equipped U-boats with Metox radar warning receivers starting in September 1942. U-boats could then submerge before being detected. However, false alarms were a problem. This countermeasure could detect radar transmissions at ranges far greater than the aircraft radar could detect the submarine. If a U-boat dove as soon as it detected a far-distant radar, it could not recharge its batteries until it resurfaced. Each time the U-boat submerged, it increased its transit time. In all of 1942 the Allies sank only four U-boats in the Bay of Biscay, though the search efforts succeeded in reducing the time U-boats spent in the shipping lanes.

From February to April 1943 the British introduced the more powerful Mark III radar, which operated in the 10-cm radio band. Allied results improved. The Germans soon recovered a Mark III unit, but it took them six months to deduce that the British would use it in their antisubmarine aircraft.

In April the U-boats tried submerging at night and running on the surface during the day. They equipped U-boats with anti-aircraft guns but that did not provide sufficient firepower to defend against aircraft. In June the U-boats with anti-aircraft guns transited the Bay of Biscay in groups to provide more firepower. The Allies countered by flying aircraft in loose formation so an aircraft spotting a submarine could summon other aircraft and they could attack in groups. With anti-aircraft guns not getting good results, the Germans tried maximum submergence tactics.

During 1943 the Germans tried several types of radar receivers. However, because the receivers were not very sensitive, they did not detect the Mark III radar. Therefore the Germans concluded that the British were *not* using the Mark III radar for antisubmarine patrols and believed the British might be using infrared detectors. They used special paint to coat the

U-boats to reduce their infrared signature, but this increased their radar signature. The British planted false confirmations about the use of infrared detectors to heighten confusion.

German losses mounted. The Allies sank six U-boats in the Bay of Biscay in May 1943, four U-boats in June, and eleven in July. U-boat skippers became convinced that the British were detecting signals emitted by the Metox receivers due to its heterodyne circuitry. A British prisoner "revealed" they could detect the Metox emissions at very long ranges (they could not). Dönitz ordered the U-boats to stop using Metox in August 1943. Because of heavy losses, Dönitz reduced U-boat operations.

In September 1943 the Germans finally concluded that the British were indeed using the Mark III radar on their antisubmarine aircraft. In November they obtained irrefutable proof with the recovery of a British aircraft carrying depth charges and equipped with a Mark III radar and a Leigh Light. As a countermeasure, the Germans introduced a marginally effective radar receiver, but it was not until April 1944 that they had effective receivers. British aircraft then countered by reducing radar power after they gained a contact (making it seem to be more distant), so the U-boats would not realize they had been detected.

In early 1944 the Germans started equipping U-boats with snorkels, which had a greatly reduced radar signature and enabled them to transit the entire Bay of Biscay underwater. However, by then it was too late. The Allies landed in Normandy in June 1944 and by August the U-boats could no longer use French ports.[21]

SURPRISE IN DEFENSE

Surprise can assist the defense in several ways. For instance, because surprise forces the enemy to confront an unanticipated situation, there is the potential for the enemy to make serious errors that leave an opening for counterattack.

In 530 CE the Persians attacked a Byzantine army at the fortress of Daras under the command of Belisarius. Despite having mostly raw recruits in his army and being outnumbered two-to-one, Belisarius chose to do battle rather than withstand a siege. He dug a deep ditch just outside the walls of the fortress. He also dug two ditches perpendicular to the walls and extending outward. These two ditches then turned outward on his flanks. Belisarius posted his unreliable infantry in the ditch near the wall, where fire from the fortress walls could support them. He posted light cavalry in the perpendicular ditches and heavy cavalry in both ditch extensions on the flanks. When the Persians approached Daras, they did not know what to make of the ditches and feared a trap. Consequently they made mistakes that afforded Belisarius a relative advantage. The Persians avoided entering the area between the ditches. This divided their force and meant that their cavalry on the wings did most of the fighting.

On Belisarius's left wing the Persians made progress until a Byzantine cavalry detachment emerged from behind a hill and attacked the Persian rear. A simultaneous attack from the light cavalry in the ditch forced the Persians back.

On Belisarius's right the Persian cavalry drove all the way to the wall of the fortress. However, this opened a gap between their attacking cavalry and the infantry in the center. Belisarius employed all his available cavalry to an attack in this gap. They first scattered the Persian cavalry, then turned on the now unprotected infantry. The Byzantines decisively defeated the Persians.[22]

Psychological Effect of Surprise

Surprise typically involves gaining the initiative by doing the unexpected, but it also tends to demoralize the enemy, thus intensifying the impact of the surprise. There are, however, circumstances in which the psychological effect of the surprise dominates the effect. If surprise so discombobulates and demoralizes the enemy commander or troops that they no longer function effectively and lose their will to fight (even temporarily), that factor by itself may lead to victory. After all, the mind of the enemy, and especially the enemy commander, is in some ways the main objective of warfare and surprise can be a *direct* assault on the enemy's mind.

In 217 BCE Hannibal crossed to the west side of the Apennine Mountains and then unexpectedly marched his army through the seemingly impenetrable marshes around the Arno River. He then moved south, devastating the land as he went. At this point Hannibal had the initiative (without any offensive action) and could shape the campaign. Hannibal knew that Caius Flaminius, the Roman commander, was impetuous and would follow him. Hannibal played Flaminius the way a matador plays a bull. As he led the Romans on, Hannibal looked for a place to engage the incautious Flaminius at a relative advantage.

The Romans thought the Carthaginians were terrified and trying to escape. Many volunteers accompanying the Roman army carried chains and fetters that they intended to use on the many Carthaginian prisoners they expected to capture. In keeping with the usual Roman practice at this time, Flaminius did not send out scouts. The Romans assumed that any substantial threat would be visible at a distance. When his army reached the area of Lake Trasimene, he had nearly caught up with Hannibal, whose camp was visible several miles ahead, just beyond a line of hills.

The next morning, 21 June 217 BCE, the Roman army left camp expecting to annihilate the Carthaginians. As the Romans entered the defile between Lake Trasimene and the line of hills, misty weather mostly shrouded the hills. When the entire Roman army was in the defile, the Carthaginians simultaneously blocked both ends of the defile and burst out of the hills along the entire length of the Roman army. Because of the mist the Roman soldiers could not see much. However, they would have heard war cries and sounds of fighting coming from all directions.

One minute the Romans were excited to be closing on a long-pursued enemy, and confident they would defeat that enemy. The next minute they were under attack by an enemy they mostly could not see. This must have been a great psychological shock. Although some of the disciplined Romans held out for three hours, many abandoned their weapons and some drowned in the lake. In the end, the Carthaginians devastated the Roman army. The Battle of Lake Trasimene was one of the greatest ambushes of all time.[23]

Use of a new weapon, sometimes even if it is ineffective, can be devastating if it has a terrorizing effect. Consider the British use of Congreve rockets at the Battle of Bladensburg in 1814. Although the rockets could explode and kill people, their flight was erratic and thus they were inaccurate. However, the rockets apparently made terrifying noises in flight. The sight must also have been frightening. When the rockets went roaring past the American troops, they panicked and fled the field, leaving the path open for the British to capture Washington DC and burn the White House.[24]

The sudden appearance of an enemy force in one's rear is often quite an effective means of sowing panic. Of course there is a relative advantage to attacking the weakly defended rear of an enemy. More importantly, the potential to cut off retreat is a psychological attack on the minds of the troops. Repeatedly we see cases throughout military history in which the appearance of even a small force in the rear of the enemy results in a rout.

Liddell Hart observed that an attack close to the rear of an army attacks the minds of the troops; an attack far in the rear of an army attacks the mind of the commander.[25]

In 425 BCE during the Peloponnesian War a Spartan force landed on the island of Sphacteria, intending to use it as a base from which to attack a nearby Athenian outpost. This proved to be a grievous error when the Athenians used their sea power to isolate the Spartans on the island. When the Athenians finally attacked Sphacteria, the Spartans resisted strenuously. Then an Athenian force took a seemingly impassible shoreline route that was unguarded and got into the rear of the Spartans. This demoralized them and they surrendered—an event that shocked the Greek world, because people believed that no Spartan would ever surrender.[26]

At the Battle of the Issus in 333 BCE, Alexander the Great fought the Persians under the command of King Darius III. Alexander's Companion heavy cavalry, personally led by Alexander, punched through the Persian lines. The decisive blow was an attack in the rear that threatened Darius and his chariot force. Thinking all was lost, Darius fled the field of battle, abandoning his wife, mother, and two daughters as well as the royal treasury. Alexander then turned on the rear of the main Persian force. Attacked in the rear and abandoned by their king, the Persian forces soon lost the will to fight.[27]

In 1356 Prince Edward, the "Black Prince," was leading the English on a raid through France. His men, hungry and exhausted from carrying their loot, had their retreat cut off by the French near Poitiers. France's King John led an army that greatly outnumbered the English. On the morning of 19 September the French attacked. As was frequently the case in medieval warfare, the attack lacked coordination. The first French division nearly overran the English, but the English threw them back with heavy losses. The second French division, led by King John's brother, the Duke of Orleans, lost heart at the sight of the disaster and joined the first division in fleeing the battlefield. As the third French division approached the exhausted English, Prince Edward, fearing his line would not hold, attacked the French. A furious battle ensued. Edward sent a small mounted force along a hidden path to the French rear. When the small force appeared in the French rear, the French overestimated the size of the force and thought all was lost. Many began to flee. The English captured thousands of prisoners, including King John.[28]

At the operational level of warfare, a prime example of the psychological effect of surprise is the Battle of France in May 1940. The attack succeeded so spectacularly because it psychologically challenged the Allies' core beliefs, made them feel trapped, and did not give them the time needed to recover psychologically.

The Maginot Line did not cover the obvious German attack route through the Low Countries—the route used during World War I—and, to avoid provoking Germany, the Netherlands and Belgium would not allow France and Britain to prepare defenses there. Therefore, France and Britain planned to rush troops forward into the Low Countries when Germany attacked in that area. Germany assumed they would do so (a case of knowing your enemy) and sent the main attack through the supposedly impenetrable Ardennes Forest. Because the French thought the Ardennes were impassable for armored forces, only a few second-rate units covered that sector. The Germans exploited the surprise first by quickly forcing their way across the Meuse River before the Allies could bring up reinforcements. Then the Germans exploited their initiative when their tanks sped toward the coast at unprecedented speeds of advance, even though that meant leaving long, vulnerable flanks. The Blitzkrieg's speed of advance did not allow the Allies time to adjust or to organize a counterattack. It also made them feel trapped because the German advance threatened to cut off the British and French forces that had advanced into Belgium.

The unexpected location of the main thrust was not the only element of surprise. The Allies failed to anticipate the massed Panzer divisions, the pace of German operations, and the close coordination of the tanks and dive-bombers. The Germans used Blitzkrieg tactics with airpower attacking obstacles and the armored units moving so rapidly that the Allies could not find time to regroup—either physically or psychologically—and counterattack. Because French training and doctrine did not account for the need to move on "tank time," every time they tried to set up a defense or a counterattack, they found that events had passed them by before the plan could be put into effect. The German flanks were long and relatively undefended. A determined Allied counterattack might have achieved success. That this did not happen was not due solely to the tactical situation and the Germans pressing their initiative.

Even after the breakthrough most of the French Army was intact and should have been capable of counterattacking the vulnerable German armored columns. The Ardennes breakthrough was a major jolt to the psyche of the French leadership because it psychologically struck at their core beliefs by overturning every assumption they had made about how the war would go. It so demoralized the French leadership that they concluded that all was lost. They were no longer capable of organizing and executing a determined counterattack that might have stopped the German thrust. The epigraph at the head of this chapter is Charles de Gaulle's recall of the situation just ten days after the attack. The psychological effect of the surprise had already defeated the French high command.[29]

For another example, consider the Tet attack. After the United States entered the Vietnam War in a major way in 1965, the situation became difficult. By the end of 1967, although the antiwar movement was gaining momentum, the Johnson administration claimed that they were making good progress. In January 1968 the Viet Cong and North Vietnamese forces broke the Tet truce and conducted major attacks throughout Vietnam. The communist forces tried to capture many cities, but overestimated their capabilities and failed in almost all cases. However, they captured Hue City and held it for nearly a month. An attack on the United States Embassy in Saigon, though unsuccessful, provided dramatic video that American television networks showed repeatedly. Overall, American and South Vietnamese forces beat the attack back decisively. The Viet Cong suffered so many casualties that they were never again a major force in the war. Tet was a major military victory for the United States. Yet, paradoxically, Tet was also the decisive political defeat that resulted in the loss of the war.

The contradiction is that although the United States won a major tactical victory on the military battlefield, the real target in war is the will of the nation and its leadership to fight. The Tet attack caused psychological dislocation of both the American public and the Johnson administration because the administration had assured the American people that the Viet Cong were beaten and no longer capable of mounting a major offensive. Tet showed that was incorrect. The psychological impact, reinforced by the media and antiwar politicians, led to the loss of the will to fight by the nation's leadership. Therefore, even though the attack was a tactical disaster for the Viet Cong, the psychological effect made it a strategic victory.[30]

The 11 September 2001 attacks by al-Qaeda achieved strategic surprise in that the attacks occurred on American soil. This altered the strategic equation because after that the American people felt vulnerable. This resulted in major changes in the lives of most Americans whenever they traveled by air or entered sensitive buildings. The 9/11 attacks were also disorienting on a number of levels. The terrorists used aircraft as weapons, not just as bargaining chips. Seizing four airplanes and attacking in two cities heightened the effect. The attacks were against the core of the country: the symbols of America's financial strength in the largest city and the symbol of America's military power in the capital city. The killing of nearly three thousand civilians was terrifying. Because large buildings seemed invulnerable, the sight of the Twin Towers crashing to the ground was profoundly shocking.

Factors that Enhance the Psychological Effect of Surprise

What is the best way to achieve the great benefits of attaining psychological surprise? Military commanders, like ordinary people, make assumptions about the situation they face. When people receive new or unexpected information, they tend to reject the new facts or rationalize and misinterpret it in a way that fits their previous mental model. At best, adjusting to the new reality takes time.

During Operation Desert Storm, the author was in the command center of the Commander, Naval Forces Central Command, on the flagship, the command-and-control ship *Blue Ridge*. On the morning of 18 February 1991, at the start of the third day of minesweeping operations in the northern Persian Gulf, the battle watch received a report that the amphibious ship *Tripoli* had struck a mine. When they plotted the reported position, the staff's immediate reaction was that was impossible because many ships had been operating in that area for the previous forty-eight hours. At that time, no one realized that the Iraqis had laid many mines much farther out in the Gulf than expected, but had incorrectly deployed the mines so that most of them would not work correctly. Being good officers, the staff quickly verified the report, notified Vice Adm. Arthur, and issued appropriate orders.[31]

Less than three hours later a report came in that the cruiser *Princeton* had struck a mine. Again, the immediate reaction was disbelief: "Surely, that's a mistake from a garbled report. They must mean the *Tripoli*, not the *Princeton*." Once again the staff quickly verified the accuracy of the report and took appropriate action. However, if they had not had the means to verify the original reports, would they have believed the "unbelievable" reports?[32]

In the narrow tactical arena, surprise attacks in the rear are often decisive because the psychological effect on the enemy is greatest when the surprise is sudden and makes the enemy feel trapped.[33] This idea is also applicable at the operational and strategic levels. You want to try to make the enemy think, at least for a crucial moment, that there is no hope. This in turn works best if the surprise hits at the enemy's core beliefs.

Thus, to achieve decisive results, you should try to do what the enemy least expects. This may be geographically the area where the enemy least expects to be attacked, but it should also be what the enemy least expects psychologically. Liddell Hart called this being "psychologically indirect" and claimed it was a necessary condition for a battle to be decisive.[34]

Doing the "impossible"—Hannibal crossing the Alps—or something that seems insane—Cochrane attacking Valdivia with 250 men—has a strong psychological impact on the enemy.

Technological innovation can create surprise. If the technology is new, then the enemy has no defense immediately available. The enemy may feel helpless and trapped. However, the psychological impact of surprise depends on the concept, not on technical issues, because human psychology does not change. For example, operational innovations considered impossible are often decisive.

Unexpected operations also have the benefit that the surprise does not have to be perfect because the enemy will often overlook or rationalize away clues. In part, this is due to the bias of the human mind, which is more willing to accept information that reinforces preconceptions than to accept information that challenges them. Thus, contradictory evidence must meet a higher standard than confirming evidence. The more critical the situation, the more likely it is that people go into denial about the contradictory information. The more strongly a person holds a belief, the harder it is for him or her to change his or her preconceptions.[35] The more ambiguous the clues are, the more likely that preconceptions will prevail.[36]

To change a person's mind about preconceived beliefs, it usually takes one significant piece of incontestable and unambiguous evidence, rather than an accumulation of many small pieces of evidence. People are less likely to reverse their assumptions and preconceptions if evidence arrives piecemeal, because that allows each piece of information to be rationalized and assimilated psychologically.[37]

One can surprise an enemy about the location, time, strength, intention, or manner of an attack. Even if it is not feasible to surprise the enemy about some of these factors, a surprise in just one aspect of an attack can be useful.

To enhance the possibility of achieving a psychological surprise, commanders should take actions that the enemy thinks are impossible, strike at the enemy's core beliefs, and make the enemy feel trapped. The primary tools to achieve surprise include deception, knowing your enemy, and imagination.

To surprise your enemy and achieve the greatest psychological impact, it helps to know what your enemy fears most, and then try to attack the enemy (psychologically) there. The better you know your enemy, the more likely it is that you can find a point at which to attack the enemy psychologically.

The most powerful tool for achieving surprise is imagination:

> History has proven that it's not the quantity of men or the quality of weapons that make the ultimate difference; it's the ability to out-think and out-imagine the enemy that always has, and always will, determine the ultimate victor (Pete Blaber).[38]

According to the 9/11 Commission Report, the attacks were, at least partially, made possible by a failure of imagination. Nobody thought of people flying airplanes into buildings.[39] Actually, Tom Clancy had. In his 1994 novel *Debt of Honor* a terrorist flies a 747 airliner into the Capitol Building while the president is addressing a joint session of Congress.[40]

Deception and surprise often are closely related but are not the same. Although deception frequently helps achieve surprise, you can achieve surprise without deception, and deception can be valuable even when it does not lead to surprise. We will discuss deception in the next chapter.

TEN

Deception

All warfare is based on deception. Therefore, when capable, feign incapacity;
when active, inactivity. When near, make it appear that you are far away;
when far away, that you are near. Offer the enemy a bait to lure him;
feign disorder and strike him.

—SUN TZU, *Art of War*

....

DECEPTION'S ROOTS GO FAR BACK IN TIME. ABOUT 1285 BCE
Pharaoh Ramses II of Egypt attacked the Hittite stronghold of Qadesh.
Two Hittite deserters came to him and offered to guide him. He accepted,
and they led him into an ambush.[1]

The Bible in Judges chapter 7 describes how Gideon and the Israelites drove off the Midianites around 900 BCE. Gideon, greatly outnumbered, equipped three hundred men with trumpets and torches inside clay
pots. They surrounded the enemy camp at night, then broke the pots to
expose the torches and blew the trumpets. They shouted, "The sword of
the Lord and of Gideon." The enemy panicked and ran.

Deception is usually included in the current principle of surprise. Certainly, deception is an excellent way to achieve surprise, but deception is
not necessary to achieve surprise. You can achieve surprise without deception if you do something the enemy does not expect. You might act faster
than the enemy thinks possible. Good knowledge of the environment
might enable you to do things the enemy does not think are possible.
You might simply be more imaginative than your enemy and able to find
unique ways to operate. If the enemy has preconceptions about where or
when you will attack, you can do something different without the need
to deceive. In many cases, however, deception will reinforce the enemy
commander's preconceptions so that he or she comes to the wrong conclusion about what you will do.

Deception has important applications in addition to achieving surprise. One possible goal of deception is deterrence. For instance, if your defenses in a particular place are weak, try to convince your enemy that they are strong so that they will not attack you there. Deception can mislead the enemy into wasting scarce resources. Deception is potentially extremely powerful because it attacks the enemy commander's mind.

New Principle of Deception

Although no contemporary military lists deception as a principle of war on its own, we think it should be included. Here is our proposed definition:

The goal of deception should be to cause the enemy to act in a specific manner that you can exploit.

Possible goals of deception include the following: (1) Induce the enemy to let down their guard and weaken their defenses. (2) Induce the enemy to misallocate their defensive assets. (3) Deter the enemy from attacking. (4) Lure the enemy into an unsound attack or ambush. (5) Induce the enemy to waste their resources. (6) Induce the enemy to misuse sensors and weapons. (7) Persuade the enemy to give up. (8) Induce the enemy to reveal information.

Deception works best when it reinforces the enemy commander's expectations. Deception should use as many information pathways as possible. Any information that is likely to reach the enemy should be susceptible to the enemy believing it supports the deception narrative, even if that is not the most likely explanation.

To avoid falling for enemy deception, commanders should suspend judgment as long as possible and periodically review all evidence to weed out discredited information and to avoid overweighting early evidence. Commanders should form alternative hypotheses and determine whether the evidence will support them. Commanders should also identify "missing" evidence that one would expect to observe if a hypothesis were correct.

The goal of deception should not be merely to mislead the enemy or to cause confusion. It should also not be to deprive the enemy of all information. In either case you cannot predict what the enemy might do. They might stumble onto an effective course of action for all the wrong reasons. Alternatively, the enemy might follow an ineffective course of action, but you would not be in a good position to exploit their mistakes. That is not to say that confusing the enemy is bad, but deception aims to achieve much more.

In early 1941 Gen. Archibald Wavell intended to conquer the northern part of Abyssinia. To facilitate his attack he wanted the Italian defenders to move some of their forces to the south beforehand so their defenses in the north would be weakened. The British successfully deceived the Italians into thinking the British attack would be in south Abyssinia. However, the Italians responded by evacuating the south and moving their troops north to more defensible positions. The lesson is that you must aim deception at what you want the enemy *to do*, not what you want the enemy *to think*.[2]

Before World War II the Germans developed the He-100 single-seat fighter. It captured the world speed record but suffered from many mechanical problems. The Germans were not happy with the plane and settled on the Me-109 as their primary fighter for the war, never using the twelve He-100s they had built in combat. Then someone got an idea to deceive the British. In spring 1940 the Germans redesignated the He-100 as the He-113. They photographed the planes with a variety of squadron insignia and announced it in the newspapers as a new high-performance fighter. The ruse fooled the British, who warned their pilots about the new fighter. Within a month, British pilots reported meeting He-113s and evaluated its performance relative to that of the Me-109 (which was what they had actually seen). Such reports continued for years. The British did not realize the Germans had duped them until after the war.[3]

We can learn two lessons from this deception. First, it illustrates how we see what we expect to see. The British pilots expected to see He-113s, so this was what they "saw." Second, the German ruse definitely induced the British to think they faced a new high-performance fighter. However, because the deception had no goal other than to confuse the British, it was actually a failure in that it did not cause them to react in any exploitable manner.

The goal of deception should be to cause the enemy to act in a *specific* manner that you can exploit. Of course, that specific act should be something that furthers your plans or stymies the enemy's plans. Ideally the primary target for deception is the enemy commander's mind. Again, it is not sufficient to make the enemy commander believe something; he must *act* on that belief in a reasonably *predictable* manner.[4]

We categorize deception according to the action we would like the enemy to take. Our proposed definition for the principle of deception lists eight categories of specific goals for deception. We discuss each of these categories in what follows.

Induce Enemy to Let Down Their Guard

A common goal of deception is to induce the enemy to let down their guard and weaken their defenses or allow them to remain weak. If a deception convinces the enemy that you are not about to attack, then your enemy may not prepare their defenses physically, or be mentally prepared for an attack. An attack at a time when a complacent enemy does not expect it can thus achieve surprise with all the benefits discussed in the previous chapter.

Perhaps the most famous deception of all time is captured in the story of the Trojan Horse. The Greeks evacuated their forces to indicate that they were giving up the war. It should have been clear to the Trojans that something was strange about the Greek gift. Certainly the Trojans should never have brought the horse inside the city, much less left it unguarded. That they did all that was folly.[5] Most likely, after many years of battle, the people of Troy must have desperately wanted the war to end. If Homer's story is true, one concludes that the Trojans' hopes for peace clouded their perception of the evidence in front of them.

Following a false routine in view of the enemy is one method of lulling the latter into a false sense of security. As related in the previous chapter, Scipio Africanus "conditioned" the Carthaginian commander at Ilipa by repeating the same actions for several days.

After the disaster of the Sicilian expedition in 415 BCE, Athens rebuilt her fleet and held off the Spartans. In 405 BCE the Spartan admiral Lysander sought to bring the Athenian fleet to decisive battle. He took his fleet to the entrance of the Dardanelles to intercept Athens' vital grain shipments from the Black Sea. Because this threatened Athens' lifeline, her entire fleet of 180 ships responded. They offered battle for four successive days, but Lysander refused. The Athenians might have camped at a safe harbor, but instead they made camp at a beach on the opposite side of the strait at Aegospotami. Because the Spartans declined to fight, the

Athenians believed they feared a battle. Hence the Athenians became complacent and careless. On the fifth day, under circumstances that are not clear to us, the Spartans rowed across the strait and caught most of the Athenian fleet on the beach without their crews onboard, capturing nearly the entire fleet. Without a fleet Athens could neither import grain nor defend her empire. In one hour Lysander effectively ended the twenty-seven-year-long Peloponnesian War.[6]

OPERATION BARBAROSSA

Operation Barbarossa—the German invasion of the Soviet Union starting on 22 June 1941—illustrates how it is possible to achieve strategic surprise through deception. The Non-Aggression Pact between Germany and the Soviet Union in August 1939 set the stage for the invasion of Poland that began World War II in Europe. On 18 December 1940 Hitler signed a directive ordering preparations for the invasion of the Soviet Union. Germany took several measures to ensure secrecy and to deceive the Soviet Union. Germany's deception should not have succeeded because the massive size of the invasion force meant that they could not hide a lot of contrary evidence. This effort succeeded in part because deceiving the Soviet Union required deceiving just one man—Joseph Stalin. Fortunately for Germany, Stalin believed that Germany would not attack the USSR (at least not in 1941), being convinced that the Germans could not attack before they defeated the British and that they would follow their previous pattern of demands and provocations before any attack.[7] Stalin's extrapolation of a supposed pattern in previous German aggressions is an example of the human bias of exaggerating the significance of small samples (fewer than five instances in this case).[8]

Germany's deception of the Soviet Union had many components. In late 1940 the German military attaché in Moscow told the Russians the troop movements to the east were simply younger men replacing older men to free the latter for war production. In a letter to Stalin in early 1941 Hitler claimed that the buildup in the east was to train troops for the invasion of Britain in a location out of reach of British bombers. Hitler tried to convince Stalin that the German build up in Poland was part of the greatest deception of all time, which was to lead to the invasion of Britain. It might well have been the greatest deception of all time, just not in the way Stalin believed. To avoid leaks, most German troops stationed in Poland were not told of the invasion of Russia until the night before. In mid-June, Josef Goebbels wrote in a Nazi periodical that the fall of Crete indicated that the invasion of Britain would come soon. To call attention to his pronouncement Goebbels immediately confiscated the entire issue as soon as it reached the foreign press.[9]

Stalin received eighty-four separate warnings that Germany would invade the Soviet Union—all of which he either overlooked, rationalized away, or dismissed. Churchill tried to pass evidence of the coming attack to Stalin, but the latter regarded this as an English attempt to disrupt the German-Soviet alliance. On 19 April Soviet intelligence gave Stalin evidence gathered in Czechoslovakia of a German attack planned for mid- or late June, in response to which Stalin scrawled on the report, "English provocation, investigate! Stalin."[10] Meanwhile, Stalin had become obsessed with the need to avoid provoking Germany, which affected the interpretation of, and reaction to, many of these warnings. When Adm. Nikolai Kuznetsov gave orders to shoot down German aircraft conducting reconnaissance flights over Soviet-controlled territory, Stalin countermanded the order because he did not want to provoke Germany. Richard Sorge, a Soviet spy in Japan, sent four warnings in May. In June 1941 Sorge provided detailed operational and tactical information on the German plans. Stalin dismissed these warnings with contempt for Sorge. An anti-Nazi German ambassador approached a Soviet trade negotiator, who rejected his warnings. When the Soviet defense minister showed Stalin evidence of large-scale rail traffic and troop concentrations by the Germans, Stalin countered with similar reports that concluded the evidence was part of the German deception for the invasion of Britain. Nearly two weeks before the invasion Germany evacuated dependents from their embassy in Moscow and then burned papers. The Soviet Western Special Military District reported German preparations just over the border. On 18 June a German deserter claimed the invasion would start on 22 June. This report reached Stalin three days later.[11]

Mere hours before the attack, Stalin finally issued a warning to the Soviet armed forces, but he still demanded that they do nothing to provoke the Germans. On Sunday, 22 June 1941, the invasion began. Many Soviet officers were away from their units and the army was quite unprepared. The Germans destroyed two thousand Soviet aircraft on the first day; most never took off. Despite a mountain of evidence to the contrary, almost to the end Stalin had clung to his belief that Germany would not attack the Soviet Union in 1941.[12]

The Egyptian deception plan prior to the Yom Kippur War in 1973 was successful, largely because it reinforced Israeli perceptions. In the run-up to their attack the goal of Egyptian deception was to slow the Israeli response by capitalizing on perceptions held by Israelis that Egyptians could not keep secrets, were inefficient militarily, and did not have the ability to plan and coordinate complex operations. To that effect they planted rumors that, due to Egyptian incompetence, Soviet equipment had been deteriorating since the Soviets left in July 1972.

The plan provided plausible alternative interpretations of the buildup of forces along the Suez Canal and Golan Heights. The Egyptians increased the "noise" with a series of false "cry wolf" alerts in May, August, and September 1973, each accompanied by bellicose rhetoric, sowing the idea that these alerts were just saber rattling for home consumption to placate hawks. In the final buildup period, when the Egyptians needed to explain movements of troops toward the border that they could not conceal, they used the usual fall training exercises as cover. They sent troops forward during the day, but only half of the newly arrived troops returned to their barracks at the end of the day. "Lazy squads" sat on the canal bank fishing, eating oranges, and soaking their feet. In the end, the Israelis managed to figure out an attack was coming just a few hours before the Egyptians attacked on 6 October 1973, but this was not sufficient to blunt most of the surprise effect. The attack led to initial gains, though Israel later regrouped and routed the Egyptians.[13]

Induce the Enemy to Misallocate Defensive Assets

Sometimes it is quite clear that an attack will take place. In that case the goal of deception may be to induce the enemy to misallocate defensive assets through deception about the location or strength of the attack. Depending on the circumstances, you might want to convince your enemy that you will attack in a particular place or that you will not attack in a specific location.

In the summer of 1863 during the American Civil War the Tennessee River prevented the Union Army of the Tennessee under Maj. Gen. William Rosecrans from capturing Chattanooga. There were various obvious reasons to cross the river upstream of Chattanooga and that is what the Confederate general Braxton Bragg expected Rosencrans to do. Rosecrans demonstrated upstream with three brigades. They lit fires in the rear of all possible crossings, threw wood chips into tributaries of the Tennessee, and pounded on barrels twenty-four hours a day to convince the Confederates that they were building boats to cross the river in that area. Consequently, Bragg reinforced the area with troops drawn from downstream; Rosecrans's army then crossed downstream and soon captured Chattanooga without a fight. The deception was successful because it reinforced the course of action that the Confederates expected Rosencrans to follow.[14]

Deception might have the aim of drawing an enemy force away from a particular location. At the Battle of Leyte Gulf in 1944 the Japanese used most of their remaining aircraft carriers as decoys. At this time the primary constraint on the Japanese Navy's air arm was the lack of trained pilots. Therefore, the carriers were no longer as valuable as they had been earlier in the war; for this battle they sailed with few aircraft onboard. Nevertheless, these were not plywood and canvas decoys, but large, seemingly important ships, indicative of the desperation of the Japanese. These decoys succeeded in drawing Adm. Bill Halsey's main force away from the landing area, opening the door to a potentially devastating Japanese attack on the American amphibious ships.

EL ALAMEIN

In the autumn of 1942 Lt. Gen. Bernard Law Montgomery prepared his Eighth Army for an offensive against Field Marshal Erwin Rommel's German and Italian forces. The sea and the impassible Qattara Depression limited his options for where to attack. Because Montgomery planned to attack in the northern sector of the front, he wanted to conceal his buildup of forces and stores in the north while simulating a large buildup in the southern sector.

Col. Dudley Clarke directed an incredibly complex deception plan. In the southern sector Clarke created bogus supply dumps, building a fake twenty-mile-long water pipeline with dummy pipes in a trench. The British also constructed three fake pump houses and fake reservoirs. The building rate of the pipeline indicated a completion date about ten days after the planned date of the attack. The British placed armor close to its starting point but disguised the tanks using covers to make them appear to be trucks. After establishing three staging areas toward the south, armor openly moved into these areas, then moved north under the truck-appearing covers. Meanwhile, fake tanks made from split palms under camouflage nets replaced the real tanks in the dummy staging areas.

The British constructed gun pits with fake artillery near the actual location of the planned attack and avoided movement around the fake guns to convince the Germans that these were in fact fake. Near the time of the attack, real guns replaced the fake guns that the Germans by this point believed were dummies. Logistics forces delivered food, ammunition, and other supplies at night. Before dawn Clarke's people stacked these supplies in the shape of three-ton trucks and covered them with nets.

The British also deployed rafts in the sea to simulate an amphibious assault behind the German lines, a feint that appealed to multiple senses. The visual cues included smoke and flares. The smell of cordite and diesel aided the deception. Recordings of confused shouting provided aural cues. Despite all the different cues, this part of the deception did not succeed.

The overall deception was successful in that the Germans detected the "buildup" in the south, but not in the north. Two German armored divisions stayed in the southern sector for four days after the real attack began. Nevertheless, the El Alamein attack still met a great deal of resistance.[15]

NORMANDY DECEPTION

The Normandy landings in France on 6 June 1944 were not a total surprise. It was common knowledge that the Allies would open a second front in the summer of 1944. The only questions were where and when. A landing near Pas de Calais involved the shortest distance across the English Channel and provided a port the invasion force could use for supplies. The short distance meant that air support for the landing would be easiest. Thus, to the Germans, Calais seemed to be the most logical landing site. Normandy, on the other hand, was farther away from England and did not offer a port. Some other factors favored Normandy, but Calais seemed most logical.

The wide-ranging deception plan for the Normandy invasion included attempts to convince the Germans that the landings might be in Norway, southern France, Greece, or anywhere else. These attempts were unsuccessful because they were not realistic.

Ordinarily, physical deceptions and radio intercepts play a major role in deceptions. However, in this case, because of the inability of the Germans to conduct effective surveillance, reports from double agents played the primary role in deceiving the Germans. The British had turned German agents into British double agents, who carefully fed the Germans information that Gen. George Patton's fictitious First U.S. Army Group would land near Calais. Physical deception efforts were simply insurance in case an unknown spy or a lucky German reconnaissance aircraft flew over the areas supposed to contain the fake forces. On the night of the invasion, bombers dropped chaff to give the impression on German radar of an invasion fleet moving toward Calais.

The deception on landing location was not wholly successful. Field Marshall Erwin Rommel, in charge of the defense of the Atlantic Wall, thought Normandy was the likely invasion site and greatly improved the defenses on the beaches of Normandy. On the morning of the invasion, he wanted to move reserves toward Normandy immediately. However, Hitler thought Normandy was a feint and that Calais would be the location of the landings, so he withheld use of the reserves. The double agents told the Germans that the real landing would occur in Calais several weeks after the "diversionary" landings in Normandy. For several weeks Hitler persisted in his belief that the Normandy landings were just a diversion, evidently rationalizing away all the evidence that the Normandy landings were the major effort because that evidence did not fit his preconceived notions. Similarly, he rejected his generals' arguments to the contrary. As a result the Germans held their Fifteenth Army near Calais for seven crucial weeks. Thus, even though the deception was only partially successful, it delayed the deployment of reserves and ultimately contributed to the success of the Allied landings.[16]

DESERT STORM

Many aspects of deception and surprise change in a world of instant broadcasting of events. Yet great deceptions are still possible. After the Iraqis invaded Kuwait in August 1990 they set up defenses along the southern border of Kuwait. They assumed that the desert terrain to the west was too difficult for a military force. They also assumed the coalition force would attempt an amphibious landing. Because the coalition forces had air supremacy, the Iraqis' primary source of information was commercial media reports. With the news media hungry for anything to broadcast, the coalition fed them accurate information about the capabilities that they wanted to emphasize; in the process, the media repackaged capabilities as intentions.

Early in the planning Gen. H. Norman Schwarzkopf ruled out an amphibious assault. However, he did not tell the U.S. Navy or Marine forces about this decision, directing Vice Adm. Hank Mauz, commander of the naval forces, to plan an amphibious operation. Nor did Schwarzkopf tell Mauz's successor, Vice Adm. Stan Arthur, his real intentions. At a meeting on the naval commander's flagship on 2 February, the naval staff briefed Schwarzkopf on the state of the plans for amphibious operations. Schwarzkopf approved nothing but told Arthur to keep the amphibious option open.[17]

Meanwhile, the presence of U.S. Marines in ships offshore was an obvious source of news. The coalition highlighted the amphibious capabilities to the media without explicitly saying that an amphibious operation was in the plans. They did, however, advertise a large amphibious exercise in late January.[18]

During the run-up to the ground attack in Desert Storm, the British provided television reporters the opportunity to film British ground forces with the Persian Gulf in the background. When the British forces moved inland, these opportunities ceased. Subsequently, British television networks accompanied their reports with library footage showing the sea in the background without explaining that it was library footage (as the British had anticipated).[19]

At the same time, when the American armored forces stationed in the desert near the coast moved far to the west, they left behind fake headquarters that continued to transmit radio signals. The VII Corps created a fake base, including missiles, vehicles, and fuel dumps, which transmitted both radio traffic and radar signals.[20]

After the air campaign began on 17 January 1991, coalition ships moved farther north in the Persian Gulf. On the night of 18 January the frigate *Nicholas*, her embarked U.S. Army helicopters, and the Kuwaiti ship *Istiqlal* captured Iraqi personnel occupying nine oil platforms in the Durra oil field. On 24 January two U.S. Navy ships seized minefield plans from two Iraqi vessels damaged by aircraft. Their embarked U.S. Army helicopters captured Qaruh Island and several oil platforms the Iraqis had occupied. On 16 February a large force of minesweepers and covering ships began sweeping an area for mines off the coast of southern Kuwait. The ground assault began on 24 February 1991. On the morning of 25 February, as part of the deception, the battleship *Missouri* bombarded the coast of Kuwait, and Marine helicopters flew toward the coast, but turned around just before reaching land. Two minutes later Iraq confirmed the success of the ruse by firing two Silkworm anti-ship missiles at the *Missouri* and its escorts.[21]

The Iraqis held substantial forces in Kuwait to guard against the amphibious assault that never came. Marine forces drove into Kuwait from the south to hold Iraqi units in place. Meanwhile, American and British armored units drove through the "impassable" desert far to the west. They went all the way into southern Iraq and then turned east to get behind most of the Republican Guard units. Unfortunately, a premature end to the war prevented capturing nearly all of the Republican Guard divisions.

Deter Enemy Attack

When threatened, it is common for animals to appear larger or more ferocious than they really are in order to deter predators. This technique also works among combatants. If one appears to be extremely strong, this may deter the enemy from attacking.

In the sixth century the Byzantine general Belisarius had fewer than twenty thousand men to stop an invasion of two hundred thousand Persians. When a Persian envoy visited Belisarius's camp to discuss possible peace, Belisarius posted contingents of his best troops along the envoy's route. To persuade the envoy that this was the outpost of a great army, Belisarius had his soldiers spread out and keep in constant motion to make them appear to be more numerous. This display, together with Belisarius's confident manner, convinced the envoy to tell his king that it was too dangerous to fight such a large army.[22]

During World War II an area south of Ireland was well-suited for a minefield to stop German submarines, but no minelayers were available. A British-controlled double agent told the Germans that a (nonexistent) minelaying friend had revealed the details of a new (fictitious) minefield off the Irish coast. The gods of chance frowned on the Germans when a U-boat hit a stray mine in that area soon after the agent sent the report.

Deception

Red Cross accounts of survivors landing in Ireland seemingly validated the minefield report. As a result of this deception, the Germans changed their operational plans. Ultra (knowledge gained from breaking Germany's Enigma code) detected a German directive instructing U-boats not to enter a thirty-six-hundred-square-mile area south of Ireland, after which point the Allies then knew they could safely pass convoys through this area.[23]

BEAUREGARD AT CORINTH

Retreating in the face of a superior enemy can be dangerous. Sometimes, to cover a retreat, the best defense is to fake an offense. In April and May 1862, in the aftermath of the Battle of Shiloh, a Union army under Maj. Gen. Henry Halleck outnumbered Gen. Pierre Beauregard's Confederate Army in Corinth, Mississippi, by more than two-to-one. In addition, Beauregard had many sick and wounded men, as well as a shortage of food. To deter a Union attack, Beauregard personally coached "deserters" he sent into the Union lines with stories of great strength and a planned offensive. He had his cavalry raid Union rear areas and spread rumors of offensives. When a train came to evacuate the sick and wounded, bands played and a regiment cheered as if welcoming reinforcements.

By the end of May, Beauregard knew he had to retreat but feared a Union attack while he was in the process of evacuating. As a deterrent, he spread the word among his troops that they would be attacking in three days, causing some timorous soldiers to desert and spread the news on the Union side. Beauregard's forces set up Quaker guns—logs positioned to look like cannons—tended by straw dummies in old uniforms. When the evacuation of frontline troops began, drummer boys kept fires going and beat reveille in the morning. A single band moved around, playing at various locations. Beauregard assigned a detachment to cheer each time a single train of empty cars moved back and forth, sounding like reinforcements were arriving. They kept this up all night.

Did the deception work? During the night Maj. Gen. John Pope reported to Halleck that "the enemy is reinforcing heavily, by trains, in my front and on my left. The cars are running constantly, and the cheering is immense every time they unload in front of me. I have no doubt, from all appearances, that I shall be attacked in heavy force at daylight." Then he braced for the attack that never came.[24]

In spring 1862 Maj. Gen. George McClellan moved his Union army to the York-James peninsula near Richmond. There were not enough Confederate forces to defend the area. The Confederates formed a thinly held defensive line across the peninsula at Yorktown. If McClellan had pushed forward vigorously, he would have encountered little resistance. Brig. Gen. John Magruder was in charge of the inadequate Confederate defenses at Yorktown, the first obstacle in McClellan's path. Magruder had few cannons and men to cover a thirteen-mile front; he thus put on a theatrical show, parading men repeatedly past the same point, which was visible from Union lines, so that his force appeared to be much larger than it was in reality. He moved his artillery back and forth to make it seem more numerous. He also used Quaker guns. This display convinced McClellan, who was always predisposed to believe that the Confederates greatly outnumbered him. Therefore, he delayed his attack until he could bring his siege train into position, causing nearly a month's delay and giving the main Confederate army time to reach the area and prepare for the defense of Richmond.[25]

DARDANELLES EVACUATION

One bright spot in the Dardanelles fiasco during World War I was the evacuation. The Allied deception plan sought to maintain a routine so the Turks would not realize anything was about to happen. Artillery firing continued at the same pace. Lighters continued to land "supplies" that were actually boxes filled with sand. The Turks were also "conditioned" in other ways. In one area, the Australians and New Zealanders began to periodically maintain total silence. If as a result the Turks suspected a position had been evacuated and started forward, they encountered heavy fire. After a few such episodes, they came to regard such quiet periods as routine. Thus when the evacuation actually began, the Turks took a long time to realize the trenches just a few yards away were empty. The Australians even devised a delayed-action rifle that would continue to fire long after they had abandoned a trench. Because of these and other deceptions, the entire force escaped in a series of evacuations without any additional losses.[26]

An important part of the United States' Maritime Strategy during the Cold War involved using deception to move aircraft carrier battle groups to areas near the Soviet Union without being detected. For example, a battle group would employ strict control of its electronic emissions during a transit, while false electronic signals indicative of a battle group indicated its presence in other areas. These operations demonstrated that the U.S. Navy could move battle groups to strike positions near the Soviet Union without being detected, and therefore played a part in deterring the Soviet Union from starting a war.[27]

SADDAM HUSSEIN

In the years leading up to 2003, Saddam Hussein's greatest fear was not the United States, but Iran. To deter and intimidate the latter, the Iraqi leader played a two-faced game in which he publicly denied having weapons of mass destruction with a sort of wink so that Iran would think he had them. As a result, he successfully convinced nearly every intelligence agency in the world that he had such weapons. He even convinced his own generals. This widespread belief led antiwar activists to argue that the United States would suffer heavy casualties from Iraqi chemical weapons if it invaded Iraq. On the eve of the 2003 invasion, the United States government and nearly every intelligence agency in the world was completely convinced that Iraq possessed weapons of mass destruction, and therefore the United States invaded Iraq.[28]

Thus, on the one hand Saddam Hussein's deception was successful in that it convinced nearly everyone and it deterred Iran. On the other hand it was so "successful" that it backfired by provoking the United States into invading Iraq and deposing Saddam Hussein. This is a cautionary case in that Iraq was defeated in part *because* it used deception successfully. Iraq's problem was that it had two enemies, each of which Iraq wanted to deceive in a different, conflicting way.

Lure Enemy into Unsound Attack or Ambush

If one feigns weakness, this may induce the enemy to attack in an unsound manner or lead the enemy into an ambush. One might feign flight to induce an unbalanced attack susceptible to counterattack. One might pretend to be weak at a particular location to tempt the enemy into attacking a strong position. Anything that convinces the enemy to attack in an unsound manner is worth considering.

Pretended flight has been a common tactical deception throughout the ages. In many cases it is not clear today whether the flight was feigned or real. In either case the pursuing enemy can become overeager and engage in an unbalanced pursuit that is susceptible to a devastating counterattack. Such a counterattack also has a strong psychological component. The enemy's apparent flight seems to indicate that the enemy thinks they are beaten. The attacking force is quick to believe they have won the battle (their hope and perhaps their expectation) and all that remains is to massacre the fleeing enemy and collect the spoils. When the counterattack comes, the pursuing force suddenly goes from the elation of apparent victory to the psychologically devastating agony of being attacked, compounded by the feeling of being trapped. Historically, cases of real or feigned flight that resulted in victorious counterattacks include the battles of Lake Trasimene and Cowpens. In 1066 the Saxon king Harold used this deception successfully at Stamford Bridge to defeat the Norwegians, but then his forces fell for the same ruse a few days later at the Battle of Hastings.[29]

Something similar occurred at the Battle of Jutland. First, Adm. David Beatty's British battle cruiser squadron pursued Vice Adm. Franz Hipper's German battle cruiser group, as the latter attempted to lead the British to the German main battle fleet. However, Beatty saw Vice Adm. Reinhard Scheer's German High Seas Fleet in time to turn away. Then the entire High Seas Fleet pursued Beatty as he led them to Adm. John Jellicoe's British Grand Fleet of battleships. The Germans did not see the British fleet of battleships until it was too late to avoid getting their "tee" crossed.

Another common deception that dates back to ancient times is the feigned defection. We have already mentioned the Battle of Qadesh in 1285 BCE; one more example here should suffice to make the point. Namely, before the Battle of Salamis in 480 BCE, Themistocles, the commander of the Greek forces, sent his servant to the Persian King Xerxes to tell him that Themistocles was going to defect and that Xerxes could capture the Greek fleet by advancing into the Strait of Salamis. Xerxes consequently ordered the Persian fleet to advance into the narrow waters that would negate their advantage of superior numbers. At the Battle of Salamis the

Deception

next day, the Greek ships' apparent flight induced the Persian fleet to advance in disorder, facilitating a successful Greek counterattack.[30]

PICKETT'S CHARGE

On the third day of battle at Gettysburg, Pickett's Charge was to begin when Col. Edward Alexander, commander of the artillery for Longstreet's Corps, judged that he had knocked out some of the Union artillery in that sector. After an hour's bombardment the Confederates' artillery ammunition was getting low. Alexander told Longstreet that if he wanted artillery support during the attack, then the attack had better start soon. Shortly thereafter, Alexander noticed that the Union artillery fire had slackened and some guns were retreating. The reason for these events was Union deception.

On Cemetery Ridge the Union chief of artillery, Brig. Gen. Henry Hunt, saw that Confederate artillery fire had not driven any of his guns or gun crews from their positions. He thought that perhaps the Confederates would not charge after all. Hunt knew that Meade hoped that they would do so and had no fear of the outcome. Prodded by a suggestion by a subordinate, Hunt stopped his guns firing—gradually rather than all at once so the Confederates would think the Union artillery had been driven off or the guns were being knocked out group by group. He also gave permission for one damaged Rhode Island battery to leave the field because it was almost out of ammunition. It was this Rhode Island battery that Alexander saw leaving the field. When he saw the Union artillery fire slacken and then stop, he sent another message to Longstreet to attack quickly.[31]

ROBIN OLDS

After the Vietnam War intensified in 1965, the U.S. Air Force deployed F-4 Phantom fighters and obsolescent F-105 Thunderchief fighter-bombers. The F-105s made most of the bombing attacks early in the war. The F-4s protected them from North Vietnamese MiG-17s with considerable success. However, in September 1966 North Vietnam started deploying MiG-21 interceptors. These aircraft had success against the F-105s. Because the MiGs only attacked under favorable conditions, covering F-4s were not able to protect the F-105s adequately. The MiG-21s shot down some F-105s and caused many to jettison their bombs to evade the MiGs. The U.S. Air Force wanted to attack the MiGs at their airfields near Hanoi and Haiphong, but the Lyndon Johnson administration would not allow them to do so. Because the North Vietnamese could stay safely at their bases and only attack when and where they chose to do so, they had the initiative. Furthermore, when the American fighters approached the MiGs, they often escaped by flying into China.

Col. Robin Olds of the Air Force devised a ruse to lure the MiGs into a position where the Americans could engage them on favorable terms. Flights of F-4s mimicked flights of F-105s, following the same airspeed, altitude, routes, formation, and refueling profiles as the F-105s. When the MiG-21s came up to intercept the Americans, the F-4s engaged them. They claimed that they destroyed seven MiG-21s, which was nearly half of all the MiG-21s that North Vietnam possessed. (The North Vietnamese admitted losing five aircraft.) A few days later, two F-4s flew close together to mimic a single reconnaissance aircraft. Again, the North Vietnamese took the bait. The F-4s claimed to have shot down two of the four MiGs. The ruses had worked so well that the MiG-21s went into a three-month standdown.[32]

SCIPIO AFRICANUS

During the Second Punic War, Scipio Africanus, a master of deception, deceived the Carthaginians at many levels in the events leading up to the battle related in this section. His culminating achievement was to have his enemies, oblivious to the nearby hostile force, respond unarmed to what they thought was a natural disaster and run into an ambush that was more a slaughter than a battle.

After Scipio defeated the Carthaginians in Spain, he invaded Africa in 204 BCE. He laid siege to Utica but did not make much progress; in the meantime, the Carthaginians and their Numidian allies approached. Hasdrubal led the Carthaginians with thirty thousand foot soldiers and three thousand horsemen, while Syphax led the Numidians with fifty thousand foot soldiers and ten thousand horsemen. Faced with this numerically superior force, Scipio lifted the siege of Utica and went into a fortified camp nearby for the winter. Hasdrubal and Syphax established separate camps about a mile apart and seven miles away from Scipio.

The first part of the deception involved obtaining information about the enemy camps. To this end Scipio entered into peace talks with Syphax.

His messengers traveled to the enemy camp and observed the conditions. They noticed that the Carthaginians had built wooden huts close together and that the Numidians constructed their huts of reeds and matting, placing them very close together. In addition, some of the huts were outside the walls of the camp. Scipio began sending expert scouts and centurions disguised as servants of the officers who acted as the messengers. They roamed both enemy camps and gathered information about the layout of the camps, the location of the gates, and the stationing and changing of the guards. To familiarize as many of his officers as possible with the camps, Scipio rotated the "servants" each time he sent ambassadors to the enemy camps.

In keeping with the ethical standards of the time, Scipio then broke off the peace talks prior to conducting an attack that exploited the information gained. Hasdrubal and Syphax also planned an attack, but Scipio was ready first. The second part of Scipio's deception diverted the enemy's attention and covered his movement into position. To focus the enemy's attention on Utica, he sent a small force to seize a hill near that city. To avoid leaks Scipio told his men they would be attacking Utica and had his men still in camp act normally until it was dark. Then he marched most of his men to a point near the enemy camps, arriving about midnight. Of course, he had previously determined the best route for the night march. Scipio then split his men, with a subordinate commanding the second group that approached Syphax's camp.

The culminating phase of the deception lured the enemy into an ambush. The Romans set fire to the huts just outside Syphax's camp. Because of the huts' flammable construction and the lack of proper spacing, the fire quickly spread throughout the camp. Because they thought the fire was accidental, Syphax's men woke and came out of their tents unarmed. Many men were burned or trampled in the panic. Meanwhile, Scipio's men waited outside the camp gates and cut down those who tried to escape the camp.

When Hasdrubal's men in the other enemy camp saw the fire, they also thought it was accidental. Many ran the mile between camps to help their allies put out the fire. Scipio attacked those exiting Hasdrubal's camp and then rushed into the camp through the unguarded gates. The Romans also set the huts in this camp on fire. Much of the scene of confusion and panic was repeated in Hasdrubal's camp, along with the slaughter of the Carthaginians trying to escape. The Romans and the flames killed perhaps forty thousand men. It was a masterpiece of deception. It also demonstrates the value of surprise, knowing your enemy, and relative advantage.[33]

Induce Enemy to Waste Resources

Any decoy designed to draw enemy fire qualifies as a deception that may cause the enemy to waste resources. Of course, decoys also can have other functions, such as revealing the enemy's position or drawing him into an unsound attack. In one peculiar case, decoys were used to obtain weapons from the enemy.

In 755 CE a Chinese rebel force besieged a loyalist walled city. The defenders prepared one thousand straw dummies, dressed them in black, and lowered them over the wall on ropes at twilight. The rebels thought the defenders were coming out to counterattack and shot them full of arrows. The defenders pulled the dummies up, along with thousands of captured arrows. The next night (or more probably after several nights of repetition to "condition" the rebel commander), real soldiers descended on ropes. The rebels, now conditioned to think they were dummies, neither shot arrows at them nor prepared to defend. The defenders became the attackers and defeated the besieging army of rebels.[34]

During the prolonged Union campaign to capture Vicksburg during the American Civil War, ironclad gunboats periodically ran past the Vicksburg batteries commanding the Mississippi River. In February 1863 Acting Rear Adm. David D. Porter used an old barge and constructed a dummy monitor. He set this adrift so that it floated past the Vicksburg batteries, thus inducing them to waste ammunition bombarding the fake warship.[35]

During World War I German submarines preferred to surface and use their deck gun to sink enemy merchant ships. This conserved their torpedoes and allowed them to stay on patrol longer. The British countered this German tactic with Q-ships, innocent-looking merchant ships fitted with hidden guns. A cargo such as cork would fill their holds so the ships could take a lot of damage without sinking. When a submarine appeared, a "panic crew" would hurriedly launch the lifeboats and try to get away. When the submarine surfaced and approached the Q-ship to finish it off with its deck gun, the real Q-ship crew would uncover their guns and turn the table on the submarine. The Q-ships were few in number and did not sink a large number of German submarines—just eleven in

World War I—but they were successful in deterring some surface attacks on merchant ships.[36] The threat of Q-ships resulted in German U-boats sometimes using torpedoes instead of surface gunfire to finish off merchant ships, causing them to expend additional torpedoes and therefore having to return from patrol sooner.

Sometimes the enemy will identify decoys as such. During the North African campaign, the British erected many dummy aircraft. Because these dummies were flimsy, the British put struts under the wings to support them. One day an officer reported to Col. Dudley Clarke that intercepted German communications indicated that they had identified dummy aircraft by the struts under their wings. We suggest the reader take thirty seconds to consider what he or she would do if presented with this situation. Barton Whaley claimed that less than one-tenth of the officer-students in his classes on deception were able to deduce the correct response. The "obvious" reaction was to strengthen the dummy aircraft so they did not need struts or to make the struts less visible. However, Clarke's immediate response was that they should put struts under all the real aircraft as well. This would protect the real aircraft from attack. Furthermore, Whaley suggested they should have used a mix of real struts on dummy aircraft, dummy struts on real aircraft, and invisible struts on dummy aircraft. Doing so would cause the Germans to think many of the dummy aircraft were real and many of the real aircraft were dummies. They would also incorrectly estimate the number of British aircraft.[37]

After World War II Jewish guerrillas sought to establish the state of Israel in the British mandate in Palestine, fighting both the British and the Arabs in the name of this cause. In a particularly effective deception, whenever they made a large-scale attack against the Arabs, the guerrillas used British equipment and wore British uniforms (contrary to the laws of war). As a result, the Arabs attacked British convoys because the Arabs thought they were Jews in disguise.[38]

In 1959, during the Algerian War for Independence, the French ambushed and killed an Algerian district commander, Colonel Lofti, and his staff. However, the French did not announce this achievement. Instead, they had a fake Colonel Lofti report via radio to the overall commander of the liberation army. Over the next several months the fake Colonel Lofti repeatedly called for reinforcements, arms, and money. When this was sent, the French intercepted all of it. The deception operation also caused distrust among the rebels.[39]

Induce Enemy to Misuse Their Sensors and Weapons

If one misleads the enemy about the effectiveness of their weapons or sensors, this might induce the enemy to misuse their assets.

In 1941 Malta was a vital strategic British base, but German aircraft from nearby Sicily attacked it relentlessly. Due to the short distance to German airfields and the limited defenses on the island, Malta relied heavily on its early-warning radar. One day Malta reported that a strong new German radar jammer had rendered their early-warning radar useless. Malta asked the British Air Ministry for advice on how to counter the jamming. R. V. Jones, the brilliant head of the British Air Ministry's scientific intelligence, directed Malta to continue to operate the early-warning radar as though nothing was wrong. After a few days the Germans stopped using their jammer because they assumed it was ineffective. After the war the German general in charge confirmed this reason to Jones.[40]

In 1944 the Germans launched many V-1 Buzz Bombs toward London. These crude cruise missiles had no aiming sensors; their range was determined by the time of fuel cutoff. The Germans adjusted the range based on their spies' reports of the time and place of impacts. Because the British controlled all the German spies in Britain, Jones had them report impact points some distance beyond the target point, which was the Tower Bridge. Over time he induced the Germans to shorten the range repeatedly, until their actual target point was four miles short of their intended target point. Ironically, this moved the actual target point to the area in which Jones's parents lived. Jones patriotically continued the ruse despite the danger to his parents.[41]

If the human mind receives information in small increments, it can be gradually acclimated (conditioned) to a major change of view. In February 1942 the German battleships *Scharnhorst* and *Gneisenau* and the heavy cruiser *Prinz Eugen* broke out of Brest and successfully dashed through the English Channel to German bases. Part of the reason for their evasion of effective British attack was that the Germans jammed the British radars, and the radar operators did not realize they were being jammed. A month prior to the breakout, the Germans started jamming British radars for a few minutes each dawn in a manner that was similar to the effects of atmospheric interference. Each day the Germans increased the duration of the jamming by a small amount. By the time of the breakout, the British radar operators expected atmospheric interference early in the day and therefore did not report it.[42]

Persuade the Enemy to Give Up

If one can persuade the enemy that their position is hopeless, then the enemy's morale and will to resist will suffer. The story of Gideon and the Midianites illustrates how convincing a greatly superior enemy that you are powerful can cause the enemy to lose hope. The ultimate deception is to convince the enemy that you are so strong that he literally gives up and surrenders to you.

THOMAS COCHRANE

When Brazil was trying to achieve independence from Portugal, the rebel Emperor Pedro hired Thomas Cochrane as the first admiral of the Brazilian navy. In July 1823 Cochrane sailed one ship, *Pedro Primiero*, into the port of São Luis. Because he flew the British flag, he was able to approach the town without opposition. When within gun range of the town, he raised the Brazilian flag and sent a message ashore that he was the forerunner of a mighty fleet with many warships and troop transports that were coming to liberate the province. The next day the local authorities came to Cochran's ship and swore allegiance to Emperor Pedro. Cochran installed a new provincial government. Concerned that his deception might be exposed, he persuaded the Portuguese commander and his troops to board ships and return to Portugal immediately.[43]

NATHAN BEDFORD FORREST

In May 1863, after a three-day running fight, a group of Confederate cavalry led by Brig. Gen. Nathan Bedford Forrest cornered a Union raiding party led by Col. Abel Streight in Alabama. They met under a flag of truce. Forrest, claiming to have a large force at hand, demanded the Union colonel surrender. Streight would not surrender without proof that he faced an overwhelming force and his position was hopeless, proof that Forrest refused to give him. While they were talking, however, Streight saw over Forrest's shoulder a large number of artillery pieces moving into position. When Streight exclaimed that he had counted fifteen guns already,

Forrest responded, "I reckon that's all that has kept up." Actually, Forrest's men moved his only two guns repeatedly across open ground that Streight could view. The Union colonel soon surrendered to a force less than half the size of his own.[44]

On 24 September 1864 Forrest attacked Union forces led by Col. Wallace Campbell at Athens, Alabama. Forrest sent Campbell a note demanding unconditional surrender to avoid useless bloodshed. They met under a flag of truce, at which point Forrest said he had ten thousand men and if he had to storm the fort, he would massacre the entire garrison. Campbell insisted that he see for himself the size of Forrest's force. When Forrest took him on a tour to see the Confederate force, Campbell was convinced that he was facing eight to ten thousand men. In fact, Forrest had shown Campbell a group of dismounted cavalry that appeared to be infantry. When he and Campbell moved on, these men mounted their horses and later in the tour appeared again, this time as cavalry. In reality Forrest had only forty-five hundred men. As before, he also moved his artillery around to exaggerate the number of guns he had. Campbell, who had been stalling in hope of Union reinforcements arriving, finally surrendered in the face of such "superior" force. When the Union reinforcements arrived a bit later, Forrest also forced their surrender.[45]

BATTLE OF THE RIVER PLATE

In December 1939 the German pocket battleship *Admiral Graf Spee*, armed with 11-inch guns, was raiding merchant ships in the South Atlantic. When it encountered three British cruisers in the Battle of the River Plate, it battered the heavy cruiser *Exeter* but sustained some damage itself. Capt. Hans Langsdorff put into Montevideo, Uruguay, to repair the *Graf Spee*. International law at that time allowed combatants to stay in a neutral port only twenty-four hours if seaworthy, seventy-two hours if not. Langsdorff requested fourteen days to complete repairs that he claimed were necessary to make the *Graf Spee* seaworthy. The British made a show of objecting strenuously, but secretly told the Uruguayans to let the *Graf Spee* stay longer than seventy-two hours. The Uruguayans allowed the German ship to stay for a total of ninety-six hours.

The nearest British heavy combatants would take at least a week to reach Montevideo. The best they could do for reinforcement was the heavy cruiser *Cumberland*. Thus, the British had only one heavy cruiser and two light cruisers on station. These should have been no match for the German pocket battleship. Nevertheless, the British started rumors about British battleships just over the horizon itching to fight the *Graf Spee*. The light cruiser *Achilles* sent radio signals as though a large British fleet was present. To ensure that the Germans could not check out the rumors themselves, the British tied up every aircraft that the Germans might have chartered to observe the situation. After the time allowed by the Uruguayans, the *Graf Spee* got underway. Rather than face what Langsdorff thought was a formidable British force, the German captain scuttled the ship.[46]

Induce Enemy to Reveal Information

Another use of deception is to obtain information from the enemy.

THOMAS COCHRAN

In the early nineteenth century, chains of semaphore stations were the fore-runners of the telegraph. In 1808 Thomas Cochran raided a semaphore station on the French coast. The French guards retreated at his approach and watched him burn the station. When they returned they were much relieved to find their partially burned codebooks in the ashes. "Obviously," the British had failed to find them. In fact, Cochrane had copied the information and deliberately left the half-burned remnants behind so the French would find them and conclude the codes were still secure. After that, any British ship close enough to read the semaphore signals could decipher them. The goal of Cochran's deception had been quite specific: to induce the French to continue using compromised codes by convincing them that their codes had not been stolen despite the British having had free run of the semaphore station for some time.[47]

MIDWAY

In early May 1942 U.S. Navy codebreakers in Hawaii deduced that the Japanese were planning a large operation with the target "AF." Cdr. Joseph Rochefort, the chief codebreaker, and Adm. Chester Nimitz both became convinced that AF was Midway Island. However, naval intelligence in Washington believed the target was either Johnston Island or in the south Pacific, or even the west coast of the United States. Rochefort and Nimitz needed evidence to convince Washington of the validity of their point of view. Therefore they had the commander on Midway send a message, both in plain language and via a code known to be compromised, that their water distillation plant had exploded and they needed water. Shortly thereafter the codebreakers decrypted a Japanese message stating that A F had reported a water shortage. This convinced naval intelligence that Rochefort and Nimitz were correct. Although Rochefort designed this ruse to convince Washington about the correct identity of the target, some historians have incorrectly interpreted the incident as being used to convince Nimitz that Midway was the target, but Nimitz already believed that to be correct. The ruse was necessary to convince people in Washington.[48]

Early in World War II the British discovered several German spies. Rather simply throwing them in prison or executing them, the British turned them into double agents, who in turn led them to other German spies in Britain. This tactic also enabled the British to catch every new spy, as whenever the Germans would send a new spy to Britain, they would ask current spies to help them. The British then used this to turn the newly arrived spies into double agents. Soon, every German spy in Britain was actually working for the British. When the Germans no longer could conduct aerial surveillance over Britain that might contradict reports from their spies, it became easy to deceive the Germans. Thus the operation facilitated deception in many different ways, one of which was obscuring the location and timing of the invasion of France.[49]

Does Deception Qualify as a Principle of War?

To determine whether we should consider deception to be a principle of war, let us consider the criteria we set forth in chapter 4. Deception is applicable to all levels of warfare and broadly applicable to many types of warfare. Deception qualifies as a fundamental element of warfare. It is not just a gimmick but also something that appears repeatedly. Because armies have used deception throughout at least thirty-two hundred years of history, it qualifies as an enduring concept. To be sure, as the flow of information becomes greater and knowledge of the enemy becomes more complete, it can be more difficult to deceive the other side. Nevertheless, as long as human beings are in control of the conduct of warfare, they

will tend to believe their preconceived ideas and thus remain susceptible to deception.

Let us now consider our four-part discrimination test. Deception has
frequently been an important contributor to victory, but rarely caused
defeat. The nearest case of causing defeat might be Saddam Hussein's successful deception that led to the United States' invasion in 2003. At the
same time, the Japanese in World War II in the Pacific, in particular in the
Battle of Midway, provide a cautionary tale about the danger of spreading
forces into multiple groups with complex plans, part of which involved
deception. During the execution phase Yamamoto did not pass intelligence to Nagumo because he wanted to maintain radio silence. Indeed,
you sacrifice operational efficiency every time you use radio silence to
avoid revealing your position. We conclude that we need to consider certain detriments to deception, but it is a stretch to say that anyone other
than Saddam Hussein was defeated *because of* the deception perpetrated.

We are confident that the failure to use deception has rarely been an
important factor in victory. However, can we say that violation of the
concept of deception frequently led to defeat? If we interpret "violation
of the concept of deception" as meaning to telegraph one's moves to the
opposition, then nearly every frontal attack qualifies as a violation of the
concept, and many telegraphed frontal attacks have led to defeat.

Is the concept important in that commanders frequently violate the concept? Again, every attack without deception qualifies as a violation of the
concept of deception. Therefore we conclude that the idea is important.

Things that Enhance Deception

The deception planner must always keep in mind that the goal of deception is not just to confuse the enemy but to cause the enemy commander
to take a particular action—or not to take a certain action. We suggest
using as a guide the list of eight types of effects listed earlier in this chapter.

The planner should know the enemy well and tailor the deception to
the latter's predilections. The deception plan should take measures to
entice the enemy into believing the deceptive information. The planner
should understand human psychological biases, as well as cognitive dissonance and those factors that enhance it. The actual operation will have
the greatest impact on the enemy's mind if it brings up his worst fears and
causes him to feel trapped.

The primary thing that enhances deception is knowledge of your enemy. You need to understand what they expect you to do, what they hope you will do, what they fear, how they gather information, and the nature of their preconceptions.

Knowing your enemy's expectations allows you to deceive your enemy into thinking you will do what they expect, but then choose a different course of action. Reinforcing your enemy's preconceptions is easier than inducing them to believe things that they think are unlikely. People tend to compare their current situation using templates formed by experience.[50] When the Germans had to estimate where the Allies would land in 1944, they recalled the difficulties they had faced four years previously in planning the invasion of Great Britain. Severe naval constraints forced them to choose the shortest possible crossing. In addition, they needed the landing beach to be within range of their fighter airfields. The Germans assumed the allies would face similar problems and choose the shortest possible route. Hence the Germans were susceptible to deception that indicated the landings would be near Calais.

Barton Whaley claims that if a deception correctly anticipates and reinforces the enemy's expectations, then the deception is absolutely secure. Even if there are security leaks or other warnings, nothing can reverse the enemy's "fatally false expectations."[51]

The connection between deception and knowing your enemy can be a two-way street. Before he burned the two camps, Scipio Africanus used deception to gain the information he needed for a successful attack. After he gained that information, he was able to use that knowledge of the enemy to include deception in his attack plan.

Having a reliable source of information about the enemy's reaction to your deception attempts is invaluable. During World War II the Allied codebreaking of the German Enigma encryption system allowed them to determine which pieces of deception the Germans believed. In the larger scheme of things, constantly reading the German message traffic allowed the Allies to determine the types of information the Germans valued most.[52]

Tailoring a deception to your enemy's hopes seems to have obvious value. In 1941 Stalin hoped Germany would not invade the Soviet Union that year. He did not want to believe evidence that indicated such a thing was about to occur. Because they knew Stalin did not want to believe such evidence, subordinates started not giving him the information or giving him an interpretation of the information that fit Stalin's preconceptions. On the other hand, Richards Heuer claims there is no evidence that wishful thinking has an effect beyond that of people's expectations.[53]

The North Vietnamese planned the Tet attack to use every available asset to induce a popular uprising and capture many cities and towns in South Vietnam. They designed a deception plan that took advantage of American expectations and hopes. When the North Vietnamese proposed a longer-than-usual truce for Tet, the Americans took it as a sign of weakness. A captured Viet Cong agent said the North had sent him to open negotiations on matters such as prisoner of war exchanges. Meanwhile, North Vietnam's foreign minister claimed they would open bilateral peace talks if the Americans stopped bombing the North. The Americans regarded this as another sign of weakness.

On the military front, the communists conducted several diversionary attacks to draw attention away from the real targets in heavily populated areas. Ten days before the Tet attack the North Vietnamese attacked the remote American firebase at Khe Sanh near the border with North Vietnam. The Americans had expected the North Vietnamese to attack an isolated base in an attempt to replicate their victory over the French at Dien Bien Phu; as a result, General Westmoreland believed the Khe Sanh attack was the main effort, even saying that he expected some attacks elsewhere to divert attention from Khe Sanh.

The Americans ignored much intelligence about an impending major attack aimed at attaining final victory. On 5 January 1968 the Americans issued a press release stating that captured documents indicated that the communists would attack and capture cities and towns, and that they wanted to induce uprisings of the local populace. However, the Americans knew the communists were not capable of capturing and holding cities. Therefore, they expected that the enemy would not try to do so. The Americans were correct in their assessment of the communists' capabilities, but wrong in their deduction. The deception worked because it fed on U.S. expectations. Although the American forces were put on alert, the communists achieved surprise in the scope of the attacks, as related previously.[54]

Successful deception relies on the application of psychological factors and a knowledge of the enemy's susceptibility to believe various things. If a deception exploits enemy psychological weaknesses, the deception need not be perfect to work. Because human beings are prone to various biases, the enemy will sometimes do much of the work by deceiving themselves.

The human brain has evolved various mechanisms to help interpret the complexity of the world. In the evolutionary past of homo sapiens, these mechanisms generally proved their value. However, some of these mechanisms result in a tendency to see and interpret evidence in a manner that distorts reality under some circumstances. Psychologists have identified many such biases that show up repeatedly and predictably. These are general guides to human behavior, but are not fixed rules. Here we discuss a few that have relevance to military deception.

One part of "confirmation bias" is the tendency of people to interpret evidence in a way that supports their beliefs. A second part is that people *look for* evidence to support their beliefs and overvalue that evidence when they find it. They are inclined to ignore or distort information that contradicts their expectations. As they gather more and more "confirming" evidence, they become ever more adamant in their beliefs.[55] The more critical the situation, the more likely it is that people go into denial about any contradictory information.

The "anchoring effect" is the tendency for the first piece of evidence people obtain to influence them excessively because this information then serves as the reference point with which they compare subsequent information.[56] It then follows that stronger evidence is required to *change* a person's perceptions than to *form* an initial perception. People form perceptions quickly but resist changing those perceptions. The more strongly people hold their beliefs, the harder it is for them to change their views. Furthermore, ambiguous evidence affects people's perceptions even after better information becomes available.

The order in which a person receives evidence matters because early information has a greater impact than that received later. The astounding thing is that *this is true even for information that is later discredited.*[57] The tendency for misinformation to continue to influence us even after the wrong information is corrected is called the "continued influence effect."[58]

"Cognitive dissonance" is the psychological discomfort you feel when new information contradicts what you know or believe. The dissonance increases with the strength of your commitment to your belief, the degree of logical inconsistency, the number of inconsistent facts, and the relative importance you place on the discrepant cognition. The theory of cognitive dissonance says that this dissonance creates unpleasant psychological tension. This is not a trivial factor; the desire to eliminate this tension is similar to the drive to eliminate hunger or thirst.[59]

When the change in beliefs necessary to reconcile the new information is significant, the psychological impact is great. People are surprisingly capable of interpreting "clearly" contradictory facts in such a way as to support their strongly held beliefs. The drive to eliminate dissonance initially manifests itself by seeking validation of our initial beliefs, by denying contradictory evidence, by rationalizing new information, by downplaying the importance of the discrepant information, and other ego-defense mechanisms.[60]

If information arrives in small increments, people rationalize each of these facts sequentially as fitting into their beliefs so that no change in those beliefs is necessary. One's beliefs change only if there is one dramatic contradictory piece of information that overcomes the resistance to changing one's beliefs. Even when one's belief changes to accommodate the new information, the process of dealing with cognitive dissonance is quite unsettling and not conducive to the exercise of good judgment. The absence of evidence often will not spoil a deception, because people tend not to notice something that is not there, unless they specifically look for it.

The deception planner should control as many information channels as possible to reduce contrary evidence reaching the enemy. Nevertheless, people's capacity to rationalize contrary evidence is often sufficient to overcome security leaks and uncontrolled information channels. A good example of this is Stalin's rationalization of the evidence of Germany's coming attack.

It is best to discredit true information in the mind of the enemy as well as to build up false information. The El Alamein deception is an example of how one can do this.

One useful deception ploy is to exploit the cry-wolf syndrome. If one side repeatedly does things that appear to portend an attack and then backs off, the other side comes to regard the moves as routine. The enemy forms a memory of these false alarms and gradually begins to ignore the events that raised the alarms. The Egyptians used the cry-wolf syndrome to good effect to achieve surprise in the Yom Kippur War.

Psychological factors tell us that deception works best when it reinforces the enemy commander's preconceptions. The enemy will inevitably receive evidence that is contrary to the deception. The deception can survive such "leaks" provided each item of evidence is small or is susceptible to the enemy interpreting it as supporting his preconceptions and hopes. If possible, deception planners should not attempt to deceive the enemy in a manner that requires the enemy to go against his preconceptions. It is much easier to convince the enemy that you will do what they expect, and change your operations plan to do the unexpected. However, if it is imperative to change the enemy's mind about his expectations, then the first goal should be to get the other side to consider the possibility of the interpretation that you would like it to adopt. The way to do this is to use strong and obvious evidence early. Otherwise the enemy is likely to rationalize the planted information that disagrees with its expectations.[61]

Getting the Enemy to Accept Deception Clues

Deception works best when you feed it to your enemy via multiple pathways. Then, the enemy will see each piece of information as confirmation of similar information from a seemingly independent source. If possible, one of the multiple pathways should be your enemy's favorite, most trusted source of information.

At the tactical level, an effective way of using multiple pathways is to appeal to multiple senses. When Rosecrans wanted the Confederates to think he intended to cross the Tennessee River upstream from Chattanooga, he fed them visual clues (fires behind potential crossing points, wood chips in the water) and aural clues (men banging on barrels to make it sound as though they were building boats).

The Allies designed a deception for the airborne landings behind the Normandy beachhead to delay the German response to the vulnerable troops arriving by glider and parachute. It accomplished this by overloading the German information inputs about the location of the airborne landings. To do this the planners used deceptions aimed at several different senses. Visual clues included dummy parachutists and flares. Aural clues came from battle simulators on the dummy parachutists that produced the sounds of small-arms fire, mortars, and grenades. In addition, gramophones put out the sounds of soldiers' voices and other battle sounds. The deception even included olfactory clues from canisters of chemicals that spewed forth the smells of combat. Because the real parachute drops missed their target landing zones and were scattered over the countryside, this further added to the confusion. Soon the operations map at German headquarters was a mass of red over the entire area, making it impossible to determine the actual areas under attack.[62]

If a deception packages false information with information the enemy knows is accurate, this makes the enemy more likely to fall for the deception. In August 1942 British forces in Egypt prepared to repel an expected German attack. The British knew the Germans had captured British "going maps" that showed the drivability of various parts of the desert in different colors. They arranged for a scout car to be blown up and abandoned in a German minefield. Inside the wrecked vehicle was a creased and worn "going map" with multiple tea stains. This map showed areas occupied by the Germans and covered by maps they already possessed, as well as areas not covered by German-held maps. Because they could verify half of the map, the Germans believed all of it.[63]

This false "going map" showed that a particular route to the British stronghold on the ridge Alam el Halfa was difficult but a second route would be easy going. In the ensuing attack, the Germans sent their tanks along the route that the map indicated was easy. As a result, the tanks bogged down in deep soft sand in an area that the false map showed to be good conditions. Due to the resulting tripling of fuel consumption and slow progress, Rommel called off the attack. Two months later a captured German general stated that they had been fortunate to capture the valuable (false) map. He had originally intended to take the first route, but changed the plan to the second route after capturing the map. The interrogator decided not to tell the German the truth.[64]

Ideally a deception planner should deny the enemy sources of information that might contradict the desired interpretation. For example, during Operation Desert Storm, allied ground forces did not move west to position themselves for the "left hook" until they had precluded Iraqi air surveillance that might detect the force movement. In the deception for the Normandy invasion, the Germans' inability to fly reconnaissance flights over Britain, failure to obtain useful intelligence from radio intercepts, and inability to obtain other sources of intelligence forced the German General Staff to rely heavily on reports from their spies in Britain, all of which the British controlled.[65]

Often, depriving the enemy of all information about real force movements is impossible. Inevitably, information leaks out. Therefore, one should plan deceptions so that any force movements or other information that is likely to leak is ambiguous. That is, when the enemy obtains the information, they can interpret it as supporting the deception scenario. It is not even necessary that the deception scenario be the most likely explanation for the data point. Provided other aspects of the deception have previously conditioned the enemy commander's mind, the enemy commander may interpret the leaked data in the way that he (and you) would like. The greater the enemy's preconceptions, the more likely the enemy is to be misled. Of course, the fewer the leaks, the less likely the enemy is to catch on. As a corollary to this, one should defer as long as possible any necessary actions that may make the deception obvious to the enemy.[66] During the preparation for the Normandy invasion, allied aircraft cut transportation links that the Germans could use to reinforce the Normandy landing area in such a manner that it would also cut links to reinforce the Calais area.

To achieve a successful deception, one must suppress other ways for the enemy to determine the truth. In particular, one needs to deal with the problem of the enemy detecting a deception by observing the construction of a decoy installation. During World War II both the British and the Germans extensively utilized decoys to protect airfields and other important installations. For example, the Germans built an elaborate fake airfield in the Netherlands, including buildings, aircraft, vehicles, and anti-aircraft guns. The Germans constructed almost all of the bogus facility out of wood. The extensive decoy required a lot of time to build, so the Allies observed it and correctly identified the installation as a decoy. As soon as the Germans finished construction, a single British aircraft "attacked" the airfield and dropped one large bomb—made entirely of wood![67]

Contrast the above situation with the British deception before El Alamein. As described previously in this chapter, the British substituted fake objects for real objects (and vice versa) to provide continuity in what the Germans observed. Second, the British made any changes at night so the German could not observe the substitution.

In addition to using deception, commanders should also avoid being deceived. A review of the preceding section on psychological factors suggests several useful approaches.

Commanders should examine their own beliefs for preconceptions the enemy might try to exploit. Commanders should avoid relying on a single channel for information. Redundant information channels are useful for detecting inconsistencies that might indicate enemy deception. When examining the historical record of the enemy's actions, avoid being misled by a small number of events (small sample size), especially if the stakes were lower in past events.[68]

Commanders should realize that the enemy might employ the "cry wolf" syndrome by triggering numerous false warnings. This is extremely difficult to overcome. One probably apocryphal anecdote contains tongue-in-cheek advice for avoiding the problem. Supposedly, when an old British Foreign Officer retired in 1950 after nearly fifty years of service, he spoke about how every year people came to him with frenzied warnings that a major war was about to break out. He dismissed the warnings every time. And, in his entire fifty-year career, he had only been wrong twice![69]

To avoid forming false perceptions that may be difficult to change due to confirmation bias, one should suspend judgment as long as possible. To avoid overweighting early evidence due to the anchoring effect and to weed out discredited information, one should periodically take a fresh look at all information. People do not change judgments as often as they should, and organizational incentives promote consistency at the cost of accuracy. One should consider other methods for avoiding the anchoring effect and confirmation bias such as using competitive analyses, devil's advocates, interdisciplinary reviews, and a counter-deception staff.[70]

In planning for the Battle of Midway, Nimitz relied heavily on the decryption of Japanese messages. General Eammons forwarded an army intelligence criticism of Nimitz's plan.[71] The critics pointed out that plans should be based on enemy *capabilities* rather than perceived enemy *intentions*. This elementary lesson in the use of intelligence annoyed some of Nimitz's staff. However, Nimitz saw it as a useful warning. He assigned a senior officer on his staff to challenge and reassess every bit of information submitted by intelligence. He also directed Edwin Layton, his intelligence officer, to review all the data. In addition, he told the officers plotting locations of the Japanese ships to compare their movements with the supposed Japanese plan.[72]

One must counter the tendency to see evidence as supporting beliefs when the evidence might also support other hypotheses. You should note what evidence you do not observe. If you do not see evidence that you would expect to see if your beliefs were correct, that should alert you to a problem with your beliefs.

One should try always to be open to alternative interpretations. You should form alternate hypotheses and identify observables for each hypothesis. You should then evaluate alternative hypotheses in view of actual evidence and expected observables that were not detected.[73]

For those readers who believe they are less susceptible to psychological bias than the average person, there is another important bias to consider. The "bias blind spot" is people's tendency to think they are less susceptible to bias than the average person. We readily recognize that other people are biased, but we refuse to believe that we too have cognitive biases. One study found that 85 percent of the people surveyed thought they were less biased than the average person.[74] The bias blind spot follows from people's desire to view themselves as rational. The first step in avoiding the various psychological biases is to admit that we have a bias blind spot and need to take active measures to control our problem.[75]

ELEVEN

Simplicity

For every complex problem, there is an answer
that is clear, simple, and wrong.

—LOUIS MENCKEN

....

ONLY THE UNITED STATES, NATO, AND ISRAEL CONSIDER SIM-
plicity to be a principle of war. The American explanation of the principle of simplicity addresses both plans and orders:

> The purpose of simplicity is to increase the probability that plans and
> operations will be executed as intended by preparing clear, uncomplicated
> plans and concise orders.
>
> Simplicity contributes to successful operations. Simple plans and clear,
> concise orders minimize misunderstanding and confusion. When other
> factors are equal, the simplest plan is preferable. Simplicity in plans allows
> better understanding and execution planning at all echelons. Simplicity
> and clarity of expression greatly facilitate mission execution in the stress,
> fatigue, fog of war, and complexities of modern combat and are especially
> critical to success in multinational operations.[1]

The NATO explanation of the principle of simplicity is similar to that of
the Americans but is simpler and more concise: "Simple plans and clear,
concise orders minimize misunderstanding and confusion."[2] In contrast,
the Israeli definition addresses plans, but not orders. It recognizes that
plans may be complex and only requires that each element of the plan
should be simple to execute.[3]

Many battles have been lost because subordinates did not implement
a plan as the commander intended. The principle of simplicity claims to
facilitate accurate execution of plans by reducing misinterpretations of
orders. Complex plans tend to be hard to understand and susceptible to

something going wrong. The more moving parts a plan has, the more things that can go wrong and doom an operation. The more coordination of elements that is required, the more scope there is for a lack of coordination. However, there is another side to the issue.

Weaknesses of Concept

Although complex plans tend to be hard to understand and susceptible to things going wrong, this does not mean we can make all plans simple. Simplicity in plans is not always desirable. After all, the simplest plan is to go straight at the enemy. Flank attacks, probing for weakness, deception, surprise, and seeking relative advantage all introduce complexity. The problem is that warfare is not simple; it is inherently complex.

History provides many examples of commanders who violated simplicity in favor of complex operations, yet won a great victory because of the plan. For example, the Allied invasion of Normandy in June 1944 involved a multinational command structure, armies from five countries, months of air attacks, elaborate deception operations, landings at five separate beaches, two artificial harbors, paratroop drops, glider landings, naval gunfire support, and numerous other complications. More than seven thousand naval vessels and eleven thousand military aircraft supported the operation. More than 155,000 men landed on the first day, 500,000 in the first two weeks. Within a month the Allies had put ashore one million men and 190,000 vehicles.[4] Yet the Normandy invasion was successful. Writers have described it in many ways, but never as being simple.

Amphibious operations in the Pacific theater during World War II were nearly as complex as the Normandy invasion. In addition, there usually was a threat from the Japanese fleet or, later in the war, kamikaze attacks. Defending against this threat required a strong naval force led by multiple aircraft carriers. If a Japanese fleet approached the landing area, the American fleet commander then had to decide whether to give priority to defending the landing force or to attacking the Japanese fleet. The Americans made numerous mistakes early in the war, but gradually learned how to conduct such complex operations with few costly mistakes. They learned how all the pieces fit together and what was required of each element. Furthermore, they tried to allow for the likelihood that some aspects of the operation would go awry by ensuring the plan would withstand some things not going right. The currently recognized principle of simplicity is an overly simplistic solution to a real problem.

Essence of Concept

The goal of the principle of simplicity is to improve the probability of executing plans as intended by minimizing errors due to misunderstandings of orders. Let us separate the two parts of this goal.

First, the American principle claims that simple plans are easier for subordinates to understand and execute. (Simple plans can also be easier for the enemy to anticipate.) Overly complex plans can be difficult to execute because they are susceptible to many individual pieces going wrong because of misunderstandings, confusion, enemy responses, or friction from various sources.

One vulnerability of complex plans is that there are many "moving parts": often, if one piece goes awry, the entire plan fails. One way to ameliorate this problem is to devise *robust* plans, so that if a few things go wrong, the operation can still be successful. To minimize individual pieces of the operation failing, the plan must incorporate *realistic* expectations.

The second part of the American principle involves minimizing the misunderstanding of orders. Its recommended solution is to make orders concise and simple. Certainly, verbose, poorly organized, poorly written, or ambiguous orders invite misinterpretations. However, simplicity is not necessarily the solution. After all, "Win the war" is a concise and clear order, but it is also useless. Commanders are most likely to minimize misunderstandings not by avoiding complexity where necessary but rather by demanding *clarity* in orders and reports. One mark of a good writer is the ability to explain complex concepts clearly.

In summary, the essence of the concepts behind the principle of simplicity is *robust* plans, *realistic* expectations, and *clarity* in orders and reports. Robust plans reduce the consequences of a few things going wrong. Incorporating realistic expectations in a plan reduces the number of things going wrong by making it more likely that subordinates can accomplish critical tasks according to the plan. Clarity in orders reduces the number of things going wrong by enhancing subordinates' understanding of their assigned tasks. Clarity in reports reduces the fog of war, allowing commanders to make better decisions that accord with the real situation. The following sections elaborate on each of these points.

Simple plans are not necessarily easy to execute, nor are complex plans necessarily bad. We have noted that historically, many operations such as the Normandy invasion were immensely complex, but necessarily so. The solution is not mere simplicity. Rather, one should root out *unnecessary* complexity. As Albert Einstein is reputed to have said, "Everything should be made as simple as possible, but not simpler."[5]

Simplicity of plans does not always lead to success. On the morning of 4 June 1942 the Japanese Kido Butai (four aircraft carriers and supporting ships) approached Midway Island. Its search plan was simple: seven aircraft would fly outbound a distance of three hundred miles (150 miles for one of them), turn left, fly 60 miles, and return to their ship. The outbound legs of the flights were twenty-three degrees apart for the five sectors to the east and northeast. Two additional sectors covered the area directly south of the Japanese force. Five of the aircraft were seaplanes; two were torpedo planes from the aircraft carriers.[6]

As is commonly the case in warfare, not everything went according to plan. Most critically, the no. 1 plane from the Japanese cruiser *Chikuma* failed to detect the American task force despite its planned search path passing close to the American fleet. It either flew off course or flew above the clouds, thus severely limiting its detection range.[7] The fate of the Japanese Kido Butai—the core of Japan's military strength—depended on a single aircraft (1) launching on time, (2) following the assigned search pattern, (3) sighting the American task force if it flew close to it, and (4) accurately reporting the location and composition of the force in a timely manner. Failure to do any one of those four things could result in disaster for the Kido Butai and the Japanese prospects in the naval war. *Chikuma's* no. 1 plane failed to do the third (or perhaps the second) of the necessary four things. This *single* failure was a vital contributing factor to the loss of four Japanese aircraft carriers.

Other search aircraft also experienced problems. The no. 4 plane from the cruiser *Tone* had mechanical problems that delayed its launch by half an hour. The search plan specified that *Tone's* no. 4 plane would search the sector immediately south of that searched by *Chikuma's* no. 1 plane. Because of its late launch, the pilot evidently cut his outbound leg short to about 220 miles.[8] By a stroke of good luck, this failure to follow the plan resulted in sighting the American task force much earlier than if the plan had been followed. The pilot, however, failed to identify any of the ships as being an aircraft carrier for a critical half hour.[9]

As a result, the Japanese search did not detect the American carriers until at least ninety minutes later than they would have done if the Japanese aircraft had followed the plan perfectly. The delay in launching *Tone*'s no. 4 aircraft could have been disastrous—for many years it was described as being the deadly error that led to the Japanese defeat—but as Parshall and Tully have shown, this was not the fatal error. It was the failure of *Chikuma*'s no. 1 plane to detect the American task force even though it evidently flew over them.

The problem with the Japanese search plan was not complexity or simplicity. The problem was the lack of *redundancy*. The Japanese assigned too few aircraft to search such a large area. They counted on decent weather and nearly perfect timing on the part of all aircraft. They should have assigned more aircraft to scouting and had overlapping sectors so that even if one aircraft failed to sight the American force, another aircraft would do so. At this stage of the war the actions of the Japanese search plan was typical of both sides when they did not think the enemy was near. Nevertheless, the plan was not robust because it had no redundancy.[10]

On the western front during World War I, machine guns, barbed wire, and artillery made offensive breakthroughs difficult. However, the robust system of multiple, widely separated lines of trenches, backed up by reserves, made breakthroughs nearly impossible by ensuring that losing the first line of trenches would not be fatal.[11]

The critical factor is not the simplicity or complexity of a plan; it is that a plan must be *robust*. Simplicity is not the same as robustness, and simplicity certainly does not guarantee that a plan will be robust.

When we describe a plan as being robust, we mean that it is able to recover from unexpected setbacks. Sometimes, this is called being "fault-tolerant." A plan will be robust if there are alternative routes to success. An operation that absolutely requires the use of route A will fail if an enemy, Mother Nature, or an accident blocks that route. However, if an operation can succeed if *either* route A or route B is passable, then success is much more likely. Better yet would be a choice of three or more routes. A rigid adherence to the false principle of simplicity would favor a plan that used only route A because that would be simpler. At the other extreme, a fragile plan might require each of three routes to be passable. Note that the "routes" in this paragraph need not be geographic paths but can be any intermediate activity necessary to reach the final objective of the operation.

The classic story about how minor events can snowball and cause disaster is the nursery rhyme:

> For the want of a nail the shoe was lost,
> For the want of a shoe the horse was lost,
> For the want of a horse the rider was lost,
> For the want of a rider the battle was lost,
> For the want of a battle the kingdom was lost,
> And all for the want of a horseshoe-nail.

One might blame such an outcome on bad luck, or one might call it friction (Clausewitz's term for small real-world problems that accumulate to derail military operations). However, if winning or losing a battle depends on a single rider arriving on time, then the battle plan has a single mode of failure and is fragile. If it is vital that a message reaches someone, then the sender should transmit that message via multiple modes. In this case, the message-sender should have used multiple riders taking multiple routes.

During the Seven Weeks' War between Prussia and Austria in 1866, the two Prussian armies were separated when Gen. Helmuth von Moltke (the Elder) learned that his First Army was about to battle the Austrian army near Sedowa. Moltke needed to order the crown prince's Second Army to march at once to their support. Time was critical. However, the telegraph line was out of commission. To be sure his order would get through, Moltke sent *two* mounted officers, traveling by *separate* routes to the headquarters.[12]

Commanders should devise plans that are robust, with few or no single modes of failure, thus making the overall plan less likely to fail. Complex plans that require precise coordination among numerous units are likely to fail in a way that some people will blame on friction, but in reality, the problem is that such plans have many modes of failure.

To make a plan robust, commanders should have redundancy in critical paths. They should avoid plans with critical failure nodes, where a single thing going wrong can derail the entire operation. If a plan contains such critical nodes, then a commander must take extra care to be as sure as possible that the assigned forces will be able to accomplish the critical task. In turn this requires that commanders have realistic expectations of the ability of their forces to accomplish critical tasks.

Realistic Expectations

Friction in war has been a problem in all forms of warfare throughout history. It is a common cause of operations running into trouble. As Clausewitz emphasized, friction is an ever-present reality of warfare. Friction is often a large number of small factors that accumulate to frustrate a commander's intent and sabotage an operation. Factors contributing to friction are things such as some minor factor delaying a unit, with the unit's tardiness causing a delay in an attack, thus upsetting the timetable for the entire battle plan. Things never seem to happen as rapidly as they should. The principle of simplicity claims that enhancing subordinates' understanding of the plan will reduce confusion and hence friction.

Indeed, complexity can magnify the effects of friction. Complexity can cause friction if subordinates do not understand what they are supposed to do and how it fits into the big picture. However, complexity is not the primary cause of friction, and simplicity does not eliminate friction. Important causes of friction include the failure to understand limitations of one's own forces, lack of appreciation for environmental difficulties, and failure to anticipate things that could go wrong.

Plans that assume perfect execution are doomed to fail. The way to minimize friction is to anticipate potential problems, to be alert to the possibility of friction interfering with plans, and to make allowances for things going wrong. Plans should be realistic and within the capability of the assigned force to execute. Commanders must not adopt plans that their forces cannot execute in the real world. Fallwell suggests that the ability of commanders to achieve the unity of effort needed to execute a plan should be a dominant factor in determining whether to accept or reject a course of action.[13]

Commanders need to know the capabilities of their own forces and know the environment. That way, they know what is possible and what is not. They must search for things that might go wrong. If a unit must travel by a particular route to reach a position at a certain time, commanders must not engage in wishful thinking about how long it will take. Rather, they must have realistic expectations and consider that things are not likely to go as smoothly as it seems they should, especially if they are comfortably sitting behind a desk.

The Japanese plan for the Battle of Leyte Gulf involved three widely separated forces reaching particular locations on a tight timetable. The plan did not allow for maneuvers necessary to avoid enemy attacks. As one result, the powerful force penetrating San Bernardino Strait did not do so at the time planned.[14]

One part of having realistic expectations is to understand the level of training of the people assigned to carry out a task. In the complexity of the modern world, ordinary people can accomplish amazing things when they each have specialized knowledge or skills and their efforts can be coordinated. In the military arena also, specialization can allow each person or unit to train on a narrowly defined role. The Israeli principle of simplicity requires that each piece of the plan must be simple to execute. Then if a plan puts all the pieces together in a robust manner, success follows.

In 1973 Egyptian forces successfully carried out a sophisticated attack on the Israeli defenses on the Suez Canal. The Egyptian troops were not well-educated or well-trained by American standards, but they trained extensively on their tiny piece of the overall attack until they could do it with near perfection. Without sufficient training, however, many men would not have properly executed their individual tasks.[15]

As related in chapter 14 on knowing yourself, because Col. Daniel Morgan knew the serious limitations of his militia troops at the Battle of Cowpens during the Revolutionary War, he devised a plan with realistic expectations of the militia's performance.

To prepare for Operation Anaconda in Afghanistan in 2002, Pete Blaber had his men conduct what he called environmental reconnaissance of the approaches to several observation points overlooking the Shahikot Valley. They experimented taking different approaches and tried different equipment. Among other things they learned the best routes to take and the speed at which they could travel over each path while avoiding detection by the enemy. They then formed plans based on realistic expectations of the times needed to reach their observation points.[16]

Clarity

Orders often must convey complex plans. However, that does not mean one must write orders unintelligibly in the bureaucratic prose that the military often uses. If orders are misunderstood, it is most likely not because the orders are not simple, but because the orders are not clear. It is not simplicity in orders and reports that leads to understanding—or absence of misunderstandings—but rather *clarity*. If plans can be made simple and yet do the task, that is to be commended. However, when plans are necessarily complex, those writing the orders must expend considerable effort to ensure that the complexities are broken into clear and specific tasks and presented with clarity. The more complex the situation, the harder the order-writer must work to be clear. Orders must be unambiguous. Similarly, reports from subordinates to commanders also must be clear and unambiguous. The common denominator in three of the following examples is that a commander issued an order, the meaning of which was clear to him but was not clear to the recipients of the order.

The United States owes its independence to an ambiguous order. During the Battle of the Capes on 5 September 1781 the British fleet under Rear Adm. Thomas Graves attempted to break the blockade of the French fleet that bottled up Cornwallis's army at Yorktown. The British flag signal system limited the orders Graves could transmit. He wanted his entire line, then sailing parallel to the French line and two miles away, to move close to the French ships while maintaining their formation. He signaled both "line ahead" and "close the enemy," hoping his captains would correctly deduce his intentions. Instead, they were confused and chose to obey the "line ahead" order. As a result, the British line formed an acute angle with the French line. While the least-capable British ships in the van engaged the most-capable French ships, the most-capable British ships remained out of range of the least-capable French ships. The British got the worst of the battle and the French maintained the blockade that led to Cornwallis's surrender.[17]

Contrast Graves's hopes that his captains would deduce his intentions with the way Horatio Nelson anticipated problems before the Battle of Trafalgar in 1805. Nelson hosted his captains at two dinners on his flagship *Victory* to give them his "commander's intentions." He explained his plan for the coming battle. Nelson knew he could not plan every aspect of the action, so he did not give his captains a long list of detailed instructions. Instead, he gave them general guidance: "In case signals can neither be seen or perfectly understood no Captain can do very wrong if he places his Ship alongside that of an Enemy."[18] Thus Nelson's captains could have no doubt about what to do when confronted with uncertainty.

Before the battle of Waterloo, Napoleon defeated Field Marshal Blücher's army at Ligny but did not destroy it. He wanted Marshal Grouchy to prevent Blücher from joining Wellington by positioning himself, with his thirty-three thousand men, between Blücher and Wellington. On the morning of the battle Napoleon sent Grouchy an order, which was ambiguous and confusing. The order directed Grouchy to follow the enemy's column on his right but also told him to go to Wavre if the enemy was already there. However, if Blücher headed for Wellington and Grouchy went to Wavre, then that would soon put him on the wrong side of Blücher. Undoubtedly confused, Grouchy chose the one clear direction in the message and proceeded to Wavre. Later, Grouchy sent a report to Napoleon that was also confusing and uninformative.[19] Grouchy's failure to prevent Blücher from joining Wellington resulted in the defeat of Napoleon at Waterloo. Ironically, Napoleon is reputed to have warned, "An order that can be misunderstood will be misunderstood."[20]

In the Battle of Balaclava on 25 October 1854 during the Crimean War, Lord Raglan, the commander in chief of the British Army, wanted the cavalry to prevent the Russians from carrying away a set of guns the enemy had just captured. Raglan sent a written order to Lt. Gen. the Earl of Lucan, commander of the cavalry, "Lord Raglan wishes the cavalry to advance rapidly to the front—follow the enemy and try to prevent the enemy from carrying away the guns."[21] This order was simple but it lacked clarity. It did not specify which guns. From Raglan's position it was obvious which guns he meant, but from Lucan's location, it appeared to be the guns at the end of the well-defended valley, which is what Raglan's aide mistakenly indicated to Lucan. Raglan's aide also told Lucan that Raglan wanted the cavalry to "attack immediately," even though the word "attack" was not in Raglan's written order. This resulted in the disastrous Charge of the Light Brigade.[22]

On 20 September 1863 Maj. Gen. William Rosecrans commanded the Union Army of the Cumberland at the Battle of Chickamauga. He was concerned with shifting forces to his threatened left, which Maj. Gen. George Thomas led. One of Thomas's aides rode to ask Rosecrans for more reinforcements. He reported to Rosecrans that, along the way, he had seen a gap in the line between the divisions of Brig. Gens. Thomas Wood and Joseph Reynolds. Rosecrans also received another similar report. Both reports were wrong. Brig. Gen. James Brannan's division filled the supposed gap, but deep woods hid their position from the two observers. Without verifying the reports, Rosecrans immediately ordered Wood to move his division to close the gap. Unfortunately, Rosecrans did not have his chief of staff, Brig. Gen. James Garfield, write the order. Garfield, who wrote every other order that day, knew there was no gap, and he would have corrected the erroneous perception if another matter had not distracted him. Instead, another aide wrote an order whose meaning was ambiguous: "The general commanding directs that you close up on Reynolds as fast as possible and support him." This order was simple but not clear. "Close up" means to move laterally to close a gap. "Support" is ambiguous in this context. Wood knew he could not slide his division laterally because Brannan's division was in the way. Therefore Wood moved his division out of the line—creating an actual gap in the Union line— and moved behind Brannan and toward Reynolds. To complicate matters further, Wood encountered Thomas on his way to request reinforcements. Thomas knew Reynolds did not need any support. Therefore Thomas ordered Wood to move his division farther away to shore up the left wing.

Meanwhile, on the Confederate side, Lt. Gen. James Longstreet had learned from Pickett's Charge. Instead of attacking on a broad front, he arranged a deep column of sixteen thousand fresh men attacking on a narrow front. In a dramatic demonstration of the importance of chance in warfare, for his point of attack, Longstreet picked the very spot in the Union line where Rosecrans had unwittingly created the gap. Furthermore, he chose a time that was ten minutes after the creation of the gap, before anyone on the Union side could realize the error and plug it. Longstreet's pile-driver attack broke through the gap and routed much of the Union army. Only a determined stand by Thomas, the "Rock of Chickamauga," saved the army from complete destruction.[23]

During the Battle of Leyte Gulf, Adm. Bill Halsey sent a message that was misinterpreted by most of those who received it. He intended it to say that he might in the future detach battleships from his carrier task forces to form Task Force 34. However, most recipients interpreted it to mean that Halsey would form Task Force 34 immediately and this force would guard San Bernardino Strait.[24] As a consequence, a strong Japanese naval force passed undetected through San Bernardino Strait and attacked the American ships near the landing area. Disaster was averted only when the Japanese commander lost his nerve in the face of determined opposition by the outgunned Americans.

Fog of War in the Battle of Cape Esperance

The fog of war permeated many of the night battles in the Solomons. The Japanese and Americans fought the Battle of Cape Esperance on the night of 11–12 October 1942 at the entrance of the strait between Savo Island and Guadalcanal. Rear Adm. Aritomo Goto led a Japanese bombardment group of three heavy cruisers and two destroyers with the mission of bombarding Henderson Field on Guadalcanal. A reinforcement group would precede him by several hours and land supplies on Guadalcanal. Goto's confidence was high. In night battles up to this point in the war, the Japanese had sunk eight American cruisers and three destroyers; the Americans had not sunk a single Japanese ship.

American aircraft detected approaching Japanese ships during the day. Rear Adm. Norman Scott led an American force of four cruisers (*San Francisco*, *Salt Lake City*, *Helena*, and *Boise*) and five destroyers. He planned to surprise the Japanese by getting to the area before them. Because the Americans had no night fighting doctrine and his ships had little opportunity to train together, Scott chose a simple column formation with destroyers ahead and behind his cruisers. The Japanese ships did not have radar, but the American ships did. Two of the cruisers, *Helena* and *Boise*, had the newly introduced SG radar that was markedly superior to its predecessor. However, due in part to secrecy, most officers were not familiar with its capabilities. Scott chose to embark on the cruiser *San Francisco*, which had the older, less-capable SC radar.

During the approach to the battle Scott ordered his four cruisers to launch float planes to search for the Japanese. *Helena* did not receive the order and jettisoned its plane as a fire hazard prior to the battle. Flares in the rear cockpit of *Salt Lake City*'s plane ignited shortly after takeoff and the plane crashed in flames. Engine problems forced *Boise*'s plane to land.[25] *San Francisco*'s aircraft spotted part of the Japanese reinforcement group and reported "One large, two small vessels, one six miles from Savo." This report was ambiguous. Did it mean all three vessels were sixteen miles from Savo Island, or did it mean that one of the three ships was six miles from Savo?[26]

The Japanese submarine I-26 sighted the American ships but chose to attempt an attack before reporting the sighting. By the time I-26 reported, it was too late to be useful. Japanese sailors ashore unloading reinforcement group ships heard American float planes, an indication of American ships nearby, but did not report this to the force at sea.

At local time 2333 Scott ordered a course change by columns, in which each ship follows in the wake of the ship ahead, with all ships turning at the same point in the ocean. All the ships in the force received the order correctly, except the flagship. The order did not travel accurately up one deck to the bridge. The flagship misinterpreted the order as a simultaneous turn. When the leading destroyer started its turn, the flagship also turned. The next ship in line followed the flagship. Each subsequent ship followed the wake of the ship ahead. This left the three destroyers in the van by themselves.

Eight minutes before this course change, *Helena's* SG radar had detected the Japanese force at a range of nearly thirty thousand yards. However, *Helena* did not report this to Scott. By the time *Helena* had range, course, and speed estimates on its contact, Scott ordered the course change. Because of the course change, *Helena* assumed the flagship must also have had the contact (it did not). *Boise* also gained contact on its SG radar but did not report it, for similar reasons.

Because experts erroneously had told Scott the Japanese could detect emissions from the older SC radar, Scott had ordered his ships not to use it. *Salt Lake City* used its SC radar anyway. Surprisingly, it picked up a contact about the same time as Helena's contact. However, the skipper of the *Salt Lake City* had little confidence in the contact. Therefore he did not report it to Scott. Thus, every American cruiser had radar contact except the one Scott was on. However, none of the three cruisers reported its contact.

Helena finally reported its contact at a range of twelve thousand yards to Scott at local time 2342, fully seventeen minutes after its initial contact. When *Boise* reported its surface contact to Scott two minutes later, its description was five "bogies" bearing sixty-five degrees. *Boise* intended to report the contact as being ships, but the term "bogies" meant aircraft. If the sixty-five-degree bearing were relative to the ship's heading, that would indicate it was the same contact as that reported by the *Helena*. However, if the bearing were true, that would indicate a second Japanese force was nearby. *Boise* soon clarified that the bearing of sixty-five degrees was relative, not true, but not all ships received the clarification. The Americans had a single radio channel for communications, and needless chatter often blocked vital communications.

Ironically, because of the mismanaged turn Scott managed to cross the "tee" of the Japanese force. Even more ironically, had the American column turned as Scott intended, it might have been the Americans who had their "tee" crossed by the Japanese. Such is the influence of the gods of chance. As it was, Scott was in an ideal situation. However, he did not realize it because he did not have a good picture of the situation and was very concerned about firing on the three destroyers that were originally in the van. On the *Boise* and *Helena*, their SG radars with its plot-like display gave them a good picture of the overall situation and they were certain their targets were not the American destroyers.

To gain a better mental picture, Scott asked the commander of the destroyer squadron whether he was taking station ahead. The destroyer commander, who did not understand why he found himself out of formation, interpreted the question as a command. He replied in the affirmative and assumed his other two destroyers would follow him. Right about that time, *Duncan*, one of the other destroyers, gained contact on its fire-control radar, pulled out of formation, and headed for the Japanese force to launch a torpedo attack. It did not tell anyone that it was doing so.

Meanwhile, as the forces approached each other, Japanese lookouts reported seeing three ships at a range of eleven thousand yards. However, Goto, the Japanese commander, thought they might be the Japanese reinforcement group and he transmitted the Japanese recognition signal.

Without a clear picture of the situation, Scott initially held off firing at the Japanese because he worried about firing at his own destroyers. *San Francisco* gained radar contact on its fire-control radar at a mere five thousand yards, but classified it as "unidentified." *Helena*'s commanding officer was certain his radar contacts were Japanese ships headed right at them, and they were getting dangerously close. Then *Helena* gained visual contact on the Japanese. A worried ensign in radar plot facetiously wondered aloud whether they intended to board the Japanese ships. Although *Helena*'s commanding officer had authorization to open fire without permission, he asked Scott for permission. In accordance with the General Signal Procedure, *Helena's* skipper said, "Interrogatory Roger." Scott replied "Roger," meaning that he had received the message. However, under the General Signal Procedure, "Roger" had two meanings. Here it meant "open fire." At time 2346 *Helena* opened fire at a range of just thirty-six hundred yards, and the other American ships soon followed.[27]

On the other side of the battle, Japanese lookouts identified the ships as American at a range of seven thousand yards. Goto remained unconvinced and repeated the recognition signal. Shortly thereafter, American shells hit Goto's bridge and mortally wounded him. To escape the dire situation Goto ordered his ships to reverse course with a starboard turn, but two of his ships evidently misread the signal and turned to port.[28]

One minute after the Americans opened fire, Scott, lacking a radar to give him a good picture of the situation, ordered a cease-fire because he was still uncertain about the identity of the targets. Either most of his ships did not hear the order or they ignored it. Four minutes later, Scott ordered gunfire to resume.

In the battle the Japanese seriously damaged one American light cruiser. The Americans sank one Japanese heavy cruiser, seriously damaged another, and sank one destroyer. In addition, both Japanese and American gunfire sank the American destroyer *Duncan* and damaged another American destroyer.[29]

The Battle of Cape Esperance illustrates the extreme confusion often present in combat, the ubiquity of chance, and the pervasiveness of the fog of war. One might describe some of the incidents as being friction. Several things contributed to the fog. People failed to report vital information to the decision-maker on at least five occasions. Commanders did not believe at least three accurate reports. In at least four cases orders did not reach their destination (or subordinates failed to carry out the order). Subordinates misunderstood at least two orders (or orders were transmitted inaccurately). At least four communications were ambiguous or used poor terminology. *None of these problems resulted from a lack of simplicity.*

Nearly all of these mistakes were avoidable if commanders had emphasized reporting all contacts, making reports clear, and using unambiguous terms in communications. If Scott had chosen as his flagship either *Boise* or *Helena*, the two light cruisers in his group equipped with SG radar, he would have had a much clearer picture of the situation. The decision-maker should ensure that he has the best information, or the person with the best information should make the decisions.

Is Simplicity a Principle of War?

The principle of simplicity claims to reduce misunderstandings and to increase the probability of executing plans as intended. As we have shown, however, the principle of simplicity does not address the real causes of the problems it aims to cure.

Although simplicity gives some practical guidance, it does not give much guidance on how to avoid oversimplifying things—other than urging commanders to use their judgment. We see numerous cases—such as the amphibious assaults during World War II—in which a plan violated the concept of simplicity yet the battle was won because of a complex plan.

We conclude that the current principle of simplicity does not merit status as a principle of war. The concept of simplicity is simply too simplistic. Keeping things simple is still an excellent idea, but not to the exclusion of all else.

Instead, commanders should concentrate on avoiding the real causes of misunderstandings and failed plans: clarity of orders and reports, robust plans, and realistic expectations that minimize the impact of friction. Despite their importance, we do not consider these three concepts to be principles of war because we do not think they are fundamental elements of warfare. However, we do classify clarity (of orders and reports), robust plans, and realistic expectations to be in a class of important second-tier concepts that we call "near-principles."

TWELVE

Other American Principles

IN THIS CHAPTER, WE CONSIDER THREE AMERICAN PRINCI-
ples of war that we have not discussed previously—economy of force,
security, and maneuver—plus three relatively recent additions to the principles of joint operations—restraint, perseverance, and legitimacy. For
each of these principles, we present the currently recognized principle
and its explanation. Then we consider whether it meets our criteria for
a principle of war.

Current Principle of Economy of Force

The United States has listed economy of force as one of the principles of
war since 1921. It emphasizes economizing the use of force in everything
but the primary effort so that commanders can employ maximum force
at the decisive place and time:

> The purpose of economy of force is to expend minimum essential com-
> bat power on secondary efforts to allocate the maximum possible com-
> bat power on primary efforts.
>
> Economy of force is the judicious employment and distribution of forces.
> It is the measured allocation of available combat power to such tasks as
> limited attacks, defense, delays, deception, or even retrograde operations
> to achieve mass elsewhere at the decisive point and time.[1]

France and India also list economy of force as a principle of war, while
Israel uses the principle of optimum utilization of forces. The British originally used economy of force but changed to economy of effort after World
War II. Other English-speaking countries use the British principle, which
does not differ markedly from the American:

> Economy of effort is central to conserving fighting power. Command-
> ers must prioritise resources between engagements, actions and activities,

and the sustainability demands of the operation as a whole. Economy of effort is best summarized as creating the right effect, in the right place, at the right time with the appropriate resources.[2]

Because resources are always limited, one must economize on forces devoted to secondary efforts to achieve mass at the critical point. Thus, the principle of economy of force is complementary to the principle of mass. Indeed, Robert Leonhard argues that economy of force underlies all other principles because war is inherently wasteful and involves managing scarce resources.[3]

Economy of force is in opposition to the current principle of security. Taken together, the two principles seem to say, "Practice economy of force, but not so much economy as to jeopardize security; practice security, but not so much security as to violate economy of force." That is not useful guidance for a commanding officer.

In their plans to defend against a German invasion in the run-up to World War II, the French practiced economy of force. They built the Maginot Line so that fewer troops per mile would be required to defend that part of the border. Because terrain would restrict an enemy advance in the Ardennes sector, the French defended the area only with a few second-class divisions. As a result of these measures, they could send the strongest possible force into Belgium to defend against the expected path of the German invasion. That was the critical place—or so they thought. The French plan was in accordance with the principle of economy of force. They thought they were allocating "minimum essential combat power to secondary efforts" so they could achieve superiority and overwhelming effects in the decisive operation on the Belgian front. Obviously the French miscalculated the "decisive point," as well as the "minimum essential combat power" necessary to defend the Ardennes sector.

In the previous chapter we described how the Japanese search plan at the Battle of Midway failed, as it employed too few aircraft to make the search robust. The reason the Japanese assigned so few aircraft to scouting was that they were practicing economy of force. By minimizing the air assets devoted to scouting, they maximized the aircraft that could be devoted to attacks at the critical place and time, which turned out to be a false economy.

Other American Principles

Regarding the relationship between deception and economy of force, the American definition of economy of force calls for the "measured allocation of available combat power to . . . deception . . . to achieve mass elsewhere at the decisive point and time." As Liddell Hart points out, sometimes it is necessary to devote a large force to deception. If you employ a large part of your force to distract your enemy and successfully do so, your enemy is likely to be much weaker at the point you have chosen for your primary attack.[4] Jomini gave a quantitative example of the potential benefits of employing a substantial part of one's force in a deception:

> If you have one hundred battalions against an equal number of the enemy's . . . bring eighty of them to the decisive point while employing the remaining twenty to observe and deceive half of the opposing army. You will thus have eighty battalions against fifty at the point where the important contest is to take place."[5]

Thus, by using a larger portion of your own force to distract the enemy, you may greatly increase your relative advantage at the critical location that you have chosen. An example of this is the German attack in May 1940. A large portion of the German forces attacked into the Netherlands and Belgium to distract the Allies. Then the strong armored force that penetrated the Ardennes had a great relative advantage when they met the French defenders at the Meuse River.

A counterexample is the British resupply mission to Malta in August 1942. As a deception plan the British ordered four merchant ships, escorted by four cruisers and thirteen destroyers, to sail from Alexandria and Haifa to fake an assault on Crete. However, the force allocated to the deception was not large enough to pose a credible operational threat. Therefore the Germans quickly recognized it as a deception and did not divert any forces from their main effort, instead concentrating on the resupply force and severely damaging it.[6]

We do not think the current principle of economy of force merits designation as a principle of war because it fails to meet three of our criteria for a principle of war. First, the concept is not sufficiently profound for us to consider it fundamental. The essence of the principle of economy of force is to accept prudent risk in selected areas to achieve what we call relative advantage in the decisive area. That is perfectly sound, but we consider the concept to be implicitly included in the new principle of relative advantage, as well as the principle of prioritized objectives. Given a fixed amount of force, how else are you going to concentrate your forces if not by economizing elsewhere?

Second, it does not provide practical guidance. There is an inherent conflict between the principles of economy of force and security. How to balance the two concerns is a matter of judgment, and the principle says only to be judicious and measured in making that judgment. A principle that says "use good judgement" is not useful.

Third, the practice of economy of force does not meet the test of frequently having resulted in victory. It is hard to find cases in which a commander won a battle *because of* following economy of force. He may have won because he kept his eye on the primary objective and concentrated as much of his force as he could to achieve a relative advantage over the enemy in a battle. In the situation described in the previous sentence, we submit that it was following the principle of relative advantage that was an important determinant of victory, not economy of force. The latter merely refers to *how* the commander went about gaining the relative advantage. Although lack of economy of force can indirectly lose a battle by not allowing one to gain a relative advantage, the primary cause of a victory is unlikely to be due to practicing economy of force.

Therefore we reject economy of force as a principle of war. It is, however, an important tool for achieving relative advantage and ultimately to reach your objective. Prioritizing objectives provides guidance for where to economize and where to concentrate to achieve a relative advantage.

Notice how the idea of using substantial forces to distract the enemy does not fit in with either the principle of economy of force or that of mass. However, it does fit naturally into the principle of relative advantage. When done well, a large force can distract a proportionately larger portion of the enemy force from the critical sector, thus giving you a better relative advantage in the forces remaining in the critical sector.

Current Principle of Security

The principle of security is nearly universal. Except for the French, all Western militaries that list principles of war include security in that list. Furthermore, it has always been included since the British first listed it in 1920. The American definition is:

The purpose of security is to prevent the enemy from acquiring unexpected advantage.

Security enhances freedom of action by reducing friendly vulnerability to hostile acts, influence, or surprise. Security results from the measures taken by commanders to protect their forces, the population, or other critical priorities. Staff planning and an understanding of enemy strategy, tactics, and doctrine will enhance security. Risk is inherent in military operations. Application of this principle includes prudent risk management, not undue caution.[7]

The British definition is similar:

Security entails balancing the likelihood of loss against achieving objectives. It demands managing risk, protecting high-value assets and resilience. Security does not imply undue caution or avoiding all risks, for bold action is essential to success. Neither does it demand over-committing our resources to guard against every threat or possibility, thereby diminishing relative fighting power.[8]

Never allowing the enemy to gain an unexpected advantage is certainly desirable. Avoiding both surprise and excessive caution is certainly desirable. "Prudent risk management" is certainly advantageous. All of this, however, is rather obvious. Does security merit designation as a principle of war?

The essence of the currently accepted principle of security is to prevent the enemy from gaining a relative advantage. Thus you must avoid being surprised or otherwise allowing the enemy to seize the initiative. Indeed, one part of the principle of surprise is to prevent the enemy from surprising you.

The concept of security does not rise to the level of a fundamental element of warfare, which our criteria require to designate a concept as a principle of war. A corollary to the new principle of relative advantage is that of preventing the enemy from gaining a relative advantage. This corollary includes the essence of the current principle of security.

As discussed in the previous section, the principle of security is in opposition to the principle of economy of force. The two principles taken together do not provide practical guidance to a commander on how to proceed. This factor also precludes us from including security in the new principles of war.

Finally, security does not qualify as a principle of war because, although there are plenty of cases in which battles and wars were lost due to poor security, it is hard to find cases in which a commander won a battle *because of* practicing security. Although lack of security can lead to defeat, the primary cause of a victory is unlikely to be the practice of good security.

Therefore we reject security as a principle of war. It is, however, an important aspect of preventing the enemy from achieving a relative advantage.

Current Principle of Maneuver

The U.S. armed forces are the only military to list maneuver as one of the principles of war. As related in chapter 1, the 1921 title of the American principle was "movement," but it was changed to "maneuver" in 1949:

> The purpose of maneuver is to place the enemy in a position of disadvantage through the flexible application of combat power.
>
> Maneuver is the movement of forces in relation to the enemy to secure or retain positional advantage, usually to deliver—or threaten delivery of—the direct and indirect fires of the maneuvering force. Effective maneuver keeps the enemy off balance and thus also protects the friendly force. It contributes materially in exploiting successes, preserving freedom of action, and reducing vulnerability by continually posing new problems for the enemy.[9]

The Israelis evidently think the concept of maneuver is too obvious to include it as a principle of war. While the British listed the similar principle of mobility as a principle of war in 1920, it was notably absent from the British revival of the principles after World War II. One can see some aspects of the principle of mobility in the British principle of flexibility introduced after World War II. However, flexibility is a much broader principle that we will discuss in chapter 13.

Maneuver is a means of achieving a positional advantage that threatens the enemy. Repeated maneuvers can keep the enemy off balance and thus allow your forces to obtain and retain the initiative. Furthermore, maneuver is often an alternative to direct combat because you can achieve the

Other American Principles

same advantage through threats, that is, maneuver might yield the same benefit as direct combat but without the risks of the latter.

An example of "rapidity of movement . . . limited only by physical endurance and the means of transportation available" occurred during the Second Punic War. Hannibal's brother Hasdrubal led an army across the Alps and entered Italy in the spring of 207 BCE, intending to join Hannibal in the south of the peninsula. In response, Caius Claudius Nero, the Roman consul commanding the Roman army shadowing Hannibal in the south, took six thousand foot soldiers and one thousand cavalry and hurried to join his fellow consul Marcus Livius Salinator in confronting Hasdrubal before he could join Hannibal. To speed the march, towns along the way provided food and supplies, including mules and carts to carry the weary. Nero's army marched 250 miles in about a week. To conceal these reinforcements Nero's men arrived at night and shared tents with Salinator's men. Nevertheless, Hasdrubal figured out the ruse and retreated to the Metaurus River.

When the two armies fought, Nero commanded the right wing of the Roman army. Making no progress against a strong position, Nero took a substantial portion of his force, marched behind the entire Roman battle line, and attacked the Carthaginians on their extreme right flank. This surprise attack resulted in routing the entire Carthaginian army. Nero then took his soldiers and marched the 250 miles back to southern Italy in six days, arriving before Hannibal discovered his absence.[10]

Nero's operational maneuver of a 250-mile rapid march gained operational surprise and the initiative for the Romans. In the actual battle, Nero's tactical maneuver of switching from one side of the battlefield to the other gained tactical surprise and a great relative advantage that yielded victory.

As seen from this example, maneuver can allow you to gain and maintain the initiative. Often there is an intermediate stage. Maneuver creates a threat, or multiple threats, and then the threat leads to the initiative, which in turn leads to a position of relative advantage. However, maneuver is a means to an end. It is a method for gaining the initiative, a relative advantage, or surprise. Maneuvering gains nothing if it does not gain an initiative, a relative advantage, or surprise, or at least prevent the enemy from doing so.

THE CONCEPT OF MANEUVER FAILS OUR DISCRIMINATION test on two counts. First, it is difficult to find many cases in which a commander failed to maneuver and as a result lost the battle. Second, it is hard to find cases in which a commander won a battle because he maneuvered. He may have won because he kept his eye on the primary objective, then used maneuver to concentrate as much of his force as he could to achieve the initiative and a relative advantage over the enemy. Maneuver may have contributed to victory, but it was not an important direct factor.

Finally, maneuver is merely a tool to reach a favorable position. It is not a fundamental concept. Because it does not meet our criteria, we conclude that maneuver does not qualify as a principle of war.

Three Principles of Joint Operations

The 2011 revision of U.S. Joint Doctrine replaced nine principles of war with twelve principles of joint operations. These principles include three new principles: restraint, perseverance, and legitimacy. The 2018 update to the joint doctrine stated that all twelve of the principles of joint operation apply to combat operations.[11] In this section we consider whether these three new principles meet our criteria as principles of war.

The definition of the principle of restraint begins,

The purpose of restraint is to prevent the unnecessary use of force.[12]

This candidate principle provides good advice for noncombat operations and for combat operations such as the recent operations in Iraq and Afghanistan. However, this book concentrates on principles of *war*. Given the goal of this candidate principle, it clearly does not qualify as a fundamental concept applicable to the conduct of war. Although it provides useful guidance for some situations, restraint does not qualify for designation as a principle of war.

The definition of the principle of perseverance begins,

The purpose of perseverance is to ensure the commitment necessary to achieve national objectives.[13]

This is applicable to the National Command Authority rather than to the military. In Western democracies the concept of civilian control of the military is deeply entrenched. If the civilian leaders order the military to continue fighting for many years, the military will obey orders.

This candidate principle provides good advice for a civilian government in a democracy. In order to persevere, the government must convince the people that a prolonged effort is necessary. However, this book concentrates on *military* principles of war. This concept does not qualify as a fundamental element applicable to the military's conduct of war. Therefore, we do not deem perseverance to be a principle of war.

The definition of the principle of legitimacy begins,

> The purpose of legitimacy is to maintain legal and moral authority in the conduct of operations.[14]

This candidate principle provides good guidance for noncombat operations and for some combat operations such as recent operations in Iraq and Afghanistan. However, this book concentrates on principles of *war*. Given the goal of this candidate principle, it clearly does not qualify as a fundamental concept applicable to the conduct of war. It provides useful guidance for some situations, but only a few. Western democracies prefer to operate legitimately, do the right thing, and act morally. In some situations, this helps achieve victory. However, throughout history, many nations have won battles and wars without the benefit of legitimacy. Therefore, legitimacy does not qualify for designation as a principle of war.

THIRTEEN

Other British Principles of War

IN CHAPTERS 5 THROUGH 12 WE CONSIDERED AS CANDIDATE principles of war the nine American principles of war and three other principles of joint operations. The ten British principles of war are more widely used than the American principles of war. Therefore, we also should consider each of them as candidate principles of war. The British principles of security and surprise are essentially the same as their American counterparts. Four British principles are reasonably close to corresponding American principles, with a change of title being the only appreciable difference: selection and maintenance of aim (objective), offensive action (offensive), concentration of force (mass), and economy of effort (economy of force). More substantively, the British replace the American principle of unity of command with cooperation. Previously in this book, we considered each of these seven British principles alongside their American counterparts. At the same time, while the British do not use the American principles of maneuver and simplicity, they add the principles of flexibility, sustainability, and maintenance of morale to their own list. Therefore, in this chapter we consider these latter three principles as candidate principles of war.

Flexibility

All English-speaking militaries other than the United States list flexibility as one of their principles of war, as does NATO. In this section we consider flexibility as a candidate principle and test it against our criteria to decide whether it merits designation as a principle of war.

Candidate Principle of Flexibility

The British definition of the principle of flexibility is:

> Flexibility comprises mental and physical aspects. Flexible organisations and cultures encourage people to think creatively, and to be resource-

ful and imaginative. Flexibility needs the physical and structural ability to allow forces to act rapidly, especially when operating in complex situations. Really flexible organisations are highly responsive. This can be measured in their speed of action/reaction or how quickly a commander seizes the initiative.[1]

Some authors advocate adding the principle of flexibility to the American principles of war.[2] In particular, Robert S. Frost has written cogently on the subject and advocates adding the need to focus on the objective and the commander's intent. We will use his proposed definitions. His short definition is:

The purpose of flexibility is to be responsive to change and adaptable to the volatility, pressures and complexities of military operations, while constantly focusing on the objective.[3]

Frost's longer definition is:

Flexibility is both a state of mind and a characteristic of effective military units. It is an antidote to surprise, uncertainty and chance. Flexibility represents the fundamental ability to avoid dogmatic rigidity and to "bend" as each situation demands—to be receptive, responsive, and adaptive—and like a flexible, resilient rod, to neither break nor lose orientation. Flexible units and leaders adapt to changing conditions in opportunistic and innovative ways, yet never lose focus on the commander's intent. Flexible leaders encourage critical, creative thinking, ensuring all realistic alternatives and possible outcomes are considered throughout mission planning and execution. Flexibility—the antithesis of rigidity—is the crucial ability to synthesize all of the principles, and to prioritize their application in each unique situation to ensure success.[4]

The core of Frost's definition of flexibility is that commanders should be receptive, responsive, and adaptive. In addition they should be open to doing things in a different way than in the past (that is, they should be innovative). Above all, commanders and organizations should maintain a flexible mindset. To prevent flexibility from turning into indecisiveness,

Frost emphasizes that all aspects of flexibility are subject to maintaining focus on the objective and the commander's intentions.[5]

BE RECEPTIVE

Commanders should anticipate that unexpected things will occur and should be receptive to information that challenges their previous beliefs. Wars and battles rarely turn out in the way either side anticipated. The element of chance plays a large role in war. Because warfare is chaotic in the mathematical sense, small changes in seemingly minor situations can have major, unpredictable effects. One reason that warfare is so complex is that the enemy constantly tries to thwart your plans and devise plans to surprise you. The enemy commander has also read the books that advise military commanders to do the unexpected. In addition, no rules constrain the enemy's attempts to thwart you. In chess each side takes turns moving, the battlefield consists of exactly sixty-four squares, and each piece can move only in a specified manner. In war each side tries to take three turns to the enemy's one, turn the enemy's flanks by moving "off" or over the battlefield, and use weapons in an unexpected manner. Hence commanders should have a flexible mindset and expect that things will not work out as planned; indeed, they should expect the enemy to do the unexpected and should seek to reduce unpleasant surprises.

When the enemy does the unexpected, commanders must reevaluate their situation and perhaps adapt their plans. To react in a timely manner it is important to detect information about unexpected conditions as soon as possible. To do this commanders should be receptive to information that challenges their prior beliefs; as a corollary, they must be wary of confirmation bias. Furthermore, they should actively seek to obtain evidence that would contradict their beliefs. They should observe carefully what the enemy is doing, gathering clues about unexpected weapons, tactics, or capabilities. Germane here is the discussion at the end of chapter 10 about methods to resist enemy deception.

Commanders should also encourage subordinates to question and criticize the interpretation of available information. One sure way for commanders to avoid getting information about unexpected developments is to discourage subordinates from bringing them bad news. Rather than blaming the messenger for unwelcome news, commanders should encourage subordinates to bring bad news and critical views.

Other British Principles

Furthermore, commanders must be alert for erroneous assumptions underlying their plan. They should seek information that might contradict their assumptions and should encourage subordinates to question the assumptions behind the plan. Indeed, assumptions can harbor three types of dangerous errors: known unknowns, unknown knowns, and unknown unknowns. Known unknowns are identified uncertainties in assumptions. These could turn out to be a problem, but at least commanders should be aware which aspects of the situation are uncertain.

Unknown knowns are things you are confident you know, when in fact your confidence is misplaced. Because you start out convinced that you know a certain aspect of the situation, figuring out that this assumption is in error is extremely difficult. In World War II, both the Germans and the Japanese "knew" their highest codes were unbreakable. Therefore, whenever they received information suggesting that their codes had been broken, they ended up explaining away the evidence as due to some other factor.[6]

Unknown unknowns are things that you never considered that turn out to be important, even critical. When the factor becomes so obvious that one can no longer deny its reality, that realization can be disorienting, even paralyzing. For example, prior to World War II the British strengthened the defenses of Singapore. As a result of these fortifications, Churchill believed it was impregnable. Although the defenses would have been strong against an attack from the sea, Singapore was vulnerable to an attack from the landward side. When the defenders realized the Japanese were coming from that direction, they were so paralyzed they did not put up as much of a fight as they might have done, even considering their disadvantages. In summary, commanders must be mentally flexible in interpreting the available evidence and seeking potentially contradictory evidence.

BE RESPONSIVE

When someone accused him of changing his mind, John Maynard Keynes is reputed to have said, "When my information changes, I alter my conclusions. What do you do, sir?" When commanders receive new information, they must have the flexibility of mind to accept any new evidence that indicates that some previous assumptions (or presumed facts) are wrong. As discussed previously, when confronted by information that contradicts our assumptions, human nature tends to ignore or explain away that information. The principle of flexibility, however, demands that commanders fight the urge to explain away information that might contradict their assumptions. Instead, they must constantly reevaluate the situation whenever they obtain fresh information.

As many people have noted, no battle plan survives first contact with the enemy. Of course, commanders should devise plans that are robust so that the plan is still worth following even if not every assumption is correct. They should also make their plans flexible, with options in case the enemy does the unexpected. The more flexible and robust the original plan, the less likely it is that changes will be necessary. Nevertheless, as circumstances change, commanders must be prepared to revise plans.

One way in which adaptation to new information might come into play is learning from earlier events in the war. Commanders must objectively determine whether their weapons and tactics are working as well as predicted. New weapons, when first introduced, are often quite effective. However, with time the enemy devises countermeasures. Commanders must observe these and be prepared to change their own tactics in response.

Prior to World War II both the U.S. Navy and the Japanese Navy regarded the battleship as the capital ship. In part due to necessity, the United States quickly realized that in fact aircraft carriers were now the capital ships in navies. The U.S. Navy in the Pacific then developed doctrines based on early battles that allowed commanders to employ all available assets in a coherent manner.[7] On the other hand, the Japanese periodically slipped backward and planned for aircraft carriers to soften the enemy up so that battleships could move in for the kill.

The Americans were far more effective than the Japanese were in learning from experience, for example, by applying lessons they learned from the Battle of Coral Sea at the Battle of Midway a month later. In contrast, the Japanese did not make any significant changes. The aviation fuel officer on the carrier *Yorktown* observed how fire consumed the aircraft carrier *Lexington* at Coral Sea. At Midway, when they expected an attack, personnel on *Yorktown* drained aviation fuel tanks on the hangar deck and purged the fuel lines with carbon dioxide. As a result, although the Japanese hit *Yorktown* several times during the Battle of Midway, major fires did not ravage it. On the other hand, quickly spreading fires were a factor in the destruction of each of the four carriers that Japan lost. Furthermore, while the Americans improved their combat air patrols and fighter direction at Midway based on the experience at Coral Sea, Japanese fighter tactics were nearly the same in both battles.[8]

Before the First Punic War Rome had been a land power and had few ships. Nevertheless, Rome displayed great flexibility by building a navy from scratch, becoming a naval power and winning control of the sea.

After his nearly fatal defeat at the Battle of Long Island, Gen. George Washington realized that he could not win by engaging in major battles with the British. Therefore he switched to a Fabian strategy that ultimately won independence for the United States.

In the Vicksburg campaign Ulysses Grant tried many avenues of approach to Vicksburg. These included conducting a direct assault, sending a force down the Yazoo River, rerouting the Mississippi River around Vicksburg, and finding an alternative water route to the Red River and then to the Mississippi south of Vicksburg. After all these attempts failed, Grant moved his army and gunboats to a location south of Vicksburg, crossing the river into Mississippi and successfully confronting Confederate forces that outnumbered him. Grant's campaign has been described as the most brilliant campaign ever conducted on American soil. On the one hand, Grant was determined and inflexible in not giving up on his objective of subduing Vicksburg. On the other hand he demonstrated flexibility by trying different ways to achieve that objective until he found one that succeeded.[9]

As a final example, the Iraqi insurgency after the 2003 invasion demonstrated the ability to adjust the priority of objectives as conditions changed. First, the insurgents identified United Nations personnel, as well as many international aid agencies, as weak links, attacking and driving them out of the country. Next the insurgency tried to destroy the political will of the American people by attacking American forces and agencies. As the Americans improved their performance, the insurgency shifted to attacks on Iraqi forces. To drive a wedge between the Iraqis and the Americans, they focused on attacking any Iraqis working with the Americans, creating distrust and suspicion among Iraqis. They attacked the Iraqi security forces to damage morale and discourage enlistment. Doing so also demonstrated to the Iraqi people that the police could not protect themselves, much less the populace. Similarly, the insurgents conducted attacks while wearing police uniforms so the people would not trust anyone wearing such attire.[10]

HAVE A FLEXIBLE MINDSET

As a situation changes, commanders must have a flexible mindset and be able to adapt rapidly to the new situation. They must be willing to innovate and should not reject suggestions just because they have not followed certain strategies in the past. In fact, they should encourage "out-of-the-box" thinking.

History provides examples of commanders who exhibited a stubborn adherence to an objective long after it should have been apparent that the situation had changed and the objective was no longer attainable at an acceptable cost. For example, at Fredericksburg in December 1862, Maj. Gen. Ambrose Burnside ordered repeated attacks on impregnable Confederate lines protected by a stone wall. In chapter 7 (regarding the offensive) we gave several other examples of foolish offensives.

STRENGTH OF CONCEPT

Due to its complexity and the role of chance, warfare is not predictable. Commanders must be sufficiently flexible to adapt quickly when things do not work out as planned. It is not only random events that make warfare unpredictable. If commanders achieve any kind of success, their enemy will be highly motivated to do something different to neutralize what they are doing or even to turn the tables. When the enemy tries something unexpected, commanders with flexibility are able to adapt quickly and effectively. Rigidity in commanders and organizations is clearly undesirable, and flexibility is the opposite of rigidity.

WEAKNESS OF CONCEPT

Commanders must be sufficiently flexible to bend in a hurricane, but they cannot bend before every light breeze. While they must be determined rather than indecisive, the line between flexibility and indecisiveness, or between determination and stubbornness, is difficult to locate.

Commanders can have so much flexibility that it makes them indecisive or causes them to give up too soon, even if they keep the objective in mind. Indeed, in some cases, stubbornness can be a virtue. Two years before inventing a usable lightbulb, Thomas Edison said, "Many of life's

Other British Principles

failures are people who did not realize how close they were to success when they gave up." Later he said, "I have not failed. I've just found ten thousand ways that won't work."[11]

Rome rose from a small city to a great empire because its people had unshakeable political will. They *never* gave up. As related in chapter 16, after multiple defeats at the hands of Hannibal, the Romans flatly refused to negotiate and continued to fight. The Roman attitude was that wars ended only when their opponent conceded total defeat. Rome possessed an unshakable determination to win every war in which it participated, no matter what the cost. Indeed, the Romans were not the least bit flexible.[12] They did not "avoid dogmatic rigidity" as the proposed principle urges.

In the dark days of the Civil War the Union cause appeared lost on numerous occasions: after bloody defeats at Manassas, on the James peninsula, at the second battle at Manassas, at Fredericksburg, and at Chancellorsville. Yet Abraham Lincoln was not flexible and continued the war.

By 28 May 1940 France knew it was beaten. German forces surrounded the British expeditionary force at Dunkirk and threatened to annihilate it. Britain was about to stand alone against a Germany that had conquered much of Europe. Few ground troops were available to defend Great Britain against invasion. Not only did the United States refuse to help, the American ambassador in London was certain that Germany would defeat Great Britain.

In the face of calamity, people with flexibility, such as Lord Halifax, wanted to explore whether Hitler would accept a peace deal that would preserve British independence.[13] After listening to Halifax's arguments for several days, on 28 May Churchill told the outer cabinet, "Of course, no matter what happens at Dunkirk, we shall fight on."[14] He finished his talk with the following exhortation: "If this long island story of ours is to end at last, let it end only when each one of us lies choking in his own blood upon the ground."[15] A week later Churchill told the House of Commons, "We shall go on to the end. . . . We shall fight on the beaches . . . we shall fight in the fields . . . we shall never surrender."[16] Winston Churchill was not flexible.

ESSENCE OF CONCEPT

According to Frost, flexibility is both a state of mind and a characteristic of effective military units. A person or organization should be flexible, receptive, responsive, adaptable, and innovative, but should not be rigid, dogmatic, stubborn, obstinate, or intransigent. All these character-istics are subject to focusing on the objective and the commander's intent. Commanders should bend without breaking as each situation demands.

Does Flexibility Meet the Criteria for a Principle of War?

Let us examine the candidate principle of flexibility against the criteria we presented in chapter 4. Because the proposed principle of flexibility depends on human nature, it qualifies as a fundamental element of war-fare and an enduring concept that will not require frequent revision. It applies to the strategic, operational, and tactical levels of warfare; land, sea, air, and space warfare; irregular warfare; and terrorism.

Commanders have violated the concept of flexibility sufficiently often to make a difference. Furthermore, violating the concept has often con-tributed to defeat.

Does the concept of flexibility give practical guidance? Yes and no. In urging commanders to seek information that contradicts their assump-tions, to absorb new information, and to be willing to change if the circum-stances warrant it is practical guidance. However, although the proposed principle emphasizes keeping focus on the objective, it does not give prac-tical guidance for how to tell when flexibility morphs into vacillation or lack of will.

The remaining question is whether, throughout history, flexibility has made a *difference* in determining whether a commander was successful. Does the candidate concept pass the four-part discrimination test? Adher-ence to the concept often has contributed to victory. However, it has also contributed to defeat. History provides many cases in which being too flexible resulted in defeat because a commander gave up too soon. Too much flexibility is bad.

Violating the concept through stubbornness or inflexibility often con-tributed to defeat. However, violation of the concept has sometimes con-tributed to victory. In many cases when a reasonable person would have given up, a commander was unreasonably stubborn and won a great vic-tory as a result.

Thus the candidate principle of flexibility fails two of the four dis-crimination tests. Commanders have been defeated because of too much flexibility, and commanders have won because they were unreasonably stubborn, rather than flexible.

Being flexible is clearly valuable, but flexibility does not meet our criteria to be a principle of war. Flexibility does not qualify as a principle of war because the line between flexibility and indecisiveness is blurry. It is easy to delude yourself about where your actions lie on that scale. You can have too much flexibility and end up giving up too soon. Even though flexibility does not qualify as a principle of war, it is an exceptionally important quality to look for in commanders and a particularly important characteristic of successful organizations.

Sustainability

There is a saying that claims amateurs talk tactics and armchair generals debate strategy, but professionals study logistics. All English-speaking militaries except the United States list logistics, sustainability, or administration as a principle of war. The British relatively recently adopted sustainability as a principle of war. Prior to that their corresponding principle was administration, which they introduced after World War II. Today New Zealand is the only other country that uses sustainability as a principle of war, although the Australian concept of sustainment is comparable. Canada and India still use the principle of administration, which includes both administration and logistics. The USSR lists various forms of supply, maintenance, and combat support as principles. For our candidate principle we consider the British principle of sustainability:

> Sustainability is a critical enabler of fighting power. Rigorously assessing logistic realities, including redeployment, is essential to operations planning. Sustainability may be the deciding factor in assessing the feasibility of a particular operation.[17]

Logistics is a limiting factor in all military operations. The best-trained, best-armed, and best-led military force is useless without fuel (for humans and machines) and ammunition, as well as many other vital items. When devising plans, commanders should consider their logistic capabilities.

During World War II several offenses across the North African desert stalled when the aggressors outran their supply line.[18] The German attack against the Soviet Union was halted not only by the weather but also by the long German supply lines over terrible roads.[19] At Stalingrad the Germans could not supply their surrounded army, despite Goering's promise to do so by air.

Because logistical deficiencies have often caused defeat in the past, we might consider a principle of logistics that would advise commanders to plan operations with logistical constraints in mind. However, the Allied invasion of northern France offers several instructive counterpoints. Despite two years of planning, the logistics plans did not work as intended. Unrealistic expectations and an overly rigid system resulted in much friction and congestion. The logistics system landed only half the planned amount of supplies in the first week. Logistical constraints caused the cancellation of several planned offensives. This emphasizes the importance of good logistics planning. However, there is a counterpoint to the admonition to pay attention to logistical constraints.

On 25 July 1944 the logisticians advised Eisenhower that an advance to the Seine by 4 September was impossible. Fortunately Patton did not listen to the logisticians. Despite their warnings, Patton advanced on 3 August. His army crossed the Seine by 24 August—eleven days before the date the logisticians had insisted was impossible. By 7 September they were two hundred miles past the Seine.[20] This does not invalidate the fact that logistics is important and constrains combat operations, but it is a cautionary tale.

Notwithstanding imperfections in logisticians' plans and advice, sustainability is relevant at all levels of warfare and for all types of warfare. It passes our four-part discrimination test. It contributed to many victories but caused few, if any, defeats. Poor logistics contributed to many defeats but few, if any, victories are attributable to poor logistics.

Sustainability is important at this time and certainly seems as though it will be important in the foreseeable future. In the past, fodder for horses was just as important as fuel for vehicles and aircraft is today. Even when armies did not depend on a supply line back to a base, living off the land constrained them by the need to keep moving and to find areas that had not yet been devastated.

In one sense the principle of sustainability certainly is a fundamental element of warfare. However, it is different in character from the other principles of war. It is the means needed to implement any strategy or tactic. It sets limits on all operations.

Does the principle of sustainability provide a practical guide for action?

The British principle of sustainability urges one to assess logistic realities. This is valid, but it does not provide practical guidance to commanders. Similarly, none of the corresponding principles from other countries provides practical guidance. Therefore we reluctantly conclude that, although sustainability (or logistics) is extremely important to success in war, it does not merit inclusion as a principle of war.

Maintenance of Morale

Vegetius, Napoleon, and Clausewitz were among those who stressed the importance of the moral factor in war. The British military adopted maintenance of morale as a principle of war after World War II. The USSR also included morale in their principles. Today NATO and all English-speaking militaries, except the United States, include some variation of this principle. For our candidate principle, we use the British definition:

> Maintenance of morale is crucial for operational success. High morale is characterised by steadfastness, courage, confidence, and sustained hope. Morale manifests itself as staying power and resolve, as well as the will to prevail in spite of provocation and adversity.[21]

The importance of morale to success in warfighting is undeniable. However, does it meet our criteria for a principle of war? Morale is relevant to all types and all levels of warfare. In addition, it is an enduring feature of successful warfare. Morale is important because fighting is ultimately done by individuals, and their performance depends in part on their morale. Ultimately a commander can follow all the principles of war, use good judgment, and make brilliant decisions, but if individual men and women do not fight hard, it is all for naught. Good morale is one reason individuals not only endure hardship and privation but risk their lives for a cause.

Can maintenance of morale pass our discrimination test? Poor morale has contributed to many defeats and likely has never been a significant cause of victory. Good morale has frequently been a contributor to victory. However, many armies have lost despite having good morale. Furthermore, there have been battles in which good morale *contributed* to defeat.

The Roman soldiers at Cannae were steadfast, courageous, and confident. Their high morale was combined with excellent training and good weapons. Partially because of their high morale, they attacked vigorously and pushed forward when the Carthaginian center seemed to give away. The Romans still sustained hope, courage, and resolve, but could not employ their skills effectively because they were caught in Hannibal's trap.[22]

The French knights at Crecy had excellent morale and high confidence. Their morale was so high that they rushed to attack before all the French forces were available. Their impetuous attack against the English in an excellent prepared defensive position resulted in a disastrous defeat.[23]

At the Battle of Tippecanoe on 7 November 1811, future president William Henry Harrison defeated Native Americans under the leadership of Chief Tecumseh's brother, known as the Prophet. The Native Americans attacked fearlessly because the Prophet told them they could not be hurt by American bullets. They most likely entered the battle with great morale.[24]

During the Boxer Rebellion in China, adherents to the cult of the Fists of Righteous Harmony believed that their mystical powers protected them from Western bullets.[25] They undoubtedly went into battle with high morale.

The Japanese at the Battle of Midway had great morale. Their pilots and aircraft were equal to or better than their American counterparts. The Japanese force of aircraft carriers had ranged from Hawaii to the Indian Ocean over the previous six months without experiencing defeat. Their morale was so great that, in retrospect, some termed it "Victory Disease." As a result of their overconfidence, the Japanese underestimated the Americans and made egregious errors in planning and executing the operation.[26]

If unchecked, good morale can result in foolish actions. To be valuable, good morale must be accompanied by good judgment, discipline, training, adequate equipment, and leadership. Therefore we conclude that although good morale is a valuable element of an effective fighting force, maintenance of morale does not qualify as a principle of war.

THE THREE BRITISH PRINCIPLES OF WAR THAT WERE DIScussed in this chapter—flexibility, sustainability, and maintenance of morale—are all excellent concepts that commanders would do well to heed. However, each has flaws that prevent it from being deemed a principle of war by our strict criteria. We classify these three important concepts as "near-principles."

FOURTEEN

Know Yourself

Know the enemy, know yourself; your victory will never be endangered. Know
the ground, know the weather; your victory will then be total.

—SUN TZU, *Art of War*

....

DESPITE SUN TZU'S ADVICE, NO MILITARY LISTS "KNOW YOUR-
self," "know your enemy," or "know the environment" as a principle of war.
We think all three concepts are worthy of consideration. Chapters 15 and
16 will discuss the latter two concepts as candidate principles of war. In
this chapter, we consider "know yourself" as a candidate principle of war.

New Principle: Know Yourself

Here is the definition of our proposed new principle "know yourself":

Taking advantage of strengths and compensating for weaknesses requires
that commanders have in-depth knowledge of themselves, their subordinates, their own forces, their weapons, and their current disposition.

Commanders should recognize their personal strengths and limitations.
They should be aware of their subordinates' strong and weak points and
know how to get them to perform at their best. Commanders must discern the morale of the fighting men and women under them. Commanders should know the capabilities and limitations of their own force, and of
their weapons and sensors, as well as understand the best way to employ
them. Knowledge of one's forces should result in realistic expectations that
in turn reduce some forms of friction. Commanders should maintain con-
tinual awareness of their own force's position and status.

First and foremost, commanders must know themselves. If they tend to
be impetuous, they must control those tendencies. If they are innately
cautious, they should be aware of, and fight against, that predisposition.

Commanders must recognize any biases they may have and endeavor to compensate for them. As related at the end of chapter 10, to avoid getting duped by enemy deception, commanders must admit that they have a "bias blind spot" and take active measures to counter it.

Knowledge of Subordinates

No matter how well educated and trained, human beings are not interchangeable parts. Commanders should know the strengths and limitations of their subordinates, as well as any personality issues among subordinates.

Robert E. Lee had a keen understanding of many of his lieutenants, though he sometimes did not apply that knowledge effectively. For example, Lee had the greatest confidence in Stonewall Jackson and knew he only had to give Jackson general instructions rather than detailed orders.[1] When Lt. Gen. Richard Ewell replaced Jackson after the latter's death, Lee knew that Ewell was used to specific orders under Jackson and discussed that with him. However, at Gettysburg Lee issued discretionary orders to Ewell, causing problems.[2] Furthermore, while Lee knew he could count on James Longstreet for solid defense, he knew he had to watch other generals more closely.

For example, even though cavalry leader Jeb Stuart had impressed Lee as a cadet when Lee was the superintendent of West Point, Lee knew Stuart had a tendency to seek adventure; consequently, during the Seven Days' Battle Lee warned Stuart against rashness.[3] Similarly, at the start of the Gettysburg campaign Lee emphasized to Stuart the need to keep him informed about the Union Army's movements and to rejoin the main body. However, his orders allowed Stuart sufficient discretion for Stuart to decide to ride around the Union Army. This gained glory but deprived Lee of knowledge of the enemy's movements at a crucial time.[4]

In another example, Lee showed poor judgment when he retained the incompetent Pendleton as his chief of artillery.[5] On the other hand, when Pres. Jefferson Davis asked Lee's opinion of whether John Bell Hood should replace Joe Johnston in command of the Confederate force in Georgia, Lee equivocated. He said that Hood was a great fighter but questioned whether he should command an army.[6] Davis made the move anyway, but Lee was correct in his reservations. Hood led the Confederates into a series of impetuous attacks that decimated his army. Thus Lee usually had good knowledge about his subordinates and often applied that knowledge effectively; however, he sometimes failed to take the appropriate action based on that knowledge.

Before Waterloo, Napoleon selected Marshal Grouchy, newly appointed a marshal, to command his right wing. He probably would have performed well had he been under Napoleon's direct supervision. However, in the context of an independent command he performed poorly when given the vital assignment of preventing Blücher from joining Wellington. Thus, in appointing him Napoleon exhibited lack of knowledge of Grouchy's capabilities.[7]

One of the strengths of American leadership in World War II was the willingness to select as commanders officers who were far down the seniority list. For example, Gen. George Marshall had long kept a notebook of exceptional officers he encountered during his career, and chose Eisenhower and Patton over many men who were more senior. Similarly, the U.S. Navy promoted Nimitz and Spruance over many officers with more seniority.

Knowledge of Capabilities and Limitations of One's Own Force

Commanders should be willing to ask of their force everything they are capable of doing but should avoid asking them to do what they cannot do, whether the limitation is due to training, equipment, or some other factor.

At the Battle of Cowpens during the Revolutionary War, Brig. Gen. Daniel Morgan faced Lt. Col. Banastre Tarleton. They each commanded about eleven hundred men; Morgan's force included some regulars but were mostly unreliable militia, while Tarleton's were all British regulars. Tarleton was known as "Butcher Tarleton" and had a reputation for not taking prisoners—"Tarleton's Quarter." Trying to catch Morgan, Tarleton relentlessly drove his men across the rough terrain of South Carolina, giving them little sleep or food. Finally, he had a chance to trap Morgan in front of a rain-swollen river. Tarleton got his men up and on the march by 0300 on 17 January 1781, and came upon Morgan at Cowpens, a clearing in generally wooded terrain. During past battles against militia he had always been successful by charging directly at them; he did so again here, holding some men in reserve and using his dragoons to guard his flanks.

For his part Morgan had realized he could not evade Tarleton much longer, so he chose the Cowpens to make his stand. Morgan knew the militia well. He had come from a similar background as his men. He knew they would run away if given a chance, so he chose a position in front of a rain-swollen river that made retreat impossible. He also knew the men in the militia were excellent sharpshooters but would not stand up to a British bayonet charge. (A charge by British regulars with gleaming bayonets was a fearful sight.) He took advantage of this capability by placing them in front of his regulars and asking them to fire just two volleys, and then fall back, knowing they were capable of carrying out these orders.

Morgan expected Tarleton to rush headlong at him. He arrayed his forces in three lines, planning for the first two to give way after inflicting some damage and slowing the British advance. The first group was the best militia rifle sharpshooters. The second line was the rest of the militia. As planned, the first line fired two volleys, giving special attention to British officers, and then withdrew to the second line of militia. The second line similarly fired two volleys and then withdrew behind the third line, which was composed of regulars. The retreat did not upset the third line because they knew the plan. Although the militia withdrew in good order, the British thought they were retreating in a panic and victory was at hand. They surged forward, only to run into Morgan's third line. After a few minutes, due to a misunderstood order, the third line withdrew in an orderly manner. The British saw this as a retreat and rushed ahead recklessly. Rather than panicking at this development, Morgan rallied the main line. He also talked to his militia and asked them to march around the lines and attack the British in their left rear. Because they had carried out their previous assignment successfully, they were amenable to this. At the same time, Morgan had his cavalry go around the lines on the other side and attack the British in the rear on that side. The result was a double envelopment worthy of being called a mini-Cannae. Caught in the trap, most of Tarleton's force surrendered. Only Tarleton and a small portion of his force escaped.[8]

Morgan succeeded because he knew his men and did not ask more than they were capable of doing. He knew the enemy commander and planned his defense to meet Tarleton's expected mode of attack. In addition, he was flexible enough to alter his plan when some orders were misunderstood. Finally, unlike Tarleton, Morgan did not go on the offensive until the undisciplined rush by the British gave him an opening.

In contrast, when commanders overestimate what their forces can do, failure results. After his brilliant victory at Chancellorsville in the spring of 1863, Robert E. Lee believed that his Army of Northern Virginia could accomplish anything he asked them to do. He asked them to do the impossible when he ordered Pickett's Charge. The men performed with consummate bravery, but they could not do the impossible. In contrast, Lt. Gen. James Longstreet knew the limitations of the Army of Northern Virginia. As he told Lee prior to Pickett's Charge in a last-ditch effort to change Lee's mind,

> General, I have been a soldier all my life. I have been with soldiers engaged in fights by couples, by squads, companies, regiments, divisions, and armies, and should know as well as anyone what soldiers can do. It is my opinion that no fifteen thousand men ever arrayed for battle can take that position.[9]

When George Meade took command of the Army of the Potomac a few days before Gettysburg, he knew the army had just suffered two major defeats and its morale was suspect. Consequently, he thought they were more likely to fight well if they were defending rather than if they were attacking.[10] Thus at Gettysburg Meade knew his men better than Lee knew his and consequently employed them more effectively than Lee did.

During the Battle of the Philippine Sea in 1944 Adm. Raymond Spruance asked Vice Adm. Willis Lee, commander of the battleship force, whether he wanted to attempt a night attack on the Japanese fleet with his battleships. Lee ordinarily would have leapt at such an opportunity. He would have ships that were more modern than the Japanese, with superior firepower and the advantage of radar. Nevertheless, Lee declined the opportunity because his battleships had not practiced for night battle.[11]

The importance of knowing yourself in wartime can be seen also in the context of guerrilla warfare. Because guerrillas usually have less training and fewer numbers than the government forces they are fighting, guerrilla commanders must know the true capabilities of their forces in order to achieve a relative advantage over the enemy. Because of the importance of the populace in this situation, guerrillas must have an accurate sense of the extent of their support among the people.

During the Chinese communist insurgency in the late 1920s, Mao Tsetung advocated operating in the countryside relying on the strength of the peasants. However, Marxist-Leninist dogma said that workers in the cities must be the basis of the revolution. Because it was dogma, this was not subject to debate. In 1930 the Soviets ordered the Chinese communists to conduct attacks in the cities; the Chinese complied, exposing the communists to the firepower of the nationalists, who repulsed them with heavy casualties. The communists had no relative advantage in the cities because the workers were weak while the nationalist forces were strong. In the countryside, however, the peasants were strong and the nationalists weak. Mao's insight that the strength of the revolution was in the peasants proved to be crucial to the eventual success of the movement.[12]

Knowledge of Own Weapon Capabilities and Limitations

Effective use of weapons requires good knowledge of their capabilities and limitations. At the beginning of World War II Germany employed massed tanks in Panzer divisions and equipped them with radios so their actions could be coordinated, unlike the mostly radio-less French tanks. In contrast, the French, with more and generally better tanks than the Germans, dispersed their tanks among infantry units so they could support the infantry. The French were then unable to deploy a large group of tanks to oppose the German Panzer divisions.

In World War II in the Pacific the Japanese developed tactics to take full advantage of their long-range, ship-killing Long Lance torpedoes. They also trained their crews extensively to reload the torpedo tubes quickly. The Japanese regularly practiced tactics, especially at night, so that the ship skippers knew how best to utilize the Long Lances.

As related in chapter 12, during the Battle of Cape Esperance the American commander had poor situational awareness because he chose as his flagship a ship not equipped with the latest SG radar. This was due to the Americans' lack of appreciation that the SG radar was markedly superior to its predecessor. Therefore they had not yet devised tactics to utilize this capability.[13]

One month later, in the Naval Battle of Guadalcanal on 13 November 1942, four of Rear Adm. Daniel Callahan's thirteen ships carried the SG radar. He should have placed one or both destroyers with SG radar in the van to pick up the Japanese force as soon as possible. Instead, none of the three leading destroyers had the SG radar. He should have used either the cruiser *Portland* or the cruiser *Helena* as his flagship because they had SG radar and could give him the best picture of the situation. Instead, Callahan chose the heavy cruiser *San Francisco*, which did not have the SG radar. As a result, he did not have a clear picture of the situation. The night battle developed into confused brawl in which there were numerous near collisions between Japanese and American ships. There were also several cases of fratricidal fire.[14] The fratricide resulted because Callahan's lack of understanding of the capabilities of his own sensors led to a lack of knowledge about the location of his own ships.

Knowledge of Own Force Position and Status

To employ forces effectively one must know their location and their status. For example, during the Battle of Britain in 1940, the British advantage was not just radar, but also the integrated air defense system that combined all knowledge of enemy raids with knowledge of their own aircraft locations and the status of aircraft not yet in the battle.

In contrast, during the Battle of Jutland, Adm. David Beatty's battle cruiser force engaged the German battle fleet and then led them toward the British main body. Because Beatty's reports were infrequent and incomplete, Adm. John Jellicoe did not know how best to deploy his battleships and had to make an educated guess when he ran out of time to make a decision.[15]

Toward the end of Operation Desert Storm the coalition forces had nearly cut off all of Iraq's Republican Guard divisions. However, General Schwarzkopf and the decision-makers in Washington erroneously thought they had completely cut off those key Iraqi forces. They called a ceasefire after one hundred hours of ground combat, to avoid the appearance of slaughtering Iraqis unnecessarily. This allowed a substantial portion of the Iraqi Republican Guard forces to escape the trap, even though their destruction had been one of the primary objectives of the war. Their survival enabled Saddam Hussein to survive and to suppress internal opposition to his rule.[16]

Fog and Friction

As Clausewitz pointed out, friction (little things adding up to derail operations) and the fog of war (factors leading to commanders having an uncertain, incorrect, or misleading view of the situation) are pervasive influences on the outcome of battles. If, due to the fog of war, a commander does not have an accurate view of the situation of his or her own forces then he or she cannot make effective decisions. One way for commanders to counter the fog of war is to insist that subordinates double-check reports. During Operations Desert Shield and Desert Storm, Vice Adm. Stan Arthur, commander of the U.S. naval forces, repeatedly warned his staff "the first report is always wrong." As a result, the staff got into the habit of immediately double-checking incoming information to confirm or deny its accuracy.[17]

Friction often results from a gap between a commander's expectations and the problems of operating in the real world where nothing is perfect. If a commander knows his force well enough to form realistic expectations of their capabilities, then this reduces the gap between expectations and reality. For example, if a commander knows that his or her army must travel on a single, muddy road, then the commander's plans should not depend on the army moving quickly. If the commander allows for a realistic speed, then the force will not be late in reaching its destination.

Does "Know Yourself" Meet the Criteria to Be a Principle of War?

To determine whether an objective is achievable, you must know the capabilities of your own forces. To achieve unity of effort, you must know your subordinates. In addition, you should be aware of whether any personality conflicts could affect unity. To gain a relative advantage, you must know the strength of your own force and the way to use it to best advantage. Thus knowing yourself is an important tool for implementing the three new principles of war of prioritized objectives, unity of effort, and relative advantage. Does know yourself also qualify as a principle of war on its own?

The concept of knowing yourself is clearly applicable at the strategic, operational, and tactical levels of warfare. It is also pertinent to ground, air, and naval warfare. Commanders have failed to heed the concept sufficiently often in military history that it is worth paying attention to it. Because this concept has been an important factor for centuries, it qualifies as an enduring concept. Whether it should be termed a "fundamental element of warfare," however, is a tougher call. We answer in the affirmative.

Does the concept of knowing yourself pass the four-part discrimination test described in chapter 4? First, adherence to the concept of knowing yourself has often contributed to victory. For example, in the case cited here, Daniel Morgan at Cowpens understood his own forces in detail and devised a battle plan to take advantage of that knowledge. Although other factors also contributed to the victory, knowledge of his own forces was an important contributor.

Second, adherence to the concept has rarely contributed to defeat. We are not likely to find many cases in which a commander had a good knowledge of his own forces and lost a battle *because* of having that knowledge.

Third, violation of the concept has often contributed to defeat. A lack of knowledge of his own forces has resulted in the defeat of many a commander. Robert E. Lee's overestimation of his army's capability led him to order Pickett's Charge.

Fourth, violation of the concept has rarely contributed to victory. It would be difficult to locate a case in which a commander's ignorance about his own forces contributed to victory.

Therefore, the concept passes all four parts of the discrimination test and we conclude that "know yourself" qualifies as a principle of war.

FIFTEEN

Know Your Enemy

There is no more precious asset for a general than a knowledge of his
opponent's guiding principles and character, and anyone who
thinks the opposite is at once blind and foolish.

—POLYBIUS, *Rise of the Roman Empire*

....

New Principle of Know Your Enemy

HOW CAN A COMMANDER EXPECT TO DEFEAT WHAT HE DOES
not understand? History provides many examples of occasions on which
knowledge of the enemy made a difference. Sun Tzu, Vegetius, and Polybius all thought such knowledge to be important. Today India includes
in its principles of war a principle of intelligence that includes some of
the factors considered here. China's principles of strategic warfighting
include a principle of information superiority. The USSR's principles
of combined arms tactics included morale, a principle which stressed
studying the enemy so one could exploit the enemy's moral-political
and psychological weaknesses. Yet the essential factor of knowing the
enemy is mostly absent from the principles of war espoused by contemporary militaries. Therefore, we propose a candidate principle of
know your enemy:

> Determining the best way to fight an enemy requires a deep under-
> standing of the enemy. It is desirable to learn as much as possible about
> the enemy's culture, history, motivations, way of thinking, political and
> military objectives, strategy, tactics, military doctrine, alliances, avail-
> able resources, weapons, sensors, the enemy commander's personality
> and biases, and the state of training and morale of the enemy. Deter-
> mining the enemy's current disposition of forces is worth expending
> substantial effort.

THE TYPES OF INFORMATION THAT COMMANDERS SHOULD strive to know about the enemy are many and varied. In what follows, we provide several examples.

Enemy Culture

A commander should be familiar with the culture of the enemy. This includes the history of the country as well as the enemy's motivations and way of thinking. Choosing objectives and identifying centers of gravity and critical points depends in part on knowing what the enemy most values.

Gen. George Patton, "Old Blood and Guts," might have seemed a crude personality, but he conducted extensive research on potential enemies and their culture. In addition to reading books about their history and culture, he read the writings of the potential enemy's leaders, in order to gain insight into their mindset. He believed that reading what a man wrote gave him a better window into the man's mind than what others wrote about him. Thus, Patton read books such as Adolf Hitler's *Mein Kampf*, as well as works by Lenin and Karl Marx. While in North Africa he also read the Quran. Patton claimed,

> I have read the memoirs of [our enemy's] generals and political leaders. I have even read his philosophers and listened to his music. I have studied in detail the accounts of every damned one of his battles. I know exactly how he will react under any given set of circumstances. He hasn't the slightest idea what I'm going to do. Therefore, when the day comes, I'm going to whip the hell out of him.[1]

Historically, many commanders have failed to understand their enemy's culture. After Pyrrhus twice defeated Rome in 280 and 279 BCE, he expected Rome to negotiate a peace treaty. In that period, "everyone" in the Hellenistic world did so after a major defeat. Yet, despite these defeats, the Romans never considered giving up. They refused to negotiate. The Roman attitude was that wars ended only when their opponent conceded total defeat. Nothing else was acceptable.[2]

From 218 to 216 BCE Hannibal crushed Roman armies at the Battles of Trebia, Lake Trasimene, and Cannae. Few nations have experienced three such devastating defeats. Their hundred thousand casualties represented more than 10 percent of the Roman population of military age. More than one-third of the Senate had died in battle. After the Battle of

Cannae, Hannibal expected to negotiate a peace treaty with the Senate. He sent one of his officers to accompany ten captured soldiers to Rome. The ten captives were to negotiate the customary payment of ransom for the eight thousand Roman citizens held by the Carthaginians. Hannibal's conditions for peace were not onerous. However, the Romans would not negotiate and refused to see Hannibal's emissary. The Senate not only refused to use public funds to pay the ransom for the captives, they also prohibited private citizens from doing so.[3] The Senate limited public mourning to thirty days and banned women from shedding tears in public. Finally, the Senate prohibited people from even speaking the word "peace."[4] To their ultimate detriment, both Pyrrhus and Hannibal failed to understand Rome's unshakable determination to win every war in which it participated, no matter what the cost.

In another example of the importance of knowing the enemy culture before launching an attack, in 1973 the Egyptians knew the significance of Yom Kippur to the Israelis and that many of their frontline forces would be on leave at that time. Therefore, the Egyptians opened their attack at that time.[5]

During World War II Germany failed to anticipate that Stalin could persuade the Russian people to fight to defend the Rodina (Motherland) and failed to appreciate the animosity of many eastern Europeans toward Russia. After Germany invaded the Soviet Union, many people in the Ukraine and the Baltic states hoped the Germans would help them gain independence from the Soviet Union. Germany might have persuaded them to cooperate, but failed to recognize their hatred of the Russians. Instead, the Nazis treated the occupied populace, comprised of what the Nazis considered to be inferior races, harshly.[6] Although the Germans persuaded many people to help them, they did not convince as many as they might have done. Soviet propaganda convinced many people to fight the invaders. The Germans harsh treatment of the occupied populations eventually caused many of those who initially supported the Germans to turn on them, thinking the Russians were the lesser of two evils.

Even though American planners knew of the mutual animosity between the Sunnis and the Shiites in Iraq, they did not adequately consider this in planning the invasion of Iraq in 2003. Nor did they really understand the importance of the many tribes and ethnic groups in Iraq. The CIA was wrong about Iraq's possession of weapons of mass destruction, failed to anticipate the dogged resistance of the paramilitary Fedayeen, incorrectly predicted units of the Iraqi military would surrender, and mistakenly forecast that cities in southern Iraq would welcome American forces.[7]

Because Osama bin Laden misunderstood American culture, he did not anticipate the American reaction to the 9/11 attacks. Bin Laden believed that al-Qaeda's various attacks on Americans overseas had precipitated talk of withdrawing U.S. forces from the Middle East—one of his major goals. He evidently believed that escalating the attacks to killing thousands of civilians on American soil would result in the American people demanding such a withdrawal. Bin Laden did not realize that sneak attacks on American soil, much less attacks on civilians, would elicit a visceral reaction that would unite the country. He did not anticipate that the American response to an attack would not be to give way, but to counterattack. Because of the relatively mild American response to previous attacks, bin Laden probably did not expect much more of a response than cruise missile attacks on his training camps. He never expected the Americans to go halfway around the world to drive al-Qaeda out of Afghanistan.[8]

Enemy Strategy

Sun Tzu advocated attacking the enemy's strategy as the best strategy.[9] To attack the enemy's strategy you need to understand that strategy. Commanders should also identify the enemy's political and military objectives.

Prior to the 1973 Yom Kippur War, Israel believed that Egypt's strategic objective in any war would be to regain territory lost in the 1967 war. Israel knew Egypt's armed forces were not capable of achieving that and therefore concluded Egypt was unlikely to attack. This perception contributed greatly to Egypt's successful deception. Anwar Sadat's political objective was to create the conditions to force Israel to negotiate the return of the Sinai Peninsula. That required only that the military seize and hold a small amount of territory on the east bank of the Suez Canal. Israel's failure to understand Egypt's strategy allowed the Egyptians to take them by surprise.[10]

During the Cold War an understanding of the Soviet Union's strategy in a potential war with NATO was key to the development of the United States' Maritime Strategy. For a long time the United States intelligence community believed that the Soviet Navy's primary goal during a war with NATO would be to attack the sea lines of communication between the United States and Europe. In the 1970s some experts in the United States correctly deduced that the primary missions of the Soviet Navy in wartime would be to protect their ballistic-missile submarines hidden in sanctuaries near the Soviet Union and to protect the Soviet homeland from attacks on the flanks by U.S. aircraft carriers. The U.S. Maritime Strategy sought to attack the Soviet strategy by moving forces forward at the start of any war to attack the Soviet forces. Furthermore, to deter the Soviets from starting a war the United States deliberately demonstrated the ability to do this in peacetime exercises.[11]

Enemy Commanders

One should learn about the personality and tendencies of enemy commanders, as well as their biases. As discussed in chapter 10, the key to a successful deception is knowledge of the enemy's expectations, preconceptions, and the type of information they trust the most. If one knows the personality, tendencies, and biases of the enemy commander, then one can reasonably predict his or her reactions to circumstances. One can use this information to form a plan to take advantage of the enemy commander's weaknesses.

Prior to the ambush at Lake Trasimene during the Second Punic War, Hannibal made inquiries about the Roman army commander, learned that Caius Flaminius was impetuous, and planned the Battle of Lake Trasimene to take advantage of his rashness. Hannibal moved his armies through flooded marshes to get ahead of the Romans. He devastated the land of Rome's allies to undermine Rome's prestige and to infuriate Caius Flaminius. Knowing that Caius Flaminius would pursue him, he led the Romans on until he found a great site for an ambush.[12]

At the Battle of Cowpens in 1781, Brig. Gen. Daniel Morgan knew his own forces well; he also knew his enemy. Though he had never met Lt. Col. Banastre Tarleton on the battlefield, Morgan had studied Tarleton's tactics and expected the enemy to rush headlong at him. Based on his knowledge of the enemy and his own troops, he devised a battle plan that resulted in a brilliant victory.[13]

Confederate general Robert E. Lee understood his Union counterparts. Some, such as George McClellan, he had served with in the Mexican War. Others he knew when they were cadets while Lee was the superintendent at West Point. When Lincoln replaced McClellan as commander of the Army of the Potomac, Lee lamented, "We had always understood each other so well. I fear they may continue to make these changes till they find someone whom I don't understand."[14] Later, when Joe Hooker replaced Ambrose Burnside as commander of the Army of the Potomac, Lee wrote about Hooker's apparent inability to determine a course of action.[15] At the Battle of Chancellorsville, Lee counted on Hooker not to react quickly at key times.

In contrast, when Lee took command of the Confederate army defending Richmond on 1 June 1862 George McLellan was happy because he thought Lee was "cautious and weak under grave responsibility . . . likely to be timid and irresolute in action."[16]

In June 1863 Maj. Gen. William Rosecrans faced the Confederate Army of Tennessee under the command of Gen. Braxton Bragg. Rosecrans knew that Bragg prided himself on detecting tactical deceptions. Therefore, Rosecrans devised a double feint. His idea was that Bragg would detect the first, transparent feint to the east, but then fall for the second deception to the west. Rosecrans's main attack would then be to the east. To maintain secrecy he did not inform higher command of his plan, nor did he tell his subordinates until just before the attack. Alas, Bragg identified both feints as such. Nevertheless, Rosecrans drove Bragg back.[17]

When Gen. Erwin Rommel went to North Africa in 1941 he carried with him a translation of *Generals and Generalship*, written by British general Archibald Wavell, his first opponent. Rommel read and reread the booklet and carried it with him. He claimed it gave him valuable insight into Wavell's way of thinking. In turn, Patton studied Rommel's memoir *The Infantry Attacks*.[18]

Prior to the Battle of Leyte Gulf in 1944, the Japanese reportedly studied Adm. Bill Halsey and decided that he was rash and aggressive. Therefore they devised a deception plan to lure him away from the landing area.[19] That part of their plan succeeded.

During the Six Day War in June 1967 Israel conducted a successful surprise attack on the Egyptians. However, they initially did not attack the Egyptian communication system. They wanted Egyptian commanders to radio back reports—which the Israelis knew would be unrealistically optimistic and so give Egyptian headquarters a false picture of the situation. Then Israel jammed the Egyptian radio communications. As a result, Gamal Abdel Nasser only heard the optimistic initial reports and did not realize the extent of the Israeli successes.[20]

Enemy Resources

Commanders should know the enemy's resources in men and materiel. Choosing objectives and ascertaining centers of gravity and critical points depends in part on identifying the resources available to the enemy. Thus, in starting the Civil War the Confederacy did not appreciate the Union's enormous advantage in productive capacity and overestimated the fighting ability of southerners relative to the northerners.

During World War I Germany underestimated the resources of the United States and the speed with which they could affect the war. Consequently, Germany took the risk of alienating the United States by waging unrestricted submarine warfare. Their hope was that, even if this action provoked the Americans into declaring war, Germany could win the war in Europe before the Americans could have an impact on the fighting.[21] In World War II Germany again underestimated the United States when Hitler made an "unforced error" by declaring war on the United States. Hitler had little knowledge of the United States. In his view the United States was just a "mongrelized mixture of races." As Hitler and his staff celebrated the Japanese attack on Pearl Harbor, it turned out that no one in the German high command knew where Pearl Harbor was located.[22]

Enemy Capabilities

In 57 BCE Caesar encountered numerically superior Belgic forces in northern Gaul. The Belgic forces were strong and reputed to be extremely brave. Therefore, rather than fighting a major battle immediately, Caesar tried to discover the enemy's capabilities. He did so by having his cavalry skirmish with their cavalry every day. Only after learning of their capabilities did Caesar risk a general engagement against the Belgae.[23]

At the beginning of World War II the United States thought most Japanese had poor eyesight and therefore would be quite poor at night fighting. The U.S. Navy did not appreciate that the Japanese had a history of fighting at night, wanted to fight at night, and had trained extensively to do so.[24] The United States paid a heavy price in numerous night battles in the Solomon Islands in which the Japanese spotted the Americans first despite the Americans having the advantage of radar.

Commanders should be familiar with the enemy's operational and tactical doctrine so they can anticipate what the enemy might try to do and devise the best way to counter the enemy. For example, in the Battle of Cannae a key part of Hannibal's plan correctly anticipated the Romans would form their forces with a strong center and would push forward aggressively.[25]

Byzantine emperor Justinian sent Belisarius to reconquer Italy in the sixth century CE. In his first battle with the Goths, Belisarius studied his opponents. He noticed they were poor at coordinating all their forces. He also saw it was easy to goad them into an ill-advised charge. In addition, he noted that the Goth cavalry carried only short-range weapons, such as swords and lances. Therefore, the Goths needed to close with their enemy. On the other hand, the Byzantine cavalry were skilled with the bow and could stay out of range of the Goth cavalry weapons and still assault the Goths with arrows. Belisarius also observed that, because they could not stand up to a cavalry charge, the Goth foot-archers tried to stay near the shelter of their own cavalry. Therefore, in several battles Belisarius induced the Goth cavalry to charge, leaving the Goth infantry behind and unprotected. He then used cavalry on his flanks to drive a wedge between the two components of the Goth army. Using this knowledge, Belisarius won many battles, even though he was greatly outnumbered.[26]

After Emperor Justinian recalled Belisarius because he feared Belisarius might become a rival emperor, the barbarians reoccupied Italy. Justinian then sent Narses to reconquer Italy again. Narses knew that the Goth army was inherently offensive and that they had contempt for the Byzantine infantry. In fact, the infantry generally did not withstand charges by Goth cavalry. At Taginae in 552 CE Narses used his knowledge of Goth tactics to set a trap. He put dismounted cavalry in the center of his battle line, so the Goths would perceive them as infantry. When the Goth cavalry charged the center, they were subjected to heavy fire from archers Narses had placed on his flanks but well forward of his center. Then the Goths unexpectedly encountered the dismounted Byzantine lancers who put up a stiff resistance. When the Goth cavalry retreated, they again ran a gauntlet of arrows. Then Narses unleashed his mounted cavalry to complete the rout of the Goths.

Later, when a Frankish army arrived to assist the Goths, Narses met them at Casilinum. Narses knew the Franks attacked in a deep column of infantry armed with short-range weapons such as spears, swords, and throwing axes. Based on this knowledge, Narses allowed the attack to drive back his center. Then his cavalry attacked the infantry on both sides of the column but stood off out of range of the Franks' weapons. The Franks could repel a cavalry charge as long as they kept in close formation. However, the Byzantine cavalry shot volleys of arrows into the close-packed ranks. Eventually the Frankish infantry broke ranks. At that point the Byzantine cavalry was able to run them down.[27]

Knowledge of the enemy is not sufficient if there is no action to exploit it. During World War I Gen. Sir Hubert Gough commanded the British Fifth Army. He faced the German Eighteenth Army commanded by Gen. Oskar von Hutier. Gough learned that Hutier had used a new tactic of infiltration, as opposed to the massive troop attacks, at Riga on the eastern front. Gough studied Hutier's techniques and knew that he needed to set up a defense in depth to counter them. Unfortunately, his superiors did not share Gough's beliefs and refused to give him the labor forces needed to construct the necessary defense in depth. As a result, when the Germans attacked, they drove the British back. Despite Gough having foreseen what his superiors did not see, in the wake of defeat the British high command needed a scapegoat and relieved Gough of command.[28]

Before World War II the French counterintelligence service widely distributed the translation of Heinz Guderian's book *Achtung Panzer*. When a French major visited many bases in 1937–38, he examined the book in every base library and found that not a single copy had been opened. Although France's higher commanders had knowledge of German tactics, the bulk of the French officer corps did not.[29]

Enemy Sensors and Weapons

Commanders should understand the enemy's sensors and weapons, as well as the capabilities and limitations of each. Knowing your enemy's capabilities aids in guarding against enemy surprises. Identifying the enemy's information collection methods allows one to ensure a deception "message" gets through to the enemy.

During World War II Germany failed to understand some of the Allied anti-submarine sensors and weapons. As discussed in chapter 9 under technological surprise, when British aircraft patrolling the Bay of Biscay began using an improved radar operating on a different wavelength, the Germans failed to perceive this change for many months. It took them even longer to develop adequate countermeasures. German failure to understand the capabilities of the British radar led to the loss of many U-boats.[30]

Similarly, during the Battle of Britain, Germany failed to understand the importance of British radar stations. They did not realize how the British had integrated the information from their radars and observers into a unified picture they used to direct fighter aircraft to intercept the incoming German aircraft. Had they pieced together this information, the Germans would have known that they should continue attacks on the radar stations.[31]

British Swordfish aircraft flying from the aircraft carrier *Illustrious* attacked ships in Italy's Taranto harbor on 11 November 1940. An attack from the British aircraft carrier *Hermes* that damaged the French battleship *Richelieu* in the Dakar harbor on 8 July 1940 should have alerted the Italians to the danger. Nevertheless, because the Italian Navy did not have an air arm, the Italians did not have a good understanding of air warfare against ships, leading them to defend Taranto poorly. The Italians believed that air-dropped torpedoes could not be used in the forty-foot deep harbor because the torpedoes would bury themselves in the mud. However, the British had solved that problem. Because the Italians greatly overestimated the minimum arming distance needed for aerial torpedoes, their placement of defensive balloons was not effective. Because the Italians did not realize British torpedoes had both contact and magnetic firing mechanisms, their torpedo nets only protected against contact torpedoes to the maximum draft of their ships; meanwhile, torpedoes with magnetic pistols could go under the nets and explode beneath the ships. In addition, the Italians had far fewer balloons and nets than they should have had. Despite flaws in the British plan (and the use of obsolete biplanes), this single attack took advantage of inadequate Italian defenses to damage four of Italy's six serviceable battleships. Because there were no plans to exploit the tactical success, however, the attack did not have as great an operational impact as it might have had.[32]

Tools to Obtain Knowledge of Enemy

One should know the ability, state of training, and morale of the enemy's men and officers. It is elementary that commanders need to determine the current disposition of enemy forces. Knowing your enemy includes gathering intelligence and conducting reconnaissance.

Emphasize gathering intelligence and conducting reconnaissance. During World War II in the Pacific the Japanese typically practiced economy of force by using few or no attack aircraft for search. That way they could concentrate as many attack aircraft as possible for a coordinated attack against the American aircraft carriers. In contrast, the Americans placed a high priority on searching for the enemy fleet and used a significant portion of their dive-bombers to do so. Gathering information is so important that it is often worth using assets that might otherwise be to deliver ordnance.

Seek information with a high degree of certainty; deny such information to potential enemies. Most information gathered about the enemy is subject to uncertainty. The more certain a piece of information is, the more valuable it is. When you must deduce everything, with no anchor, the potential to go wrong is enormous. You must rely on small pieces of information and guess how they fit together to form an overall picture. With nothing to pull you back to reality, mistakes are nearly inevitable. Therefore, information that is certainly true, or nearly so, is extremely valuable because it is an information multiplier. The known information serves as an anchor to test the validity and the interpretation of less-well-known pieces of information. It allows you to put things into context. Deductions are much more likely to be correct.

Thus one should try to keep the enemy from knowing things with a high degree of certainty. One needs to create uncertainty to multiply the enemy's probability of making mistakes. Make them guess about nearly everything.[33] Open societies are at a distinct disadvantage here. Routinely released information is useful to the enemy in itself and, in addition, it removes uncertainty, allowing enemies to understand other information better. Furthermore, leaks, often of vital capabilities, make the situation worse. One way to counter this is to leak false information, planting false ideas and making it harder for enemies to evaluate and interpret valid pieces of information.

One must be careful not to overestimate the certainty of information that appears to be highly reliable. As discussed previously in this book, in the run-up to the invasion of Iraq in 2003, nearly all intelligence agencies in the world believed that Iraq possessed weapons of mass destruction.

However, they were wrong. One must avoid confirmation bias. Instead, one should seek to obtain information that contradicts one's preliminary conclusions and should consider alternative explanations for previously obtained material.

Does "Know Your Enemy" Qualify as a Principle of War?

Because knowledge of the enemy has played an important role in military history for thousands of years, it seems to be an enduring concept and a fundamental element of warfare. This concept is valid at the strategic, operational, and tactical levels, as well as for all types of warfare.

It must also pass the test of history as being a concept that makes a difference. Certainly numerous commanders have met with disaster due to ignorance of their enemy. We have given several illustrations of how adherence to the concept contributed to victory. There have been cases in which a commander was so overawed by the enemy's capability that he gave up. However, it seems to be stretching a point to claim that accurate knowledge of the enemy led to defeat. On the other hand, ignorance of the enemy has frequently led to defeat. Finally, it is hard to find examples of cases in which a commander was ignorant of the enemy and *as a result* was victorious. Thus we judge that know your enemy passes the test of history as a concept that makes a difference. We conclude that know your enemy deserves designation as one of the principles of war.

SIXTEEN

Environment

In a fight between a bear and an alligator, it is
the terrain which determines who wins.

—JIM BARKSDALE, *Sayings*

....

New Principle of Environment

SUN TZU, VEGETIUS, AND MANY OTHERS URGED COMMAND-
ers to study the environment carefully. Although no military lists knowledge
of the environment as a principle of war, we believe it deserves consideration. Our proposed definition is:

> Superior knowledge of the environment and effective use of it can gain a
> relative advantage over the enemy.
>
> The environment may contain constraints and opportunities for both
> sides. Commanders should know the type of environment that gives them
> the greatest relative advantage over the enemy.
>
> Various aspects of the environment affect people, equipment, weapons,
> and movement in different manners. Detailed knowledge of the environ-
> ment leads to more realistic expectations that in turn result in reducing
> some types of friction.
>
> Environment includes topography, nature of terrain, condition of roads,
> weather, climate, tides, acoustic and electromagnetic propagation condi-
> tions, and any other feature of the battleground that might allow either
> side to gain an advantage.

ENVIRONMENT INCLUDES SOPHISTICATED THINGS SUCH AS
radar ducting, as well as simple things such as whether a river is fordable.

Often, the environment affects the battle through its effects on friction and the fog of war.

Knowledge of the environmental conditions in which one's own force has the best relative advantage over the enemy's force allows better choices for when and where to give battle. For instance, the U.S. military often prefers to fight at night when their technical capabilities give them the greatest advantage. This is not a new concept. More than two millennia ago Kautilya wrote, "During the daytime the crow kills the owl, and . . . at night the owl the crow."[1]

Knowledge of the environment is especially crucial for guerrillas. They use local knowledge of terrain and weather to gain a relative advantage. They may employ mountains, jungle, desert, swamp, or urban terrain to negate the larger numbers and better mobility of the enemy. They can use the terrain to conceal themselves. Many types of terrain favor the guerrillas because it limits the enemy's mobility, as well as the enemy's ability to employ large numbers. Thus, terrain can reduce the better-trained, better-equipped, and better-armed regular force to a level equal to the guerrillas. This has not changed since the time of the Romans.[2] Similarly, counter-guerrilla forces must learn about the terrain and study ways to take advantage of it.

We now present some examples in which knowledge of the environment led to success and other examples in which ignorance of it resulted in failure. For convenience, we loosely divide environmental factors into terrain, weather, and other.

Terrain

One way a commander's knowledge of the terrain can contribute to victory is in allowing him to find a way around an obstacle. Another way is through having an accurate evaluation of whether one's forces can cross the terrain in a timely manner. Yet another way is when a commander evaluates the terrain for its suitability as a battle site and learns how best to take advantage of the terrain in a battle.

Locating a Path

Locating a path through, around, or over obstacles such as mountain passes, swamps, or rivers can make the difference between victory and defeat. At

Thermopylae the Spartans were able to hold off vastly greater numbers of Persians because the narrow pass restricted numbers and favored the Spartans' weapons. The Persians broke the impasse when a Greek traitor showed them a path through the mountains and around the Spartan position. When the Persians suddenly appeared in the rear of the Spartans, victory was inevitable.[3]

As related in chapter 9 on surprise, Scipio Africanus made a surprise attack on New Carthage in 209 BCE. One key to his victory was learning that the protective lagoon was in fact fordable under certain conditions.[4]

In 1346 the French appeared to have trapped the badly outnumbered English army by destroying all the bridges across the Somme River. However, a peasant showed the English a ford across the river that they could use only at low tide. The English were able to use this route to escape the French and soon located an excellent place near Crecy at which to fight the French.[5]

During Justinian's attempt to reconquer Italy in 535 CE, the Goth defenses at Naples stymied Belisarius. Then one of his men became curious and explored the aqueduct into the city that Belisarius had cut. He found that if they made it wider at one spot, it could provide a covered path into the city. A group of his men snuck into the city through the narrow tunnel. They attacked the defenders in the rear while the rest of Belisarius's force attacked the walls. The city soon fell.[6]

During the 1948 war the Israelis located an ancient Roman road running through the desert. Israeli engineers prepared the road so that tanks could use it. While the Egyptians at Bir-Asluj guarded the modern road, the Israelis used the Roman road to get into the Egyptians' rear.[7]

During the Battle of Antietam in September 1862 Maj. Gen. Ambrose Burnside commanded four divisions on the left wing of the Union Army. Maj. Gen. George McClellan ordered Burnside to cross Antietam Creek on the morning of the battle. To attack the Confederates on the other side of the creek, he repeatedly attempted to cross the narrow stone Rohrbach Bridge, now known as the "Burnside Bridge." Despite the Union attackers outnumbering them by twenty to one, the Confederates had the bridge well-covered and drove back multiple attacks. By the time Burnside mounted a serious attack that was about to succeed, Maj. Gen. A. P. Hill and three thousand men arrived to reinforce the Confederate position and save the day. There was a shallow ford about a mile down from the bridge.[8] Burnside had sent men to try to locate a ford in that area, but it took them half of the day to find it. Knowledge of the ford would have enabled a timely attack that would have devastated the Confederate position. Burnside should have explored the area and located the ford on the previous day.[9] The failure to gain knowledge of the ford and effectively exploit that knowledge cost the Union a chance for an overwhelming victory.

Evaluating Difficulty of Route

One factor in avoiding unnecessary friction is to have an accurate assessment of a route's passability and the time needed to travel it. The career of Hannibal provides several examples of knowing the terrain. In the most spectacular example, the Carthaginian general correctly judged that he could move his entire force, including his elephants, across the Alps, thus achieving strategic surprise on the Romans when he appeared in northern Italy.

In another case prior to the Battle of Lake Trasimene in the spring of 217 BCE, Hannibal had a choice of several routes to take. Most were long and expected by the Romans. One route through the marshes along the Arno River was shorter but thought to be impassible in the spring. Hannibal took that route, even though the marshes were flooded and the route underwater. The Carthaginians marched through the water and mud for four days, with no dry place to sleep. Many soldiers and animals became casualties. Hannibal contracted an infection that cost him an eye. Nevertheless, the army made it through the marsh and thereby gained the initiative, which soon resulted in the ambush at Lake Trasimene.[10]

In the Saratoga campaign during the Revolutionary War, British general John Burgoyne initially underestimated the difficulty of getting supplies to his forces across miles of wilderness. Later he had to decide which route to take to get to the Hudson River from Skenesborough. Back in London, Burgoyne had determined that the army would travel by water down Lake George. Instead, he now decided to take an overland route along Wood Creek, underestimating the difficulty of his army moving along a narrow trail through the wilderness. To add to his problems, American troops wielding axes blocked the trail by felling trees every few feet, with their branches tangled up with each other. They destroyed bridges, rolled boulders onto the trail, and dammed streams. To add to the misery of the British, July 1777 was unusually rainy.[11]

During World War II Germany and France came to different conclusions about whether armored forces could pass through the Ardennes Forest. Because the French deemed the Ardennes impassable for armor, they defended the sector thinly with units that were not their best. Germany proved armored forces could push their way through the Ardennes. As a result, Germany achieved operational surprise and split the Allied forces.[12]

In Operation Desert Storm the U.S. Army ascertained that its armored forces could cross the undefended desert far to the west of the Iraqi defenses in and near Kuwait. Consequently, the armored forces made a long left-hook around the Iraqi position in Kuwait, swept into southern Iraq, and then turned east to trap many Iraqi divisions inside Kuwait.[13]

Knowledge of Battleground

Before the Battle of Trebia in December 218 BCE Hannibal studied the ground carefully. The terrain seemed to be very flat, which would give the Romans confidence that they could not be surprised. However, off to the side of the expected battlefield, Hannibal discovered a sunken watercourse. The night before the battle Hannibal moved one thousand infantry and one thousand cavalry, under the command of his brother Mago, into the ravine. The next day the Romans attacked as Hannibal expected. When they engaged the center of his line, the hidden forces emerged from the ravine and struck the Romans in the rear. As usual, the attack in the rear proved to be decisive.[14]

A year before the Battle of Waterloo, Wellington had studied the terrain and decided it would make excellent defensive ground and would be the best place to give battle if he ever had to defend Brussels.[15]

During the Civil War the Confederates erected Fort Henry in a location that was subject to flooding. When Brig. Gen. Ulysses Grant attacked the fort in February 1862, heavy rains had raised the water level of the Tennessee River so that some of the guns were underwater. Therefore the fort surrendered soon after Union gunboats attacked. After the fort surrendered, the cutter carrying Union officers to accept the surrender went into the fort through a flooded sally port. It was the first important Union victory in the Civil War.[16]

When Maj. Gen. William Sherman was outmaneuvering Gen. Joe Johnston in northern Georgia in 1864, he had the benefit of having studied the terrain in detail while stationed at Marietta twenty years previously.

While his fellow young officers spent their off days playing cards or reading novels, Sherman used his to explore the hills and valleys of the surrounding area. Indeed, Sherman claimed he won his battles because he knew the terrain better than the rebels did.[17]

Weather and Climate

The weather is generally the same for both sides. The difference comes from superior forecasting and better knowledge of the effects of the weather. In June 1944 the Germans braced for the coming Allied invasion of the continent to open a second front, but they did not know the timing. The Germans observed that the Allies had made their previous amphibious landings only in extended periods of good weather (making the mistake of relying on a very small sample size). When a storm moved into the area, the Germans concluded that landings in France would be impossible for a few days. Rommel, commander of the Germans in the west, traveled to Germany to be with his family for a few days. The Allies, however, benefitted from more weather stations to the west (where the weather fronts came from) and were able to forecast a short break in the weather. Based on a weather forecast more uncertain than most, Gen. Dwight Eisenhower decided to land on 6 June. This surprise on the timing of the invasion caught Rommel out of the area and delayed decisions.[18]

Two similarly disastrous strategic decisions 129 years apart resulted from a lack of appreciation of the severity and impact of winter weather in Russia. In 1812 Napoleon led the French Army into Russia and captured Moscow. The subsequent retreat, however, turned into a nightmare between the harassing Russian attacks and the frigid weather. In 1941 Hitler invaded the Soviet Union. The German Army evidently expected to be victorious before winter hit and did not prepare adequately for the severe cold they ended up encountering.

The role of wind and the most advantageous position for ships relative to it has varied over the years. During the age of sail, the amount of wind obviously was important. In a battle at sea, the possessor of the weather gauge—being upwind of one's opponent—had the initiative. Because sailing ships had difficulty sailing upwind, a ship downwind of its enemy did not have the option of closing with the enemy. On the other hand, the possessor of the weather gauge had the choice to either close for battle, exchange fire at long range, or avoid battle. In the opening stages of the

Battle of the Armada in 1588, because the English ships could sail much closer to the wind than the Spanish ships could, the English gained the weather gauge and therefore the initiative.[19]

In World War II aircraft carrier battles, being downwind of the enemy was an advantage for offense because one could close with the enemy while launching and recovering aircraft. If an aircraft carrier was upwind of an enemy, it would lose ground every time it turned into the wind to launch or recover aircraft.

An example from the air war in the Pacific theater during World War II illustrates how unexpected aspects of the weather can affect military operations. U.S. B-29s flying west to bomb Japan sometimes encountered headwinds well in excess of one hundred miles per hour. This was the then-unknown Jet Stream. Not knowing the vertical or horizontal extent of the jet stream, they did not know how to escape its clutches. In danger of running out of fuel, they sometimes had to jettison their bombs and return to base.

Aircraft often encountered strong winds during the Thousand-Mile War in the Aleutians. In addition, heavy fog, constant cloud cover, and mountainous seas plagued operations in the area. The Aleutians have the reputation of being the only place on earth where fog and high-speed winds occur at the same time. As a result, although maps indicated that the Aleutians were the shortest route to Japan, that path was just not practical.[20]

Rain and mud have figured prominently in many battles and campaigns throughout history. In 1415, during the Hundred Years War, the French, led by Constable of France Charles d'Albret, pursued King Henry V's English invaders. Near Agincourt the tired and hungry English army of about six thousand marching through a torrential rain found their retreat blocked by the French army of at least thirty thousand, perhaps fifty thousand, men. D'Albret paused until the next day and then waited for the English to attack him—as they would have to do eventually because they were the ones who were trapped.

When the English moved forward partway toward the French lines and fired arrows into their ranks, d'Albret was unable to control his impetuous and undisciplined nobles. He reluctantly ordered an advance. However, the French had chosen a very unfavorable battleground. The torrential rain had turned the recently plowed field into a sea of knee-deep mud. As the dismounted knights in their heavy armor slowly trudged through the mud toward the English, they quickly became exhausted. As the two lines of knights clashed, the English archers dropped their bows and attacked the knights. The unarmored archers, much more agile than the ponderous French knights, danced around them, stabbed them through joints in the armor, hit them with axes, or simply bowled them over. Once on the ground, the knights were helpless, and the English simply stabbed them in the face through their visors. Most drowned in the mud or suffocated. When the second wave of French knights attacked, the English were standing on a mound of corpses. After the English defeated the second wave, the layers of bodies stood more than six feet deep in places. French losses were at least five thousand and perhaps ten thousand men. The English lost a few hundred men.[21]

In a similar vein, one finds in the official records an aborted campaign that earned the moniker of "The Mud March." After his devastating defeat at the Battle of Fredericksburg in December 1862, Maj. Gen. Ambrose Burnside planned to cross the Rappahannock River upstream from Fredericksburg to get around Lee's left flank. On the morning of 20 January 1863, the Army of the Potomac began to march. That afternoon, it began to rain. The rain continued all night. The area became a sea of mud and the roads were rivers of churned-up muck. The rain continued all the next day. Whole regiments tried to pull the cannons. If they did not put logs under the axle when they paused, the gun sank into the mud until only the muzzle was visible, at which point the soldiers had to dig it out of the mire. Dead mules and caissons became stuck in viscous mud and littered the roads. Nevertheless, the continued heavy rain did not deter Burnside. He ordered the march to resume the next morning. To raise morale, he ordered a whiskey ration for the troops. This resulted in regiments from Pennsylvania and Maryland getting into a fight. When some men from Maine tried to break up the fight, it turned into "the biggest three-sided fist fight in the history of the world." Eventually, Burnside recognized the futility of continuing the march and ordered the army back to camp.[22]

Fighting effectively in severe environmental conditions requires periodically exercising in those conditions. For example, when a U.S. Navy carrier battle group operated in the northern Norwegian Sea for the first time in a generation in 2018, many lessons about operating in the brutal conditions had to be relearned.[23]

Other Environmental Factors

Other environmental factors of importance include the tide and the angle of the sun.

The Battle of the Dunes took place on 14 June 1658. Turenne led a French and English army against a Spanish army on the dunes near Dunkirk. He moved forward just before low tide with his cavalry divided between his flanks. As the tide changed, Turenne used cavalry no longer needed to guard his seaward flank to swing around and envelop the Spanish on the landward flank, a utilization of the environment that gained him a great victory.[24]

The most famous instance of the tide affecting a battle is the "tide that failed" at the Battle of Tarawa on 20 November 1943. Because the invasion of Tarawa was one of the first opposed amphibious landings during World War II, the landing force made many mistakes. One of these involved the complex tides of Tarawa. Exactly what the planners knew about the tide is not clear even today, but they knew the predictions were shaky at best. When the tide was lower than the planners hoped, many of the landing craft could not get over the reef to land the U.S. Marines near the beach. Thousands of Marines, heavily laden with weapons and other gear, had to get out of the landing craft and wade through chest-deep water for six to eight hundred yards, all the while under withering fire from the Japanese defenders.[25]

Another aspect of the Normandy invasion illustrates the often-close relationship between knowing the environment and knowing your enemy. The Germans believed that the Allies would choose to land at high tide when they would have the shortest stretch of beach to cross. The Allies, however, decided to land at a low tide that would expose many of the German obstacles so that landing craft could avoid them. Both sides had the same facts but came to different conclusions.[26]

During the Battle of Jutland, haze and fleet placement in relation to the setting sun severely disadvantaged the German High Seas Fleet. At the time of the climactic phase of the battle, the British battle line was north and east of the German battle line. The Germans, looking toward the darkest part of the horizon, combined with the haze and smoke, could see the British gun flashes across a huge swath of the horizon but could rarely see individual British ships in the battle line. Because of the highly variable visibility, the British also had great difficulty seeing the German battle line, but at least the German ships stood out against the bright sky in the west.[27]

During the preparation for surveillance of the Shahikot Valley in Afghanistan during Operation Anaconda in 2002, Pete Blaber and his men conducted environmental reconnaissance. This included locating the best routes, as well as developing realistic expectations for the times needed to travel those routes. They did so under the worst imaginable conditions of cold, wind, altitude, precipitation, and difficult terrain. They learned how best to survive in the inhospitable mountains and how to infiltrate the area without the enemy detecting them. In addition, they determined the best clothing and camouflage. They also gained much other knowledge necessary to understand and utilize the environment. One more benefit of these recons was that they became acclimated to high altitude. Thus they took specific actions to learn about the environment and then formed plans to make the best use of the environment.[28]

Is Environment a Principle of War?

Knowledge and clever utilization of the environment has certainly contributed to many victories over the ages. To decide whether environment qualifies as a principle of war, let us consider the criteria set forth in chapter 4.

Knowing and using the environment is clearly important at the strategic, operational, and tactical levels of warfare. It is also useful in the different types of warfare. Because it has been an important factor for centuries, it qualifies as an enduring concept. Whether it should be termed a fundamental element of warfare, however, is a borderline call.

Does the concept of knowing and using the environment pass the discrimination test described in chapter 4? First, a lack of knowledge of the environment has resulted in the defeat of many a commander. Second, it would be difficult to locate a case in which a commander won a battle because he lacked knowledge of the environment or failed to take advantage of it. Third, we are not likely to find many cases in which a commander had a good knowledge of the environment and lost a battle *as a result*. Fourth, commanders have often won battles *because* they knew the environment and took advantage of it. Numerous times, a commander devised a battle plan to take advantage of the environment. Other factors, such as knowledge of his enemy and judicious use of the initiative, may have contributed to his victory, but we judge knowledge and use of the environment to have been a critical factor in many of these victories. Therefore, we conclude that the principles of war should include knowing the environment.

SEVENTEEN

Summary

AS DISCUSSED IN CHAPTERS 1 AND 2 OF THIS BOOK, MILLEN-
nia of debate about the principles of war have evolved to the point where
most military organizations, even though they do not agree on which
specific principles to include, generally agree on the following features
of the principles of war:

Origin. The principles of war come from a study of military history.

Permanence. The principles of war are enduring, but not immutable.

Rigidity. On the law-rule-guide-suggestion spectrum, the principles
are guides (but not a straitjacket).

Application. Because every principle has exceptions, commanders
must use judgment in applying the principles to each specific situation.
Commanders may violate the principles provided they have good reasons for doing so. However, blatant disregard of the principles increases
the chances for failure.

We define a principle of war as a concept that distills enduring lessons
of military history to provide practical guidance about fundamental elements of warfare that often determine the difference between victory and
defeat. We believe the principles of war should stimulate thinking and
provide a frame of reference for studying military history and organizing thoughts. In addition, when executing an operation or formulating
a plan, principles are important considerations to take into account and
weight appropriately for the specific situation.

The principles of war are a shorthand way of referring to complex
concepts. This enhances communications between military professionals *provided* they have similar understandings of the principles and the
concepts behind them.

ANY LIST OF PRINCIPLES OF WAR WILL HAVE DEFICIENCIES that we must recognize. First, the principles of war *cannot* form a comprehensive theory of warfare because they do not include things that change with technology and social conditions.

Second, the lessons of history are complex. A thorough rendering of those lessons must contain many caveats and limitations. However, paragraphs of complex sentences with many intricate conditional clauses would be neither elegant nor memorable. At the same time, when we reduce history's lessons to principles expressed in a word or simple sentence, we lose precision. Principles must thread the needle between useful generality and inaccurate simplicity.

Using single words or short phrases to denote the principles leads to ambiguity. Some terms have multiple meanings. Various services and countries interpret terms such as unity of command differently. Inevitably, people stretch the principles far beyond their original meaning. In some cases, the principles seem to be an inkblot in which each person sees whatever he or she is predisposed to see. The net result is that instead of providing a common shorthand vocabulary, the principles can lead to misunderstandings. This is one of the reasons for sharpening the thought behind each principle.

It is the concept behind the principle that is important, not the terse title. A title is just an aid to memory and a shorthand term for communicating complex ideas.

Finally, when robots and artificial intelligence become widespread, some principles of war may no longer apply because some of the principles we have selected depend on human nature.

New Principles of War

The preceding chapters present our proposed nine new principles of war and the full definition and explanation of each. Here we just list the title of the new principles, along with a few sentences of justification, without listing the definitions or rehashing the full arguments for them.

Prioritized objectives. The currently recognized principle of the objective (or aim as the British say) is excellent for an idealized situation. However, we must recognize that in the real world, multiple objectives, some disguised as unwritten constraints, are the norm. Because the objectives guide all military operations, commanders should put a lot of careful thought into selecting and prioritizing them. Commanders and their subordinates should understand how various objectives might conflict. Commanders should either explicitly prioritize objectives or provide subordinates guidance on how to resolve conflicts among them.

Relative advantage. Commanders should seek battle under conditions of the greatest possible relative advantage. The current principle of mass or concentration gives poor guidance because it ignores half the problem. The goal should be to concentrate the bulk of your effective force against a portion of the enemy's effective force in a manner that achieves the objective and allows exploitation of success. Effective force includes all combat effects and all factors (weapons, training, environment, morale, surprise, and so forth) that might affect the effectiveness of both forces. One aspect of this principle is the need to practice economy of force in non-vital areas to maximize the force available in the most important area.

Sustained initiative. Commanders should seek to keep enemies so busy meeting threats that they have no time for their own schemes. The offensive is often the best way to gain a sustained initiative but is neither necessary nor sufficient to do so. The offensive is only good when it seizes, maintains, or exploits the initiative. The currently recognized principle of the offensive errs when it advises that the purpose of the offensive is to gain the initiative. The initiative is generally only good if it can be sustained. One can always gain a short-term initiative by attacking, but if it does not lead to a sustained initiative, then that attack is probably a bad decision. Because the initiative is a wasting asset, one must eventually convert it to a durable advantage. Other methods can gain or maintain the initiative, but the offensive is usually the only good way to exploit it.

Unity of effort. Unity of effort is the integration of forces and activities to employ all resources in an effective manner to attain the objective. In every organization, an ultimate coordinating authority must reside somewhere if there is to be a truly coordinated effort. Unity of command is usually the best way to achieve unity of effort, but unity of command is neither necessary nor sufficient to achieve unity of effort. Unity of command is how we attain unity of effort, which is the factor necessary for success. Commanders should take active measures to assure everyone understands the overall purpose of an operation. They should demand subordinates cooperate with each other to attain unity of effort.

Surprise. Surprise can gain the initiative and a relative advantage, but the strongest effect of surprise is often its moral effect on the enemy. We have revised the definition of the current principle of surprise to emphasize the importance of doing things that are psychologically unexpected and upsetting to the enemy's mental equilibrium. Plans employing surprise must include the means of exploiting the advantage gained by surprise. We like the Israeli principle of stratagem (their substitute for the principle of surprise), which includes not only surprise but also its exploitation to achieve decisive results. Of course, commanders should anticipate and prevent enemy surprises (usually included in the current principle of security).

Deception. We classify deception as a principle of war because it has many more uses than just to gain surprise. Our proposed definition includes eight other possible aims of deception. The goal of deception should be to cause the enemy to *act* in a specific manner that you can exploit. Deception works best when it reinforces the enemy's expectations. This in turn requires good knowledge of the enemy. Countering enemy deception requires recognizing one's own psychological biases. We suggest specific actions commanders can take to avoid being duped by enemy deception.

Know yourself. Commanders need to know the strengths and weaknesses of themselves, their subordinates, and their weapons, and how best to employ each of these. All forces should have good situational awareness of the location of friendly forces.

Know your enemy. To defeat your enemy, you need to learn as much as you can about your enemy. This includes background information such as the enemy's culture, the enemy commander's expectations about your actions, and the enemy's modes of operation. Information about the enemy commander's personality and past operations is also helpful. Knowing your enemy also includes learning about the enemy's current strength and location.

Environment. The environment is often a decisive factor. Commanders should know all relevant aspects of the environment and fight in a way that best takes advantage of it.

Near-Principles

In forming the list of principles of war proposed here, we used a rather restrictive set of criteria. As a result, our list excluded many important concepts that did not quite meet our criteria for various reasons. Let us call these concepts "near-principles." Each of these contains valuable guidance provided we keep its flaws and limitations in mind.

Here, we list the titles of those concepts, with a few words about their limitations.

Offensive. One can obtain a sustained initiative by means other than the offensive, but the offensive is the most common way to achieve a sustained initiative. However, before going on the offensive, commanders should always ask whether the offensive is likely to seize or increase a sustained initiative. Although an offensive nearly always gains a short-term initiative, the goal must be a long-term, sustained initiative. Usually, a commander can exploit a sustained initiative only by going on the offensive at some point. The offensive is not a principle because it is neither necessary nor sufficient to gain a sustained initiative. Failed offensives, and even some successful offenses, can lose the initiative.

Unity of command. This is not a principle of war because it tells how to achieve unity of effort, which is the real goal. Unity of command is neither necessary nor sufficient to achieve unity of effort. Nevertheless, unity of command is highly desirable. Unity of command does not guarantee effective cooperation among subordinates. Therefore, even when they have unity of command, commanders must demand that subordinates cooperate fully with any other unit involved in an operation.

Clarity of orders and reports. Commanders must emphasize clarity. If an operation must be complex to achieve an objective, planners must ensure that the operation is broken down into easily understandable pieces. Commanders should take great care to avoid ambiguity in orders. Similarly, reports up the chain of command must be clear and complete. These measures should reduce friction and the fog of war.

This and the following two near-principles replace the overly simplistic current principle of simplicity. We did not designate any of these three near-principles as principles of war because we do not consider them to be fundamental elements of warfare.

Realistic expectations. Plans that fail to account for real world conditions are destined to miscarry. To avoid having operations fail when a subordinate does not complete his or her task on time, commanders should have realistic expectations. When constructing timelines, they should consider the current state of training of subordinates, as well as the details of weather, road conditions, terrain, and so forth. Planning with realistic expectations will reduce many types of friction.

Robust plans. Plans that require perfect execution in every aspect are doomed to fail. Commanders should avoid plans that will fail overall if any single piece fails. If possible, plans should have built-in redundancies that make the plan fault tolerant. Then, when a few things inevitably go wrong, this will not jeopardize the entire operation. Robust plans reduce the impact of many types of friction.

Flexibility. Commanders and military organizations should be receptive, responsive, adaptive, and innovative. In war, one frequently encounters unanticipated situations. Commanders must be receptive to information that challenges their prior beliefs and should actively seek to obtain such information. They should adapt plans to changing conditions, yet never lose focus on the commander's intent. Above all, they must have a flexible mindset and encourage critical, creative thinking. We do not consider flexibility to be a principle of war only because the lines between flexibility and indecisiveness, between determination and stubbornness, are often difficult to recognize in real time.

Logistics. Availability of food, water, ammunition, fuel, spare parts, and thousands of other items sets the limits for any military operation. Commanders must base their plans on what their logistics can support. Attacking the enemy's logistical support can be very effective. We did not classify logistics (or sustainability) as a principle of war because we did not identify any practical guidance for commanders. We did not deem things such as "pay attention to logistics" or "base plans on what logistics can support" to be sufficiently practical.

Maintenance of morale. Ultimately, individuals do the fighting, and their performance depends in part on their morale. Although morale is an extremely valuable component of an effective fighting force, good morale must be accompanied by good judgment, discipline, training, adequate equipment, and leadership.

WE ARE NOT SO NAÏVE AS TO EXPECT EVERYONE READING this book to agree with everything we have written. At best, we hope to generate a long-overdue critical look at the currently recognized principles of war. We believe such a review is necessary because the currently recognized principles of war have become ossified. We look forward to a vigorous debate.

Epilogue

Gettysburg and the New Principles of War

THE PROLOGUE TO THIS BOOK DESCRIBED HOW GEN. ROBERT E. Lee, leading the Confederate Army of Northern Virginia, mainly followed today's currently recognized principles of war during the Battle of Gettysburg. In particular, he was on the offensive during almost the entire battle and massed his forces for Pickett's Charge in what turned out to be the critical point and time of the battle and of the entire Civil War. Yet he lost the battle. This suggests a serious deficiency in the current principles of war.

We will now examine briefly Pickett's Charge and the Battle of Gettysburg using our proposed new principles of war. First, we will consider the Confederate point of view using the proposed principles. Then we will examine the battle from the Union point of view using both the currently recognized and the proposed new principles of war.

Confederate Point of View

Prioritized objectives. The Confederacy's political objectives for Lee's invasion of the North in 1863 included encouraging the northern peace movement and gaining European support for the Confederacy. Strategic military objectives included threatening northern cities, relieving the pressure of the Union Army on Richmond, and perhaps even inducing the Union to move forces away from Vicksburg. Operational objectives included getting the Union Army out of its strong defensive position on the Rappahannock River, relieving Virginia's farmers of the burden of supplying two armies, gathering and stockpiling supplies from the farms of the North, and defeating the Army of the Potomac on Northern soil. Some of these military objectives required different dispositions of the Confederate Army during its invasion. For example, gathering supplies from Northern farms required dispersion of his forces, but defeating the Army of the Potomac required a concentration of the Army of Northern Virginia. Lee should have analyzed the inconsistent requirements imposed by all these objectives and explicitly prioritized his objectives.[1]

In addition to the objective of defeating the Union army on Northern soil, Lee also had an unspoken secondary objective of conserving his own force. He could not replace losses, while the Union could draw on a much larger manpower base. It was not an either-or situation. Lee could afford heavy casualties if he gained a significant victory, but large losses in a near-draw would be quite bad, and they would be disastrous in a defeat. In this situation, he should have followed something like Nimitz's calculated risk guidance.

Relative advantage. When Lee saw that the numerically superior Union Army had occupied a strong defensive position, he should have disengaged and sought better ground to fight with a significant relative advantage. This was his fatal error. For Lee, the critical place and time should not have been Gettysburg in July 1863. The Confederates fought the battle without any relative advantage. In fact, the Union had a relative advantage with greater numbers and good defensive positions. Pickett's Charge gave the Confederates a numerical advantage in men on the front line in that sector, but the Union's good defensive position offset that advantage, and Union reserves in the area provided approximate equality in numbers.

Sustained initiative. Lee's attacks on each of the three days gave him a short-term initiative each day. However, because he carried out the attacks with no relative advantage, he had little chance of a breakthrough that would expand his initiative. Thus, his short-term initiatives dissipated when the Union repulsed each attack. At the end of the third day of the battle, the Union had the initiative. Because Lee had no reserves left, his only realistic option was to return to Virginia. On the other hand, Meade had the choice of (1) doing nothing, (2) directly attacking the Confederate Army, or (3) attempting to get between Lee and the Potomac River, cutting off his retreat.[2] If Meade had pursued Lee more vigorously, he might have had a sustained initiative.

Unity of effort. Although the Confederates had unity of command, they did not have unity of effort. Before and during the first two days of the battle, Stuart's cavalry was off on its own quest instead of scouting out the Union Army's location. Lee wanted an attack on his left wing to accompany Pickett's Charge, but this was not coordinated. The Confederate artillery was not coordinated as well as Lee intended because Lee's chief of artillery was incompetent. The actual charge was somewhat ragged because Longstreet did not adequately coordinate the force assigned to him for the attack. Throughout the three days of battle, the Confederates repeatedly failed to coordinate their forces effectively.

Surprise. Lee did not employ surprise. The long artillery barrage telegraphed the point of the attack. Everyone in the Union line could watch the charge as it developed.

Deception. Lee did not employ deception except that he repositioned some of his artillery at night to avoid tipping his plans to the Union. In addition, Pickett's division hid from Union view before the charge.

Know yourself. Because of his army's previous brilliant successes, Lee incorrectly believed that his men could break through the Union line. He ignored Longstreet's contrary opinion. In addition, Lee retained a chief of artillery who was not good at the job. He gave discretionary orders to Ewell even though Lee knew the latter was accustomed to orders that were more specific. Furthermore, Lee's orders allowed Stuart to ride around the Union Army to the detriment of Lee's situational awareness.

Know your enemy. Because Stuart was not available for surveillance, Lee did not know the Union's position at the start of the battle. One reason Lee attacked on the second day was that he thought that not all the Union forces were on the scene. In addition, he underestimated the fighting ability of the Army of the Potomac at this stage of the war.

Environment. Lee fought the battle at a disadvantage in the terrain. Lee undoubtedly knew the terrain was unfavorable to him but probably did not realize the seriousness of the disadvantage.

THE CONFEDERATES MADE ERRORS IN APPLYING THE PRINCIPLES of prioritized objectives, unity of effort, and know your enemy. Lee did not use significant surprise or deception. However, these shortcomings were not necessarily fatal errors.

Lee's failures to follow the principles of relative advantage, sustained initiative, know yourself, and environment were serious mistakes. The lack of a sustained initiative, surprise, or deception, combined with the failure to use the terrain, meant Lee had no relative advantage when he attacked. The failure to gain a tactical relative advantage was the second most serious error. Lee's most serious error was the operational decision to give battle at Gettysburg without a relative advantage, rather than disengaging until he could fight with a significant relative advantage.

Union Point of View

In terms of the current principles of war, Maj. Gen. George Meade generally complied with the principles of objective, unity of command, simplicity, economy of force, mass, and security. He did not follow the principles of offensive, maneuver, and surprise. Critics might claim that Meade violated the principle of mass by failing to get all his forces into the battle. He used the entire Sixth Corps as a floating reserve to backup various threatened areas. Despite Lincoln's urging to get all the forces into the battle, we regard this use as reserves to be prudent and not a violation of the principle of mass. The Union's most egregious violation of the currently recognized principles of war was the lack of offense. Exceptions included the opening engagement on the first day, Joshua Chamberlain's bluff charge with bayonets after his men ran out of ammunition, and the attack on Lee's left wing at Culp's Hill on the morning of the third day. However, the bulk of the Union forces were defensive for the entire battle, especially on the third day. Nevertheless, despite the failure to follow the currently recognized principle of the offensive, the Union won the battle.

Now consider the Army of the Potomac's actions during the three-day battle using the nine proposed new principles of war.

Prioritized objectives. Lincoln and Halleck assigned Meade the primary objective of preventing the Confederates from attacking Baltimore or Washington by staying between Lee and those cities. This seriously restricted Meade's operational moves, especially after the battle. Meade also had the two objectives of defending Northern territory and defeating the invading Confederate army. These three objectives might have come into conflict if Lee had tried to evade Meade, but he did not do so. To defeat the Army of Northern Virginia, Meade's operational objective was to give battle on favorable terms. If given the opportunity, then the objective would become annihilating the Army of Northern Virginia with the likely result of ending the Civil War. After the battle Lincoln stated that Meade's prime objective should be to defeat the Army of Northern Virginia, rather than driving them out of the North.

Relative advantage. Meade went into the battle with the advantage of greater numbers. The Union then took advantage of terrain to improve its relative advantage. In addition, because Lee was willing to attack the Union, Meade kept the Union in its favorable defensive position.

Sustained initiative. The Union occupied a good defensive position, and then for the most part passively awaited the Confederate attacks. This yielded a short-term initiative to the Confederates. However, despite being on the defensive for most of the battle, the Union had the initiative at the end of the battle. Furthermore, the Confederates had somewhat more casualties than the Union, even though the Confederates had fewer men engaged in the battle. This enhanced the Union's relative numerical advantage. In addition, the Union's Sixth Corps had not participated heavily in the fighting and its soldiers were relatively fresh. Therefore, the Union was in position to sustain the initiative. They did so by pursuing the Confederates back into Virginia, but they did not follow up their victory as vigorously as they might have done.

Unity of effort. The Union had unity of command and mostly achieved unity of effort. Brig. Gen. Henry Hunt exercised unified command of the artillery. As a result, the Union artillery was much more effective than the Confederate artillery. There was little backbiting among the Union generals, which previously had been common in the Union army. Cooperation among the various units was excellent throughout the battle, with assistance given to other units in danger, often without formal orders. As a result, on the third day units from several corps were mixed together on the line when the Confederates attacked, but on the defense, this did not matter. One exception to unity of effort was Maj. Gen. Dan Sickles's move forward, which was contrary to Meade's orders.

Surprise. Meade did not employ any significant surprise in the battle.

Deception. For the most part, the Union did not use deception except for one minor, but important, incident. Just before Pickett's Charge, the Union artillery gradually stopped firing to deceive the Confederates into thinking they had knocked it out, thus encouraging them to attack.

Know yourself. Meade knew that the Army of the Potomac had endured several devastating defeats. However, he correctly believed that they could fight effectively on the defensive.

Halleck gave Meade the authority to select commanders based on ability rather than seniority, and Meade did so wisely in several instances. He chose Maj. Gen. John Reynolds as the scene-of-action commander on the first day and Maj. Gen. Winfield Scott Hancock to organize the defenses on the second day. Meade frequently rode the lines to learn the condition of all his forces. He knew the Union could hold against an attack on the third day. On the previous night, a war council of top generals unanimously voted to stay in place.

Meade did not pursue the Confederates after the battle in part because his two best and most offensive-minded corps commanders, Reynolds and Hancock, were dead and injured, respectively. He knew the remaining corps commanders either were new to the job or were not aggressive commanders.

Know your enemy. Meade knew the enemy well. His intelligence service told him all the Confederate units that were present. Going into the third day, he knew Pickett's division was the only fresh division that Lee had and therefore was likely to spearhead any attack. The night before, he anticipated that the attack on the third day would be against his center. Although he knew that Lee usually did not make frontal attacks, Meade positioned reserve units behind the threatened sector. Counting those, he actually had numerical equality in that sector.

Environment. The Union took advantage of the high ground. Early in the battle, they occupied Cemetery Ridge and then Little Round Top. Hunt positioned artillery with clear fields of fire to shoot into both flanks of Pickett's Charge.

In summary, the Union mostly followed the principles of prioritized objectives, relative advantage, unity of effort, know yourself, know the enemy, and environment. It followed the principle of the sustained initiative by giving up the short-term initiative to gain a sustainable initiative, but then failed to exploit fully that opportunity. The Union did not follow the principle of surprise. The Union employed the principle of deception in a minor way, but one that might have been significant.

The key to victory was superior numbers, unity of effort, the minor use of deception, the relative advantage due to using the environment to their advantage, and the proper use of the principle of sustained initiative.

THUS, THE PROPOSED NEW PRINCIPLES OF WAR EXPLAIN THE outcome of the Battle of Gettysburg much better than the current principles of war do.

Preface

1. Kitfield, *Prodigal Soldiers*.
2. Burnod, "Military Maxims of Napoleon," 432.
3. Fallwell, "Principles of War."
4. MacDougall, *Theory of War*.

Prologue

1. Sears, *Gettysburg*, 15–18.

1. History of Principles of War

1. Sun Tzu, *Art of War*, 72.
2. Sun Tzu, *Art of War*, 63–67.
3. Sun Tzu, *Art of War*, 77.
4. Sun Tzu, *Art of War*, 134.
5. Sun Tzu, *Art of War*, 69.
6. Sun Tzu, *Art of War*, 98.
7. Sun Tzu, *Art of War*, 129.
8. Sun Tzu, *Art of War*, 100.
9. Sun Tzu, *Art of War*, 100–101.
10. Sun Tzu, *Art of War*, ix, 179.
11. Mallick, *Principles of War*, 3; Kautilya, *Arthashastra*.
12. Kautilya, *Arthashastra*, 521.
13. Kautilya, *Arthashastra*, 513.
14. Kautilya, *Arthashastra*, 527.
15. Kautilya, *Arthashastra*, 490.
16. Kautilya, *Arthashastra*, 526.
17. Frontinus, *Strategemata*.
18. Phillips, *Roots of Strategy*, 67–69.
19. Vegetius, "Military Institutions of the Romans," 75–96.
20. Vegetius, "Military Institutions of the Romans," 99–121.
21. Phillips, *Roots of Strategy*, 72.
22. Vegetius, "Military Institutions of the Romans," 123–75.
23. Vegetius, "Military Institutions of the Romans," 172.
24. Phillips, *Roots of Strategy*, 67; Alger, *Quest for Victory*, 6.
25. Machiavelli, *Art of War*, 84.
26. Gilbert, "Machiavelli."
27. Machiavelli, *Art of War*, 94.

28. Alger, *Quest for Victory*, 9.

29. Quoted in Alger, *Quest for Victory*, 199–200.

30. Alger, *Quest for Victory*, 9–10.

31. de Saxe, "My Reveries," 202.

32. de Saxe, "My Reveries," 272.

33. de Saxe, "My Reveries," 192.

34. Burnod, "Military Maxims of Napoleon," 421.

35. Burnod, "Military Maxims of Napoleon," 401–41.

36. Clausewitz, *On War*, 70.

37. Clausewitz, *On War*, 89.

38. Clausewitz, *Principles of War*.

39. Clausewitz, *Principles of War*, 12.

40. Clausewitz, *Principles of War*, 16.

41. Clausewitz, *Principles of War*, 19.

42. Clausewitz, *Principles of War*, 30.

43. Clausewitz, *Principles of War*, 26.

44. Clausewitz, *Principles of War*, 33.

45. Clausewitz, *Principles of War*, 45–48.

46. Clausewitz, *Principles of War*, 11.

47. Clausewitz, *On War*, 65–66.

48. Clausewitz, *On War*.

49. Clausewitz, *On War*, 141.

50. Clausewitz, *On War*, 57–58.

51. Clausewitz, *On War*, 152.

52. Clausewitz, *On War*, 152.

53. Clausewitz, *On War*, 204.

54. Clausewitz, *On War*, 213.

55. Alger, *Quest for Victory*, 28–31.

56. Searle, "Inter-Service Debate," 7.

57. Alger, *Quest for Victory*, 18–27.

58. Jomini, *Art of War*, 321–23.

59. Jomini, *Art of War*, 328.

60. Jomini, *Art of War*, 70.

61. Jomini, *Art of War*, 330.

62. Jomini, *Art of War*, 329.

63. Jomini, *Art of War*, 72–73.

64. Williams, "Jomini, Antoine Henri"; Alger, *Quest for Victory*, 42–55.

65. MacDougall, *Theory of War*, 51.

66. MacDougall, *Theory of War*, 107–8.

67. MacDougall, *Theory of War*, 169.

68. Alger, *Quest for Victory*, 55–56.

69. Holborn, "Prusso-German School"; Rothenburg, "Moltke, Schlieffen"; Vandergriff, "How the Germans Defined Auftragstaktik."

70. Alger, *Quest for Victory*, 55–58.

71. Foch, *Principles of War*, 8.

72. Alger, *Quest for Victory*, 66–72; Searle, "Inter-Service Debate," 11.

73. Mahan, *Influence of Sea Power upon History*.

74. Alger, *Quest for Victory*, 88–94; Livezey, *Mahan on Sea Power*, 257–62.

75. Adams, *If Mahan Ran the Great Pacific War*, 4–9; Mahan, *Influence of Sea Power upon History*; Crowl, "Alfred Thayer Mahan"; Livezey, *Mahan on Sea Power*.

76. Alger, *Quest for Victory*, 93–94.

77. Fallwell, "Principles of War," 50–53.

78. Alger, *Quest for Victory*, 106–17.

79. War Office, *Field Service Regulations Part I* (1909), 13.

80. War Office, *Field Service Regulations Part I* (1909), 87, 112, 126–27, 132.

81. Fuller, *Foundations of the Science of War*, 14–15; Fuller, *Memoirs of an Unconventional Soldier*, 28–29, 388–89.

82. Alger, *Quest for Victory*, 121–23.

83. War Office, *Field Service Regulations Vol. II Operations (Provisional), 1920*, quoted in Alger, *Quest for Victory*, 240–41.

84. Alger, *Quest for Victory*, 122–23.

85. Searle, "Inter-Service Debate," 12–31.

86. Fuller, *Foundations of the Science of War*.

87. Searle, "Inter-Service Debate," 17–26.

88. Alger, *Quest for Victory*, 150–52.

89. UK Ministry of Defence, *Joint Doctrine Publication 0–01*, 2014.

90. Alger, *Quest for Victory*, 135–45, 241; Fallwell, "Principles of War," 50–53.

91. United States Army, *Field Service Regulations 1923*, 10, 77–78.

92. Department of the Army, *Field Service Regulations: Operations*, 1949, 21–23.

93. Alger, *Quest for Victory*, 160–68, 253–70.

94. Joint Chiefs of Staff, *Joint Publication 1*, 1991, 21–24.

95. U.S. Department of the Army, *Field Manual 100–5, Operations*, 1993, ch. 13.

96. U.S. Department of the Army, *Field Manual 100–5, Operations*, 1993, 13:4.

97. U.S. Department of the Army, *Field Manual 100–1, The Army*, 1994, 31–32.

98. Joint Chiefs of Staff, *Joint Publication 3–07*, 1995.

99. Glenn, "No More Principles of War?"

100. Joint Chiefs of Staff, *Joint Publication 3–0*, 2006.

101. Joint Chiefs of Staff, *Joint Publication 3–0*, 2018.

102. Liddell Hart, *Strategy*, 334–36.

103. Leonhard, *Principles of War for the Information Age*, 243–64.

104. Alger, *Quest for Victory*.

105. Savkin, *Basic Principles of Operational Art and Tactics*.

106. Hughes, *Fleet Tactics: Theory and Practice*; Hughes, *Fleet Tactics and Coastal Combat*; Hughes and Girrier, *Fleet Tactics: Theory and Naval Operations*.

2. Principles of War Abroad

1. Alger, *Quest for Victory*, 127–31.

2. Alger, *Quest for Victory*, 152–53.

3. French Defence Staff, *Capstone Concept for Military Operations*, 18–19.

4. French Defence Staff, *Capstone Concept for Military Operations*, 7.

5. French Defence Staff, *Capstone Concept for Military Operations*, 7–8, 19–21.

6. Armée de Terre, *General Tactics*, 30–34, 43.

7. Armée de Terre, *Employment of Land Forces in Joint Operations*, 11, 26.

8. Alger, *Quest for Victory*, 131–33.

9. Alger, *Quest for Victory*, 153–54.

10. Alger, *Quest for Victory*, 154–55.

11. Amidror, *Winning Counterinsurgency War*, 29.

12. Amidror, *Winning Counterinsurgency War*, 30.

13. Amidror, *Winning Counterinsurgency War*, 30–38.

14. Amidror, *Winning Counterinsurgency War*, 33.

15. UK Ministry of Defence, *Joint Doctrine Publication 0–01* (2014), 28.

16. UK Ministry of Defence, *Joint Doctrine Publication 0–01* (2014), 50.

17. Canadian Forces Experimentation Centre, *Canadian Military Doctrine*, 2:4–6.

18. Australian Department of Defence, *Foundations of Australian Military Doctrine*, ch. 6.

19. Australian Army, *Fundamentals of Land Power*, 16–18.

20. Royal Australian Air Force, *Air Power Manual*, 151–64.

21. Royal Australian Navy, *Australian Maritime Doctrine*, 62–67.

22. New Zealand Defence Force, *New Zealand Defence Doctrine*, x, 36, 44–49.

23. Indian Army Training Command, *Indian Army Doctrine*, 23–24; Mallick, *Principles of War*, 9–19.

24. Indian Air Force, *Basic Doctrine of the Indian Air Force*, 17.

25. NATO Standardization Office, *Allied Joint Doctrine*, 1:12–15.

26. West, *Principles of War*, foreword, 1–3.

27. West, *Principles of War*, 19.

28. West, *Principles of War*, 17.

29. West, *Principles of War*, 29–30.

30. Sun Tzu, *Art of War*, xi, 169–93.

31. Chase and Medeiros, "China's Evolving Nuclear Calculus," 131n241.

32. Straker, "36 Stratagems, in Detail."

33. Alger, *Quest for Victory*, 158–59, 247–48, 258–59.

34. Cliff et al., *Shaking the Heavens*, 7–9.

35. Fravel, "Evolution of China's Military Strategy," 92–95.

36. Fravel, "Evolution of China's Military Strategy," 92–95.

37. Cliff et al., *Entering the Dragon's Lair*, 27–43; Yang, "Military of the People's Republic of China," 191; Cliff et al., *Shaking the Heavens*, 189.

38. This seems consistent with Chinese occupation of islands in the South China Sea starting in the 1990s and building and militarizing artificial islands there beginning in 2014.

39. Alger, *Quest for Victory*, 133.

40. Savkin, *Basic Principles of Operational Art and Tactics*, 30–33.

41. Department of the Army, *Soviet Army: Operations and Tactics*, 2:3.

42. Alger *Quest for Victory*, 133–35; Savkin, *Basic Principles of Operational Art and Tactics*, 39–46.

43. Baxter, *Soviet AirLand Battle Tactics*, 6–11.

44. Savkin, *Basic Principles of Operational Art and Tactics*.

45. Baxter, *Soviet AirLand Battle Tactics*, 9–10.

46. Baxter, *Soviet AirLand Battle Tactics*; U.S. Department of the Army, *Soviet Army: Operations and Tactics*.

47. Baxter, *Soviet AirLand Battle Tactics*, 22–26.

48. Baxter, *Soviet AirLand Battle Tactics*, 24.

49. Baxter, *Soviet AirLand Battle Tactics*, 28.

50. Savkin, *Basic Principles of Operational Art and Tactics*, v–vi.

51. Savkin, *Basic Principles of Operational Art and Tactics*, 165–277.

52. Reznichenko, *Tactics: A Soviet View*, 41–58; Baxter, *Soviet AirLand Battle Tactics*, 29–31.

53. Reznichenko, *Tactics: A Soviet View*, 56.

54. Department of the Army, *Soviet Army: Operations and Tactics*, 2:3–4.

55. Stolfi, *Soviet Naval Operational Art*, ch. 1, 10–12.

3. Principles of War in U.S.

1. U.S. Department of the Navy, *Naval Doctrine Publication 1*, 33–44.

2. Swartz, *U.S. Navy Capstone Strategies and Concepts*, 32. Exceptions to the U.S. Navy's lack of attention to the principles of war include the foreword by Vice Adm. Arthur Cebrowski and the preface by Vice Adm. John Morgan and Anthony Mc Ivor in Mc Ivor, *Rethinking the Principles of War*; and Morgan and Mc Ivor's article of the same name in the *U.S. Naval Institute Proceedings*. Additionally, the Department of the Navy cosponsored an essay contest on the principles of war. The three winning essays appeared in *U.S. Naval Institute Proceedings* 131, no. 10 (October 2005): 38–55.

3. Hughes, *Fleet Tactics and Coastal Combat*, 170; Hughes and Girrier, *Fleet Tactics: Theory and Naval Operations*, 164.

4. U.S. Department of the Army, *Doctine Primer*, 4:3.

5. U.S. Department of the Army, *Doctine Primer*, 4:7.

6. U.S. Department of the Army, *Army Doctrine Reference Publication 3–0*, ch. 3.

7. U.S. Department of the Army, *Army Doctrine Reference Publication 3–0*, ch. 3.

8. Brown, *New Conception of Warfare*, xv–xlvi.

9. United States Marine Corps, *Warfighting: Fleet Marine Force Manual 1*, Foreword.

10. United States Marine Corps, *Warfighting: Fleet Marine Force Manual 1*, 31–32. In physics, momentum equals mass times velocity.

11. Quoted in United States Marine Corps, *Warfighting: Fleet Marine Force Manual 1*, 55. The source listed in that document is William Slim, *Defeat into Victory* (London: Cassell, 1956) 550–55. In the American edition, this quotation has a trivial change of wording. Slim, *Defeat into Victory*, 550–51.

12. United States Marine Corps, *Warfighting: Marine Corps Doctrinal Publication 1*, 40–42.

13. United States Marine Corps, *Marine Corps Operations*, 2001, appendix B.

14. United States Marine Corps, *Marine Corps Operations*, 2001, 2:15–16.

15. United States Marine Corps, *Marine Corps Operations*, 2001, 7:3.

16. United States Marine Corps, *Marine Corps Operations*, 2011, appendix A.

17. United States Marine Corps, *Marine Corps Operations*, 2011, 1:4.

18. United States Marine Corps, *Marine Corps Operations*, 2011, 1:4.

19. United States Marine Corps, *Marine Corps Operations*, 2011, 5:10, 9:2. In 2017 the Marine Corps published change 1 to the 2011 version of *Marine Corps Operations*, but the changes did not affect any of the discussion in the text. United States Marine Corps, *Marine Corps Operations*, 2017.

20. LeMay Center for Doctrine, *Air Force Doctrine*, ch. 4.

21. LeMay Center for Doctrine, *Air Force Doctrine*, ch. 4.

22. LeMay Center for Doctrine, *Air Force Doctrine*, ch. 4.

23. LeMay Center for Doctrine, *Air Force Doctrine*, ch. 4.

24. LeMay Center for Doctrine, *Air Force Doctrine*, ch. 5.

25. LeMay Center for Doctrine, *Air Force Doctrine*, ch. 5.

4. Criteria for Principles of War

1. Sun Tzu, *Art of War*, 120.

2. Hughes, *Fleet Tactics and Coastal Combat*, 143; Hughes and Girrier, *Fleet Tactics: Theory and Naval Operations*, 165.

3. Hughes, *Fleet Tactics and Coastal Combat*, 142; Hughes and Girrier, *Fleet Tactics: Theory and Naval Operations*, 165.

5. Prioritized Objectives

1. Joint Chiefs of Staff, *Joint Publication 3–0*, 2018, A:1.

2. UK Ministry of Defence, *Joint Doctrine Publication 0–01* (2014), 30.

3. Alger, *Quest for Victory*, 240.

4. Foote, *Fredericksburg to Meridian*, 529–30.

5. Heather, *Fall of the Roman Empire*, 227–29; Goldsworthy, *How Rome Fell*, 299–302.

6. Smith, *Utility of Force*, 217–18; Kissinger, *Years of Upheaval*, 460.

7. Nagl, *Learning to Eat Soup with a Knife*, 28, 72, 116.

8. Mitchell and Creasy, *Twenty Decisive Battles*, 63–76; Dupuy and Dupuy, *Harper Encyclopedia of Military History*, 76.

9. Foote, *Red River to Appomattox*, 531–38; McPherson, *Battle Cry of Freedom*, 758–60; McFeely, *Grant: A Biography*, 177–79; Catton, *Grant Takes Command*, 308–9, 318–25.

10. Fleming, *Operation Sea Lion*, 217–37.

11. Fleming, *Operation Sea Lion*, 223–37.

12. Morison, *Two-Ocean War*, 356–62.

13. Blaber, *Mission, the Men, and Me*, 228.

14. Blaber, *Mission, the Men, and Me*, 228, 259–60.

15. Strassler, *Landmark Thucydides*, 219–24; Kagan, *Archidamian War*, 219–21.

16. Strassler, *Landmark Thucydides*, 224–31; Kagan, *Archidamian War*, 221–34.

17. Corbett, *Principles of Maritime Strategy*, 177–78.

18. Spector, *Eagle against the Sun*, 301–12; Murray and Millett, *War to be Won*, 353–59.

19. In Halsey's defense, Nimitz's operation plan stated that if Halsey had a chance to inflict major damage to Japanese naval forces, then that would become his primary task. Hone, "U.S. Navy Surface Battle Doctrine"; Vego, *Battle for Leyte*, 126–27, 145–46. We discuss this in chapter 8.

20. Spector, *Eagle Against the Sun*, 430–42; Vego, *Battle for Leyte*, 255–59.

21. Mitchell and Creasy, *Twenty Decisive Battles*, 283–98.

22. van Creveld, *Supplying War*, 113–41.

23. Morse and Kimball, *Methods of Operations Research*, 52–53.

24. Parshall and Tully, *Shattered Sword*, 19–69.

25. Bush, "National Security Directive 54."

26. Bush, "National Security Directive 54."

27. Bush, "National Security Directive 54."

28. U.S. Department of Defense, *Conduct of the Persian Gulf War*, 96–101.

29. Based on the author's personal experience on the staff of Commander, U.S. Naval Forces, Central Command, in the theater during Operations Desert Shield and Desert Storm. See Pokrant, *Desert Shield at Sea*; Pokrant, *Desert Storm at Sea*. In addition, in a 1993 interview with Adam Meyerson, former secretary of defense Dick Cheney stated, "We liberated Kuwait, and we destroyed Saddam's offensive capability. Those were the two objectives we talked about repeatedly in the run-up to the war, and once we achieved those objectives, we stopped operations." Myerson, "Calm after Desert Storm (Interview with Dick Cheney)."

30. Naylor, *Not a Good Day to Die*, 10–21.

31. Naylor, *Not a Good Day to Die*, 10–21.

32. Corbett, *Principles of Maritime Strategy*, 238.

33. Corbett, *Principles of Maritime Strategy*, 238.

34. McPherson, *Battle Cry of Freedom*, 353–58; Sandberg, *Abraham Lincoln*, 568.

35. Sandberg, *Abraham Lincoln*, 567.

36. Sandberg, *Abraham Lincoln*, 582–90.

37. Steele, *Nimitz Graybook*, 490, 507, 520. In his message, King also urged Nimitz to concentrate his forces in the Hawaiian area because King expected the Japanese to raid Oahu prior to the Midway operation.

38. Steele, *Nimitz Graybook*, 520. It seems that the attrition tactics did not apply to a major battle but rather that attacks by submarines and land-based aircraft would have to whittle down the Japanese force *before* a decisive fleet versus fleet battle. This differs from the usual interpretation.

39. CinC U.S. Pacific Fleet, "Operation Plan No. 29–42."

40. At the Battle of Midway, of the seven aircraft carriers in the main action, five were sunk. In the Battle of the Eastern Solomons in August 1942, one of five carriers was sunk, and one was damaged. In the Battle of the Santa Cruz Islands in October 1942, one of six carriers participating was sunk and three were damaged. The cumulative score for the four carrier vs. carrier battles of 1942: twenty-three carriers participated, nine were sunk, and six were damaged.

41. Buell, *Quiet Warrior*, 136–37.

42. Nimitz, "Letter of Instructions."

43. Winnefeld and Johnson, *Joint Air Operations*, 18.

44. Robert Rubel analyzes Nimitz's Letter of Instructions and the concept of calculated risk. He criticizes Nimitz for not having any "Plan B" if the intelligence turned out to be wrong and the Japanese fleet did not appear off Midway when expected. He argues that the concept of calculated risk was useless in this case because the tactical decision point was at dawn on 4 June. Rubel, "Deconstructing Nimitz's Principle of Calculated Risk."

6. Mass vs. Relative Advantage

1. U.S. Joint Chiefs of Staff, *Joint Publication 3–0*, 2018, A:2.

2. UK Ministry of Defence, *Joint Doctrine Publication 0–01*, 2014, 31.

3. Glenn, "No More Principles of War?"

4. U.S. Department of the Army, *Field Manual 3–0*, 2008, A:2.

5. A series of books by Newt Gingrich, William R. Forstchen, and Albert S. Hanser explore a fictional alternative that might have ensued if Lee had disengaged at Gettysburg. The first book in the series is Gingrich, Forstchen, and Hanser, *Gettysburg: A Novel of the Civil War*.

6. Mitchell and Creasy, *Twenty Decisive Battles*, 1–21; Carey, Allfree, and Cairns, *Warfare in the Ancient World*, 42–45.

7. Sun Tzu, *Art of War*, 89.

8. Jomini, *Art of War*, 328.

9. Latimer, *Deception in War*, 274–76.

10. Mao, *On Guerrilla Warfare*, 46.

11. Ferrill, *Origins of War*, 194–99; Carey, Allfree, and Cairns, *Warfare in the Ancient World*, 68–71.

12. Caesar, *Conquest of Gaul*, 84–85.

13. Hughes, *Fleet Tactics: Theory and Practice*, 34–39.

14. Heather, *Fall of the Roman Empire*, 176–81; Carey, Allfree, and Cairns, *Warfare in the Ancient World*, 139–43.

15. Seward, *Hundred Years War*, 61–68; Dupuy and Dupuy, *Harper Encyclopedia of Military History*, 384–86.

16. Parshall and Tully, *Shattered Sword*, 161–75; Isom, *Midway Inquest*, 12–18, 154–81, 272–74.

17. Kagan, *Peloponnesian War*, 267–323.

18. McPherson, *Battle Cry of Freedom*, 426–27.

19. Sears, *Landscape Turned Red*.

20. Liddell Hart, *Strategy*, 259–78; Murray and Millett, *War to be Won*, 263–73.

21. van Creveld, *Supplying War*, 196–201.

22. Murray and Millett, *War to be Won*, 271.

23. Betts, *Surprise Attack*, 135.

24. Lundstrom, *First Team*, 477–85.

25. Mitchell and Creasy, *Twenty Decisive Battles*, 206–7; Ketchum, *Saratoga*, 355–72.

26. Miller, *Broadsides*, 202–10.

27. Keegan, *Price of Admiralty*, 1–107; Miller, *Broadsides*, 289–99.

28. Carey, Allfree, and Cairns, *Warfare in the Ancient World*, 113–16; Plutarch, *Lives of the Noble Romans*.

29. Carey, Allfree, and Cairns, *Warfare in the Ancient World*, 125–30; Creasy, *Decisive Battles*, 115–28.

30. Gordon and Trainor, *Generals' War*, 407–8.

31. Pope, *Battle of the River Platte*, 115–19.

32. Carey, Allfree, and Cairns, *Warfare in the Ancient World*, 46.

33. Vego, *Battle for Leyte*, 288; Fallwell, "Principles of War," 57–58.

34. Arthur and Pokrant, "Desert Storm at Sea."

35. Leonhard, *Principles of War*, 67–75.

7. Offensive vs. Initiative

1. Reznichenko, *Tactics: A Soviet View*, 42–44.

2. U.S. Joint Chiefs of Staff, *Joint Publication 3–0*, 2018, A:1–2.

3. UK Ministry of Defence, *Joint Doctrine Publication 0–01*, 2014, 30.

4. Catton, *Stillness at Appomattox*, 104–6; Foote, *Red River to Appomattox*, 190–91.

5. Fallwell, "Principles of War," 56–57.

6. David, *Military Blunders*, 13–24.

7. Murray and Hsieh, *Savage War*, 240–46.

8. McPherson, *Battle Cry of Freedom*, 733–36.

9. Liddell Hart, *Strategy*, 147.

10. Foote, *Fredericksburg to Meridian*, 520–69.

11. Dupuy and Dupuy, *Harper Encyclopedia of Military History*, 1050–52; Marshall, *World War I*, 244–49.

12. Plutarch, *Lives of the Noble Grecians and Romans*, Locations 10272–10432; Dodge, *Hannibal*, 107–11.

13. Dodge, *Hannibal*, 112–19; Plutarch, *Lives of the Noble Grecians and Romans*, Locations 10432–10452.

14. Margiotta, *Brassey's Encyclopedia of Military History and Biography*, 443.

15. Liddell Hart, *Strategy*, 84–85; Dupuy and Dupuy, *Harper Encyclopedia of Military History*, 659.

16. Leckie, *George Washington's War*, 239–41.

17. Mattingly, *Armada*, 274–77.

18. Goldsworthy, *Fall of Carthage*, 181–90.

19. Foote, *Red River to Appomattox*, 318–424, 472–92, 519–30; McPherson, *Battle Cry of Freedom*, 743–55.

20. Liddell Hart, *Strategy*, 39–40, 52–53.

21. Liddell Hart, *Strategy*, 147.

22. Spector, *Eagle against the Sun*, 305–12.

23. Nagl, *Learning to Eat Soup with a Knife*, 163.

24. Corbett, *Principles of Maritime Strategy*, 66.

25. Liddell Hart, *Strategy*, 134.

26. Strassler, *Landmark Thucydides*, 224–46.

27. Foote, *Red River to Appomattox*, 840–45.

28. Goldsworthy, *Fall of Carthage*, 190–96.

29. Jellicoe, *Jutland: The Unfinished Battle*, 199–201.

30. Sears, *Chancellorsville*, 199–281.

31. Foster, *Hit the Beach*, 28–43; Marshall, *World War I*, 177–81.

32. Frank, *Guadalcanal*, 115–21; Warner and Warner, *Disaster in the Pacific*, 258–59.

33. Vego, *Battle for Leyte*, 266–75.

8. Unity of Effort

1. Australian Department of Defence, *Foundations of Australian Military Doctrine*, 6:7.

2. Baxter, *Soviet AirLand Battle Tactics*, 22–31.

3. Savkin, *Basic Principles of Operational Art and Tactics*, 273–77.

4. Reznichenko, *Tactics: A Soviet View*, 46–47.

5. U.S. Joint Chiefs of Staff, *Joint Publication 3-0*, 2018, A:2–3.

6. UK Ministry of Defence, *Joint Doctrine Publication 0-01*, 2014, 30–31.

7. NATO Standardization Office, *NATO Standard AJP-01*, 1:13–14.

8. Carey, Allfree, and Cairns, *Warfare in the Ancient World*, 53.

9. Kagan, *Fall of the Athenian Empire*, 388–93.

10. Goldsworthy, *Fall of Carthage*, 49, 51.

11. Corbett, *Principles of Maritime Strategy*, 306.

12. LeMay Center for Doctrine, *Air Force Doctrine*, vol. 1; Winnefeld and Johnson, *Joint Air Operations*, 7–8.

13. LeMay Center for Doctrine, *Air Force Doctrine*, vol. 1, ch. 2.

14. Winnefeld and Johnson, *Joint Air Operations*, 8–10.

15. Winnefeld and Johnson, *Joint Air Operations*.

16. Pokrant, *Desert Storm at Sea*, 281–93.

17. The term harmony of action comes from Dr. Peter Perla, who in turn got the term from General William Sherman.

18. Goldsworthy, *Fall of Carthage*, 181–85.

19. Heather, *Fall of the Roman Empire*, 176–81; Carey, Allfree, and Cairns, *Warfare in the Ancient World*, 139–43.

20. Strassler, *Landmark Thucydides*, 366–402; Kagan, *Peace of Nicias and the Sicilian Expedition*, 170–243.

21. Prange, *Pearl Harbor*, 375–463.

22. Parshall and Tully, *Shattered Sword*, 405–6.

23. Vego, *Battle for Leyte*, 289–96, 335–38.

24. Vego, *Battle for Leyte*, 344.

25. Vego, *Battle for Leyte*, 290.

26. Vego, *Battle for Leyte*, 266–96, 335–38.

27. Anderson, *By Sea and By River*, 61–66.

28. Anderson, *By Sea and By River*, 42, 91–100.

29. Anderson, *By Sea and By River*, 137–52.

30. Winnefeld and Johnson, *Joint Air Operations*, 23–38; Morison, *Struggle for Guadalcanal*; Frank, *Guadalcanal*; Clubb, *Cactus Airpower at Guadalcanal*.

31. Leonhard, *Principles of War*, 194–95.

32. Cornwell, *Waterloo*, 50, 106–7, 126–27, 330; Mitchell and Creasy, *Twenty Decisive Battles*, 224–39.

33. Clapp and Southby-Tailyour, *Amphibious Assault Falklands*, 50–51.

34. Gatchel, *At the Water's Edge*, 191.

35. Khalid bin Sultan, *Desert Warrior*, 31, 244–47; Schwarzkopf, *Autobiography*, 434–35.

36. Pokrant, *Desert Shield at Sea*, 45–46, 121–22; Pokrant, *Desert Storm at Sea*, 267–68.

37. Pokrant, *Desert Storm at Sea*, 267–68.

38. Naylor, *Not a Good Day to Die*, 91–92.

39. Blaber, *Mission, the Men, and Me*, 203–15.

40. Naylor, *Not a Good Day to Die*, 101.

41. Sears, *Gettysburg*, 103–6, 502–4.

42. Schwarzkopf, *Autobiography*, 289–90.

43. Schwarzkopf, *Autobiography*, 293–94.

44. Gordon and Trainor, *Cobra II*, 344–45, 424.

45. Glenn, "No More Principles of War?"

46. Pokrant, *Desert Storm at Sea*, 274–78, 282.

47. Caesar, *Conquest of Gaul*, 66–70.

48. Foote, *Fredericksburg to Meridian*, 560–61.

49. Cook, *Battle of Cape Esperance*, 51.

50. Miller, *Broadsides*, 226–34.

51. Ketchum, *Saratoga*, 394.

52. Ketchum, *Saratoga*, 394–404; Mitchell and Creasy, *Twenty Decisive Battles*, 202–9.

53. Clapp and Southby-Tailyour, *Amphibious Assault Falklands*, 60.

54. U.S. Joint Chiefs of Staff, *Joint Publication 1*, 1991, 68.

55. Manchester and Reid, *Last Lion*, 639; Eisenhower, *Eisenhower: At War*, 159.

56. van Creveld, *Supplying War*, 221.

57. Gordon and Trainor, *Generals' War*, 362, 473.

58. Carroll, "Botched Air Strike on Lebanon"; George C. Wilson, "The Day We Fouled Up the Bombing of Lebanon," *Washington Post*, September 7, 1986, www.washingtonpost.com/archive/opinions/1986/09/07/the-day-we-fouled-up-the-bombing-of-lebanon/90b613a9-8fdd-4a8f-9261-40e53f4183c1/?utm_term=.bfe8a85b043b.

59. Naylor, *Not a Good Day to Die*, 149–58, 303–68; Blaber, *Mission, the Men, and Me*, 149–297.

60. Naylor, *Not a Good Day to Die*, 184–88; Blaber, *Mission, the Men, and Me*, 262–63.

9. Surprise

1. U.S. Joint Chiefs of Staff, *Joint Publication 3–0*, 2018, A:3.

2. UK Ministry of Defence, *Joint Doctrine Publication 0–01*, 2014, 30–31.

3. Cordingly, *Cochrane*, 280–81.

4. Miller, *Memoirs of General Miller*, 243.

5. Cordingly, *Cochrane*, 280–84.

6. Frederick the Great, "Instruction," 364.

7. Liddell Hart, *Strategy*.

8. Liddell Hart, *Scipio Africanus*, 20–43; Polybius, *Rise of the Roman Empire*, book 10.8–16; Gabriel, *Scipio Africanus*, 85–98; Goldsworthy, *Fall of Carthage*, 271–77.

9. Alexander, *Storm Landings*, 79–85; Gatchel, *At the Water's Edge*, 142–43; Whaley, *Practise to Deceive*, 136–37.

10. Gatchel, *At the Water's Edge*, 175–79; Langley, *Inchon Landing*, 39–44; Whaley, *Practise to Deceive*, 139–42.

11. Carey, Allfree, and Cairns, *Warfare in the Ancient World*, 60–62.

12. Whaley, *Practise to Deceive*, 68–69; Marshall, *World War I*, 157.

13. Whaley, *Practise to Deceive*, 83–84.

14. Liddell Hart, *Scipio Africanus*, 56–63; Gabriel, *Scipio Africanus*, 118–23; Goldsworthy, *Fall of Carthage*, 279–85.

15. Dupuy and Dupuy, *Harper Encyclopedia of Military History*, 1040–42.

16. Marshall, *World War I*, 167–69.

17. Gilbert, *First World War*, 145, 217.

18. T. N. Dupuy, *Evolution of Weapons and Warfare*, 221–23; Whaley, *Practise to Deceive*, 15.

19. Frank, *Guadalcanal*; Crenshaw, *Battle of Tassafaronga*, 111–14, 203–4.

20. United States Naval Academy, *Naval Operations Analysis*, 292–99; Tidman, *Operations Evaluation Group*, 1–92.

21. McCue, *U-Boats in the Bay of Biscay*; Sternhill and Thorndike, *Antisubmarine Warfare in World War II*.

22. Liddell Hart, *Strategy*, 41–42; Prokopios, *Wars of Justinian*, 29–37.

23. Goldsworthy, *Fall of Carthage*, 181–90.

24. Lord, *Dawn's Early Light*, 90–94.

25. Liddell Hart, *Strategy*, 331.

26. Strassler, *Landmark Thucydides*, 240–44.

27. Carey, Allfree, and Cairns, *Warfare in the Ancient World*, 71–76.

28. Seward, *Hundred Years War*, 86–93.

29. Betts, *Surprise Attack*, 28–34; Cohen and Gooch, *Military Misfortunes*, 197–230.

30. Hughes-Wilson, *Military Intelligence Blunders*, 165–217; Karnow, *Vietnam*, 515–66; Latimer, *Deception in War*, 287–90; Nagl, *Learning to Eat Soup with a Knife*, 167–68; Hammes, *Sling and the Stone*, 65.

31. Pokrant, *Desert Storm at Sea*, 151–53.

32. Pokrant, *Desert Storm at Sea*, 152–53.

33. Liddell Hart, *Strategy*, 326–27.

34. Liddell Hart, *Strategy*, 274.

35. Betts, *Surprise Attack*, 123; Central Intelligence Agency, *Deception Maxims*.

36. Betts, *Surprise Attack*, 103.

37. Central Intelligence Agency, *Deception Maxims*, 5–15.

38. Blaber, *Mission, the Men, and Me*, 58–59.

39. National Commission on Terrorist Attacks upon the United States, *9/11 Commission Report*, 339.

40. Clancy, *Debt of Honor*.

10. Deception

1. Carey, Allfree, and Cairns, *Warfare in the Ancient World*, 16–22.

2. Central Intelligence Agency, *Deception Maxims*, 40.

3. Dobbins, "Expectation and Observation."

4. Latimer, *Deception in War*, 54, 60–62.

5. Tuchman, *March of Folly*, 35–49.

6. Kagan, *Fall of the Athenian Empire*, 385–94.

7. Latimer, *Deception in War*, 132–44.

8. Central Intelligence Agency, *Deception Maxims*, 10–12.

9. Latimer, *Deception in War*, 132–44; Betts, *Surprise Attack*, 36–42; Stolfi, "Barbarossa"; Whaley, *Practise to Deceive*, 211–14.

10. Latimer, *Deception in War*, 139.

11. Latimer, *Deception in War*, 132–44; Betts, *Surprise Attack*, 36–42; Stolfi, "Barbarossa."

12. Latimer, *Deception in War*, 144; Betts, *Surprise Attack*, 36–42; Stolfi, "Barbarossa."

13. Latimer, *Deception in War*, 85–88; Betts, *Surprise Attack*, 68–80; Amos, "Deception and the 1973 Middle East War."

14. Foote, *Fredericksburg to Meridian*, 685–87.

15. Latimer, *Deception in War*, 80–85.

16. Latimer, *Deception in War*, 205–38; Hunt, "Eyewitness Report of the Fortitude Deception"; Hesketh, "Excerpt from Fortitude."

17. Pokrant, *Desert Storm at Sea*, 117–26.

18. Pokrant, *Desert Shield at Sea*; Pokrant, *Desert Storm at Sea*.

19. Smith, *Utility of Force*, 289–90.

20. Atkinson, *Untold Story of the Persian Gulf War*, 330–33; Latimer, *Deception in War*, 296–302.

21. Pokrant, *Desert Storm at Sea*, 174–79.

22. Liddell Hart, *Strategy*, 48–49.

23. Lewin, *Ultra Goes to War*, 304.

24. Foote, *Fort Sumter to Perryville*, 375–86.

25. Foote, *Fort Sumter to Perryville*, 396–410; Latimer, *Deception in War*, 27–29.

26. Latimer, *Deception in War*, 113–19.

27. Lehman, *Oceans Ventured*, 70–89; Solomon, "Maritime Deception and Concealment," 106–7.

28. Gordon and Trainor, *Cobra II*, 74–76.

29. Creasy, *Decisive Battles*, 178–79, 194–95.

30. Carey, Allfree, and Cairns, *Warfare in the Ancient World*, 47.

31. Foote, *Fredericksburg to Meridian*, 545–51; Hess, *Pickett's Charge*, 150–51.

32. Latimer, *Deception in War*, 96–99; Whaley, *Practise to Deceive*, 116–25.

33. Liddell Hart, *Scipio Africanus*, 130–37; Gabriel, *Scipio Africanus*, 163–65; Goldsworthy, *Fall of Carthage*, 292–94.

34. Whaley, *Turnabout and Deception*, part 2, 20–21.

35. Anderson, *By Sea and By River*, 142.

36. Latimer, *Deception in War*, 167.

37. Whaley, *Turnabout and Deception*, part 1, 31–32.

38. Latimer, *Deception in War*, 276–77.

39. Latimer, *Deception in War*, 283.

40. Whaley, *Turnabout and Deception*, part 1, 30–31.

41. Whaley, *Practise to Deceive*, 104–5.

42. Central Intelligence Agency, *Deception Maxims*, 11–14.

43. Cordingly, *Cochrane*, 304–5.

44. Foote, *Fredericksburg to Meridian*, 181–86.

45. Latimer, *Deception in War*, 31.

46. Pope, *Battle of the River Platte*, 191–229; van der Vat, *Atlantic Campaign*, 93–96.

47. Latimer, *Deception in War*, 44.

48. Layton, *And I Was There*, 410–22.

49. Lewin, *Ultra Goes to War*, 301–22; Keegan, *Intelligence in War*, 289–92; Hunt, "Eyewitness Report of the Fortitude Deception"; Hesketh, "Excerpt from Fortitude."

50. Latimer, *Deception in War*, 60.

51. Whaley, *Turnabout and Deception*, part 2, 82.

52. Lewin, *Ultra Goes to War*, 306–7.

53. Heuer, "Cognitive Factors in Deception and Counterdeception," 35.

54. Latimer, *Deception in War*, 287–90; Hughes-Wilson, *Military Intelligence Blunders*, 165–217; Karnow, *Vietnam*, 515–66.

55. Holm, *25 Cognitive Biases*, ch. 23–24; Tavris and Aronson, *Mistakes Were Made*, 17–20.

56. Holm, *25 Cognitive Biases*, ch. 23.

57. Heuer, "Cognitive Factors in Deception and Counterdeception," 53.

58. *100 Most Popular Cognitive Biases*, 2018, location 1066.

59. Tavris and Aronson, *Mistakes Were Made*, 11–39.

60. Tavris and Aronson, *Mistakes Were Made*, 11–39.

61. Heuer, "Cognitive Factors in Deception and Counterdeception," 42.

62. Latimer, *Deception in War*, 234–35.

63. Whaley, *Practise to Deceive*, 93–96.

64. Central Intelligence Agency, *Deception Maxims*, 42–43.

65. Hunt, "Eyewitness Report of the Fortitude Deception," 228–29.

66. Central Intelligence Agency, *Deception Maxims*, 32–33.

67. Latimer, *Deception in War*, 188.

68. Central Intelligence Agency, *Deception Maxims*, 46.

69. Central Intelligence Agency, *Deception Maxims*, 16–20.

70. Heuer, "Cognitive Factors in Deception and Counterdeception," 40–41, 66.

71. Steele, *Nimitz Graybook*, 542–44.

72. Potter, *Nimitz*, 79–83.

73. Heuer, "Cognitive Factors in Deception and Counterdeception," 51–53, 63, 66.

74. *100 Most Popular Cognitive Biases*, 2018, locations 33333–33467.

75. Holm, *25 Cognitive Biases*, ch. 25.

11. Simplicity

1. Joint Chiefs of Staff, *Joint Publication 3–0*, 2018, A:3.

2. NATO Standardization Office, *Allied Joint Doctrine*, 2017, 1:14.

3. Amidror, *Winning Counterinsurgency War*, 30–33.

4. Murray and Millett, *War to be Won*, 420–25.

5. Brainy Quotes, 2019, https://www.brainyquote.com/quotes/albert_einstein_103652.

6. Parshall and Tully, *Shattered Sword*, 107–12.

7. Parshall and Tully, *Shattered Sword*, 146–50; Isom, *Midway Inquest*, 113–15.

8. Parshall and Tully, *Shattered Sword*, 146–50.

9. Parshall and Tully, *Shattered Sword*, 159–61, 175, 183.

10. Parshall and Tully, *Shattered Sword*, 400–402; Tully and Yu, "Question of Estimates." The Japanese search plan implicitly assumed a static target. If the object of a search is moving, the search plan needs to be more complex to achieve a reasonable probability of detection.

11. Marshall, *World War I*, 138–41, 225–27, 344.

12. Mitchell and Creasy, *Twenty Decisive Battles*, 272–75.

13. Fallwell, "Principles of War," 59.

14. Vego, *Battle for Leyte*, 346–51.

15. Amos, "Deception and the 1973 Middle East War," 329.

16. Blaber, *Mission, the Men, and Me,* 228–29, 236–37.

17. Symonds, *Decision at Sea,* 7–19.

18. Miller, *Broadsides,* 365.

19. Cornwell, *Waterloo,* 146, 173.

20. Aura Quotes, 2017, http://auraquotes.com/Keyword/Military-Leadership-Quotes/1198.

21. David, *Military Blunders,* 16.

22. David, *Military Blunders,* 13–24.

23. Bowers, *Chickamauga and Chattanooga,* 113–38.

24. Vego, *Battle for Leyte,* 260–61.

25. Frank, *Guadalcanal,* 298–99.

26. Cook, *Battle of Cape Esperance,* 38–39.

27. Frank, *Guadalcanal,* 292–312; Cook, *Battle of Cape Esperance.* Morison's account of the battle differs in several places from Frank's account. Morison, *Two-Ocean War,* 147–71.

28. Dull, *Battle History of the Imperial Japanese Navy,* 225–30.

29. Frank, *Guadalcanal,* 292–312.

12. Other American Principles

1. Joint Chiefs of Staff, *Joint Publication 3–0,* 2018, A:2.

2. UK Ministry of Defence, *Joint Doctrine Publication 0–01* (2014), 30–31.

3. Leonhard, *Principles of War,* 124–28.

4. Liddell Hart, *Strategy,* 329.

5. Jomini, *Art of War,* 329.

6. Vego, "Major Convoy Operation to Malta," 141, 144, 148.

7. Joint Chiefs of Staff, *Joint Publication 3–0,* 2018, A:3.

8. UK Ministry of Defence, *Joint Doctrine Publication 0–01* (2014), 30–31.

9. Joint Chiefs of Staff, *Joint Publication 3–0,* 2018, A:2.

10. Goldsworthy, *Fall of Carthage,* 238–43; Mitchell and Creasy, *Twenty Decisive Battles,* 56–76.

11. Joint Chiefs of Staff, *Joint Publication 3–0,* 2018, A:1.

12. Joint Chiefs of Staff, *Joint Publication 3–0,* 2018, A:3–4.

13. Joint Chiefs of Staff, *Joint Publication 3–0,* 2018, A:4.

14. Joint Chiefs of Staff, *Joint Publication 3–0,* 2018, A:4.

13. Other British Principles

1. UK Ministry of Defence, *Joint Doctrine Publication 0–01* (2014), 30–31.

2. Frost, *Growing Imperative*; Van Avery, "12 New Principles of Warfare."

3. Frost, *Growing Imperative,* 16.

4. Frost, *Growing Imperative,* 16–17.

5. Frost, *Growing Imperative.*

6. Murray and Millett, *War to be Won,* 356.

7. Hone, "Replacing Battleships with Aircraft Carriers."

8. Hodge, "Key to Midway."

9. Catton, *Grant Moves South,* 367–485.

10. Hammes, *Sling and the Stone,* 174–80.

11. Thomas A. Edison Quotes, 2018, https://www.goodreads.com/author/quotes /3091287.Thomas_A_Edison.

12. Goldsworthy, *Fall of Carthage*, 38–39, 92, 126–27, 215–19, 315; Dodge, *Hannibal*, 386–90; Livy, *History of Rome*, book 22.

13. Kelly, *Never Surrender*, 229–31; Lukacs, *Five Days in London*, 123–28, 146–66.

14. Lukacs, *Five Days in London*, 3.

15. Gilbert, *Churchill*, 651.

16. Gilbert, *Churchill*, 656.

17. UK Ministry of Defence, *Joint Doctrine Publication 0–01* (2014), 30–31

18. van Creveld, *Supplying War*, 181–201.

19. van Creveld, *Supplying War*, 172–80.

20. van Creveld, *Supplying War*, 204–23.

21. UK Ministry of Defence, *Joint Doctrine Publication 0–01* (2014), 30.

22. Goldsworthy, *Fall of Carthage*, 198–206.

23. Seward, *Hundred Years War*, 61–68.

24. Symonds, *Decision at Sea*, 33.

25. Leonhard, *Principles of War*, 214.

26. Fuchida and Okumiya, *Midway*, 209–11.

14. Know Yourself

1. Freeman, *Lee*, 292.

2. Sears, *Gettysburg*, 45, 52, 227–29.

3. Foote, *Fort Sumter to Perryville*, 471.

4. Foote, *Fredericksburg to Meridian*, 441.

5. Sears, *Gettysburg*, 55, 379–82, 407.

6. Foote, *Red River to Appomattox*, 417.

7. Cornwell, *Waterloo*, 26–27, 145–47, 174–75; Mitchell and Creasy, *Twenty Decisive Battles*, 228–32.

8. Fleming, *Story of Cowpens*; Bearss, *Battle of Cowpens*; Leckie, *George Washington's War*, 600–602.

9. Foote, *Fredericksburg to Meridian*, 529–30.

10. Sears, *Gettysburg*, 151.

11. Spector, *Eagle against the Sun*, 308.

12. Hammes, *Sling and the Stone*, 44–50.

13. Frank, *Guadalcanal*, 294; Morison, *Struggle for Guadalcanal*, 154.

14. Grace, *Naval Battle of Guadalcanal*, 43–90.

15. Massie, *Castles of Steel*, 609–11.

16. Gordon and Trainor, *Generals' War*, 400–432.

17. Pokrant, *Desert Storm at Sea*, 213–14.

15. Know Your Enemy

1. McKay and McKay, *Libraries of Famous Men*, https://www.artofmanliness.com /articles/george-patton-reading-list/.

2. Goldsworthy, *Fall of Carthage*, 38–39, 92, 218.

3. Goldsworthy, *Fall of Carthage*, 92, 126–27, 215–19, 315; Livy, *History of Rome*, book 22.

4. Dodge, *Hannibal*, 386–90.

5. Betts, *Surprise Attack*, 72.

6. Murray and Millett, *War to be Won*, 356.

7. Gordon and Trainor, *Cobra II*, 572–73.

8. Hammes, *Sling and the Stone*, 148–53.

9. Sun Tzu, *Art of War*, 77. Sun Tzu said the second-best strategy is to disrupt the enemy's alliances. Sun Tzu, *Art of War*, 78. You must be aware of your enemy's alliances and the tensions between the allies in order to attack the alliance and split the allies.

10. Whaley, *Practise to Deceive*, 115–16.

11. Hattendorf, *Evolution of the U.S. Navy's Maritime Strategy*; Lehman, *Oceans Ventured*; Ford and Rosenberg, *The Admirals' Advantage*, 77–108. James McConnell, Robert Weinland, and Bradford Dismukes of the Center for Naval Analyses deduced the Soviet strategy based primarily on a careful analysis of unclassified literature. For many years, however, these analysts were "prophets without honor" because the intelligence community disagreed with their conclusion. Years later, highly classified sources confirmed the analysts' conclusions. Hattendorf, *Evolution of the U.S. Navy's Maritime Strategy*, 23–36; Ford and Rosenberg, *The Admirals' Advantage*, 79–81, 94.

12. Goldsworthy, *Fall of Carthage*, 184–85.

13. Fleming, *Story of Cowpens*; Bearss, *Battle of Cowpens*.

14. Freeman, *Lee*, 268.

15. Freeman, *Lee*, 285.

16. Foote, *Fort Sumter to Perryville*, 465.

17. Foote, *Fredericksburg to Meridian*, 665–71.

18. Whaley, *Practise to Deceive*, 91, 202.

19. Vego, *Battle for Leyte*, 347.

20. Betts, *Surprise Attack*, 68.

21. Gilbert, *First World War*, 306.

22. Murray and Millett, *War to be Won*, 135–36.

23. Caesar, *Conquest of Gaul*, 61.

24. Warner and Warner, *Disaster in the Pacific*, 55.

25. Carey, Allfree, and Cairns, *Warfare in the Ancient World*, 96.

26. Liddell Hart, *Strategy*, 52–53.

27. Liddell Hart, *Strategy*, 50–53.

28. Marshall, *World War I*, 341, 349–58.

29. Cohen and Gooch, *Military Misfortunes*, 216–17.

30. McCue, *U-Boats in the Bay of Biscay*; Sternhill and Thorndike, *Antisubmarine Warfare in World War II*.

31. Murray and Millett, *War to be Won*, 85–87.

32. Caravaggio, "Attack at Taranto."

33. There are a few situations in which one might not want to deny information to a possible enemy. For example, in peacetime, one might want to deter a potential enemy by signaling strength, as the United States frequently did during the Cold War, especially with the Maritime Strategy. In such cases, commanders must balance the need to advertise a capability to deter war with the need to deceive to conceal capabilities that would

be vital if war would break out. Hattendorf, *Evolution of the U.S. Navy's Maritime Strategy*, 54; Solomon, "Maritime Deception and Concealment," 106–7.

16. Environment

1. Kautilya, *Arthashastra*, 492.

2. Nagl, *Learning to Eat Soup with a Knife*, 16.

3. Dupuy and Dupuy, *Harper Encyclopedia of Military History*, 30.

4. Liddell Hart, *Scipio Africanus*, 27–37; Polybius, *Rise of the Roman Empire*, book 10.8–16; Gabriel, *Scipio Africanu*, 85–98; Goldsworthy, *Fall of Carthage*, 271–79.

5. Seward, *Hundred Years War*, 60.

6. Prokopios, *Wars of Justinian*, 272–76 [book 5, chs. 8–10].

7. Yadin, "Strategical Analysis," 398.

8. Some claim that one could wade across the creek in most places, but that was probably not true on the day of the battle. Sears, *Landscape Turned Red*, 439–40.

9. Sears, *Landscape Turned Red*, 187–91, 260, 287–96, 387–92, 439–40; Foote, *Fort Sumter to Perryville*, 695–700.

10. Liddell Hart, *Strategy*, 25; Goldsworthy, *Fall of Carthage*, 184.

11. Ketchum, *Saratoga*, 239–51.

12. Cohen and Gooch, *Military Misfortunes*, 220.

13. Gordon and Trainor, *Generals' War*, 146–47, 384–85.

14. Goldsworthy, *Fall of Carthage*, 176–81.

15. Creasy, *Decisive Battles*, 363.

16. Foote, *Fort Sumter to Perryville*, 181–91.

17. Foote, *Red River to Appomattox*, 345.

18. Latimer, *Deception in War*, 231–32.

19. Mattingly, *Armada*, 275–300.

20. Garfield, *Thousand-Mile War*, 62.

21. Seward, *Hundred Years War*, 162–69; Dupuy and Dupuy, *Harper Encyclopedia of Military History*, 445–50; Tuchman, *Distant Mirror*, 583–85.

22. Foote, *Fredericksburg to Meridian*, 127–30.

23. Black and Couture, "Dynamic. Distributed. Deadly."

24. Dupuy and Dupuy, *Harper Encyclopedia of Military History*, 615–16.

25. Alexander, *Utmost Savagery*, 79–84; McKiernan, "Tarawa: The Tide that Failed."

26. Latimer, *Deception in War*, 231.

27. Jellicoe, *Jutland: The Unfinished Battle*, 174–90; Hough, *Great War at Sea*, 220–32, 246–56; Massie, *Castles of Steel*, 621.

28. Blaber, *Mission, the Men, and Me*, 228–29, 236–37. Later, when a general far from the scene ordered new teams into the area, he refused to listen to the need for the other teams to familiarize themselves with the area and to acclimatize to the conditions. Blaber, *Mission, the Men, and Me*, 273–74.

Epilogue

1. Grant, "Operational Art."

2. Grant, "Operational Art."

Adams, John A. *If Mahan Ran the Great Pacific War: An Analysis of World War II Naval Strategy*. Bloomington: Indiana University Press, 2008.

Alexander, Joseph H. *Storm Landings: Epic Amphibious Battles in the Central Pacific*. Annapolis MD: Naval Institute Press, 1997.

———. *Utmost Savagery: The Three Days of Tarawa*. New York: Ballantine, 1997.

Alger, John I. *The Quest for Victory: The History of the Principles of War*. Westport CT: Greenwood, 1982.

Amidror, Yaakov. *Winning Counterinsurgency War: The Israeli Experience*. Jerusalem: Jerusalem Center for Public Affairs, 2010.

Amos, John. "Deception and the 1973 Middle East War." In *Strategic Military Deception*, edited by Donald C. Daniel and Katherine L. Herbig, 317–34. New York: Pergamon, 1982.

Anderson, Bern. *By Sea and By River: The Naval History of the Civil War*. New York: Da Capo, 1989.

Armée de Terre. *Employment of Land Forces in Joint Operations*. Paris: Armée de Terre, 2015.

———. *General Tactics: FT-02 (ENG)*. Paris: Armée de Terre, 2010.

Arthur, Stanley R., and Marvin Pokrant. "Desert Storm at Sea." *U.S. Naval Institute Proceedings* 117, no. 5 (May 1991): 82–87.

Assistant Chief of Staff for Intelligence. *The Soviet Battlefield Development Plan*. Washington DC: Office of the Assistant Chief of Staff for Intelligence, 1982.

Atkinson, Rick. *Crusade: The Untold Story of the Persian Gulf War*. Boston: Houghton Mifflin, 1993.

Australian Army. *The Fundamentals of Land Power: Land Warfare Doctrine 1*. N.p.: Australian Army, 2017.

Australian Department of Defence. *Foundations of Australian Military Doctrine (ADDP-D)*. 3rd ed. Canberra: Australian Department of Defence, 2012.

Bartlett, Merrill L., ed. *Assault from the Sea: Essays on the History of Amphibious Warfare*. Annapolis MD: Naval Institute Press, 1993.

Baxter, William. *Soviet AirLand Battle Tactics*. Novato CA: Presidio, 1986.

Bearss, Edwin C. *The Battle of Cowpens: A Documented Narrative and Troop Movement Maps*. Johnson City TN: Overmountain, 1996.

Betts, Richard K. *Surprise Attack: Lessons for Defense Planning*. Washington DC: Brookings Institution, 1982.

Blaber, Pete. *The Mission, the Men, and Me: Lessons from a Former Delta Force Commander*. New York: Berkley Caliber, 2008.

Black, Eugene H., and Jennifer S. Couture. "Dynamic. Distributed. Deadly." *U.S. Naval Institute Proceedings* 146, no. 2 (February 2020): 24–29.

Bowers, John. *Chickamauga and Chattanooga: The Battles That Doomed the Confederacy.* New York: Avon, 1995.

Brown, Ian T. *A New Conception of Warfare: John Boyd, the U.S. Marine Corps, and Maneuver Warfare.* Quantico VA: Marine Corps University Press, 2018.

Buell, Thomas B. *The Quiet Warrior: A Biography of Admiral Raymond A. Spruance.* Annapolis MD: Naval Institute Press, 2009.

Burnod. "Military Maxims of Napoleon." In *Roots of Strategy: The 5 Greatest Military Classics of All Time,* edited by Thomas R. Phillips, 401–41. Harrisburg PA: Stackpole, 1985.

Bush, George H. W. "National Security Directive 54: Responding to Iraqi Aggression in the Gulf." Washington DC: The White House, January 15, 1991.

Caesar, Julius. *The Conquest of Gaul.* Translated by S. A. Handford. Revised with a new introduction by Jane F. Gardner. London: Penguin, 1982.

Canadian Forces Experimentation Centre. *Canadian Military Doctrine, CFJP 01.* Ottawa: Department of National Defence, 2009.

Caravaggio, Angelo N. "The Attack at Taranto: Tactical Success, Operational Failure." *Naval War College Review* 59, no. 3 (2006): 103–27.

Carey, Brian Todd, Joshua B. Allfree, and John Cairns. *Warfare in the Ancient World.* Barnsley, South Yorkshire: Pen & Sword Military, 2007.

Carroll, Ward. "This Botched Air Strike on Lebanon Changed Naval Aviation Forever." *wearethemighty* (blog). August 10, 2015. https://www.wearethemighty.com/articles/this-botched-air-strike-on-lebanon-changed-naval-aviation-forever.

Catton, Bruce. *Grant Moves South.* With maps by Samuel H. Bryant. Boston: Little, Brown, 1960.

———. *Grant Takes Command.* With maps by Samuel H. Bryant. Boston: Little, Brown, 1969.

———. *A Stillness at Appomattox.* New York: Pocket, 1958.

———. *This Hallowed Ground: The Story of the Union Side of the Civil War.* New York: Pocket, 1961.

Central Intelligence Agency. *Deception Maxims: Fact and Folklore.* Deception Research Program. Washington DC: Central Intelligence Agency, 1981.

Chase, Michael S., and Evan Medeiros. "China's Evolving Nuclear Calculus: Modernization and Doctrinal Debate." In *China's Revolution in Doctrinal Affairs: Emerging Trends in the Operational Art of the Chinese People's Liberation Army,* edited by James Mulvenon and David Finkelstein, 119–57. Alexandria VA: CNA, 2005.

CinC U.S. Pacific Fleet. "Operation Plan No. 29–42." *Midway1942.com.* 1942. http://www.midway1942.com/docs/usn_doc_00.shtml, accessed March 16, 2019.

Clancy, Tom. *Debt of Honor.* New York: G. P. Putnam's Sons, 1994.

Clapp, Michael, and Ewen Southby-Tailyour. *Amphibious Assault Falklands: The Battle of San Carlos Water.* Annapolis MD: Naval Institute Press, 1996.

Clausewitz, Carl von. *On War.* Edited and translated by Michael Howard and Peter Paret. Princeton NJ: Princeton University Press, 1984.

————. *Principles of War*. Edited and translated by Hans W. Gatze. Unabridged. Mineola NY: Dover, 2003.

Cliff, Roger, Mark Burles, Michael S. Chase, Derek Eaton, and Kevin L. Pollpeter. *Entering the Dragon's Lair: Chinese Antiaccess Strategies and Their Implications for the United States*. Santa Monica CA: RAND Corporation, 2007.

Cliff, Roger, John Fei, Jeff Hagen, Elizabeth Hague, Eric Heginbotham, and John Stillion. *Shaking the Heavens and Splitting the Earth: Chinese Air Force Employment Concepts in the 21st Century*. Santa Monica CA: RAND Corporation, 2011.

Clubb, Timothy L. *Cactus Airpower at Guadalcanal*. Fort Leavenworth KS: U.S. Army Command and General Staff College, 1996.

Cohen, Eliot A., and John Gooch. *Military Misfortunes: The Anatomy of Failure in War*. New York: Vintage, 1991.

Cook, Charles. *The Battle of Cape Esperance: Encounter at Guadalcanal*. Annapolis MD: Naval Institute Press, 1992.

Corbett, Julian S. *Principles of Maritime Strategy*. Mineola NY: Dover, 2004.

Cordingly, David. *Cochrane: The Real Master and Commander*. New York: Bloomsbury, 2008.

Cornwell, Bernard. *Waterloo: The History of Four Days, Three Armies, and Three Battles*. New York: HarperCollins, 2015.

Creasy, Edward Shepherd. *Decisive Battles of the World*. Rev. ed. New York: Colonial, 1899.

Crenshaw, Russell S. *The Battle of Tassafaronga*. Baltimore MD: Nautical & Aviation, 1995.

Crowl, Philip A. "Alfred Thayer Mahan: The Naval Historian." In *Makers of Modern Strategy: From Machiavelli to the Nuclear Age*, edited by Peter Paret with the collaboration of Gordon A. Craig and Felix Gilbert, 444–77. Princeton NJ: Princeton University Press, 1986.

Daniel, Donald C., and Katherine L. Herbig, eds. *Strategic Military Deception*. New York: Pergamon, 1982.

David, Saul. *Military Blunders: The How and Why of Military Failure*. New York: Carroll & Graf, 1998.

de Gaulle, Charles. *War Memoirs*, Vol. 1, *The Call to Honour 1940–1942*. Translated by Jonathan Griffin. London: Collins, 1955.

de Saxe, Maurice. 1732. "My Reveries upon the Art of War." In *Roots of Strategy: The 5 Greatest Military Classics of All Time*, edited by Thomas R. Phillips, 177–300. Harrisburg PA: Stackpole, 1985.

Dobbins, Thomas A. "Expectation and Observation." *Sky & Telescope* 137, no. 3 (2019): 52–53.

Dodge, Theodore Ayrault. 1891. *Hannibal: A History of the Art of War among the Carthaginians and Romans down to the Battle of Pydna, 168 B.C., with a Detailed Account of the Second Punic War*. Tales End, 2012. Kindle.

Dull, Paul S. *A Battle History of the Imperial Japanese Navy (1941–1945)*. Annapolis MD: Naval Institute Press, 1978.

Dupuy, R. Ernest, and Trevor N. Dupuy. *The Harper Encyclopedia of Military History: From 3500 B.C. to the Present*. 4th ed. New York: HarperCollins, 1993.

Dupuy, Trevor N. *The Evolution of Weapons and Warfare*. Fairfax VA: Hero, 1984.

Eisenhower, David. *Eisenhower: At War: 1943–1945*. New York: Vintage, 1987.

Fallwell, Marshall L. "The Principles of War and the Solution of Military Problems." *Military Review* 35, no. 2 (1955): 48–62.

Ferrill, Arther. *The Origins of War: From the Stone Age to Alexander the Great*. New York: Thames & Hudson, 1986.

Fleming, Peter. *Operation Sea Lion: The Projected Invasion of England in 1940—An Account of the German Preparations and the British Countermeasures*. New York: Simon & Shuster, 1957.

Fleming, Thomas J. *"Downright Fighting": The Story of Cowpens*. Washington DC: National Park Service, 1988.

Foch, Ferdinand. *The Principles of War*. Translated by Hilaire Belloc. New York: Henry Holt, 1903. Digitized by the Internet Archive in 2011.

Foote, Shelby. *The Civil War: A Narrative: Fort Sumter to Perryville*. New York: Vintage, 1986.

———. *The Civil War: A Narrative: Fredericksburg to Meridian*. New York: Vintage, 1986.

———. *The Civil War: A Narrative: Red River to Appomattox*. New York: Vintage, 1986.

Ford, Christopher A., with David A. Rosenberg and assistance from Randy Carol Goguen. *The Admirals' Advantage: U.S. Navy Operational Intelligence in World War II and the Cold War*. Annapolis MD: Naval Institute Press, 2014.

Foster, Simon. *Hit the Beach: Amphibious Warfare from the Plains of Abraham to San Carlos Water*. London: Arms & Armour, 1995.

Frank, Richard B. *Guadalcanal: The Definitive Account of the Landmark Battle*. New York: Penguin, 1992.

Fravel, M. Taylor. "The Evolution of China's Military Strategy: Comparing the 1987 and 1999 Editions of Zhanluexue." In *China's Revolution in Doctrinal Affairs: Emerging Trends in the Operational Art of the Chinese People's Liberation Army*, edited by James Mulvenon and David Finkelstein, 79–99. Alexandria VA: CNA, 2005.

Frederick the Great. "The Instruction of Frederick the Great for His Generals, 1747." In *Roots of Strategy: The 5 Greatest Military Classics of All Time*, edited by Thomas R. Phillips, 301–400. Harrisburg PA: Stackpole, 1985.

Freeman, Douglas Southall (abridgement by Richard Harwell). *Lee: An Abridgement in One Volume of the Four-volume R. E. Lee*. New York: Charles Scribner's Sons, 1961.

French Defence Staff. *Capstone Concept for Military Operations*. Paris: (FRA) Joint Centre for Concepts, Doctrine and Experimentation, 2013.

Frontinus, Sextus Julius. *The Strategemata*. Translated by Charles E. Bennet. Edited by Mary B. McElwain. Seattle WA: Praetorian, 2012. Kindle.

Frost, Robert S. *The Growing Imperative to Adopt "Flexibility" as an American Principle of War*. Carlisle PA: Strategic Studies Institute, U.S. Army War College, 1999.

Fuchida, Mitsuo, and Masatake Okumiya. *Midway: The Battle that Doomed Japan: The Japanese Navy's Story*. Edited by Clarke H. Kawakami and Roger Pineau. New York: Ballantine, 1958.

Fuller, J. F. C. *The Foundations of the Science of War*. London: Hutchinson, 1926.

———. *Memoirs of an Unconventional Soldier*. London: Ivor Nicholson & Watson, 1936.

Gabriel, Richard A. *Scipio Africanus: Rome's Greatest General*. Dulles VA: Potomac, 2008.

Garfield, Brian. *The Thousand-Mile War: World War II in Alaska and the Aleutians*. New York: Bantam, 1982.

Gatchel, Theodore L. *At the Water's Edge: Defending against the Modern Amphibious Assault*. Annapolis MD: Naval Institute Press, 1996.

Gilbert, Felix. "Machiavelli: The Renaisance of the Art of War." In *Makers of Modern Strategy: From Machiavelli to the Nuclear Age,* edited by Peter Paret with the collaboration of Gordon A. Craig and Felix Gilbert, 11–31. Princeton NJ: Princeton University Press, 1986.

Gilbert, Martin. *Churchill: A Life*. New York: Owl, 1991.

———. *The First World War: A Complete History*. New York: Owl, 1996.

Gingrich, Newt, William R. Forstchen, and Albert S. Hanser. *Gettysburg: A Novel of the Civil War*. New York: Thomas Dunne, 2003.

Glenn, Russell W. "No More Principles of War?" *Parameters, US Army War College Quarterly*. (Spring 1998): 48–66.

Goldsworthy, Adrian. *The Fall of Carthage: The Punic Wars 265–146 BC*. London: Phoenix, 2006.

———. *How Rome Fell: Death of a Superpower*. New Haven CT: Yale University Press, 2009.

Gordon, Michael R., and Bernard E. Trainor. *Cobra II: The Inside Story of the Invasion and Occupation of Iraq*. New York: Vintage, 2007.

———. *The Generals' War: The Inside Story of the Conflict in the Gulf*. Boston: Little, Brown, 1995.

Grace, James W. *The Naval Battle of Guadalcanal: Night Action, 13 November 1942*. Annapolis MD: Naval Institute Press, 1999.

Grant, Arthur V. "Operational Art and the Gettysburg Campaign." In *Historical Perspectives of the Operational Art*, edited by Michael D. Krause and R. Cody Phillips, 349–91. Washington DC: Center of Military History, United States Army, 2005.

Hammes, Thomas X. *The Sling and the Stone: On War in the 21st Century*. Minneapolis MN: Zenith, 2006.

Hattendorf, John B., *The Evolution of the U.S. Navy's Maritime Strategy, 1977–1986*. Newport RI: Naval War College Press, 2004.

Heather, Peter. *The Fall of the Roman Empire: A New History of Rome and the Barbarians*. Oxford: Oxford University Press, 2007.

Hesketh, Roger Fleetwood. "Excerpt from Fortitude: A History of Strategic Deception in North Western Europe, April 1943 to May 1945." In *Strategic Military Deception*, edited by Donald C. Daniel and Katherine L. Herbig, 233–42. New York: Pergamon, 1982.

Hess, Earl J. *Pickett's Charge—The Last Attack at Gettysburg*. Chapel Hill: University of North Carolina Press, 2001.

Heuer, Richards J., Jr. "Cognitive Factors in Deception and Counterdeception." In *Strategic Military Deception*, edited by Donald C. Daniel and Katherine L. Herbig, 31–69. New York: Pergamon, 1982.

Hodge, Carl Cavanagh. "The Key to Midway: Coral Sea and a Culture of Learning." *Naval War College Review* 68, no. 1 (2015): 119–27.

Holborn, Hajo. "The Prusso-German School: Moltke and the Rise of the General Staff." In *Makers of Modern Strategy: From Machiavelli to the Nuclear Age,* edited by Peter

Paret with the collaboration of Gordon A. Craig and Felix Gilbert, 281–95. Princeton NJ: Princeton University Press, 1986.

Holm, Charles. *The 25 Cognitive Biases: Uncovering The Myth of Rational Thinking*. Amazon Digital Services, 2015. Kindle.

Hone, Thomas C. "Replacing Battleships with Aircraft Carriers in the Pacific in World War II." *Naval War College Review* 66, no. 1 (2013): 56–76.

Hone, Trent. "U.S. Navy Surface Battle Doctrine and Victory in the Pacific." *Naval War College Review* 62, no. 1 (2009): 67–105.

Hough, Richard. *The Great War at Sea: 1914–1918*. New York: Oxford University Press, 1989.

Hughes, Wayne P., Jr. *Fleet Tactics: Theory and Practice*. Annapolis MD: Naval Institute Press, 1987.

———. *Fleet Tactics and Coastal Combat*. Annapolis MD: Naval Institute Press, 2000.

Hughes, Wayne P., Jr., and Robert P. Girrier, foreword by John Richardson. *Fleet Tactics: Theory and Naval Operations*. Annapolis MD: Naval Institute Press, 2018.

Hughes-Wilson, John. *Military Intelligence Blunders*. New York: Carroll & Graf, 1999.

Hunt, Barry D. "An Eyewitness Report of the Fortitude Deception: Editorial Introduction to R. F. Hesketh's Manuscript." In *Strategic Military Deception*, edited by Donald C. Daniel and Katherine L. Herbig, 224–32. New York: Pergamon, 1982.

Indian Air Force. *Basic Doctrine of the Indian Air Force 2012*. New Delhi: Indian Air Force, 2012.

Indian Army Training Command. *Indian Army Doctrine*. Shimla, India: Headquarters Army Training Command, 2004.

Isom, Dallas Woodbury. *Midway Inquest: Why the Japanese Lost the Battle of Midway*. Bloomington: Indiana University Press, 2007.

Jellicoe, Nicholas. *Jutland: The Unfinished Battle*. Annapolis MD: Naval Institute Press, 2016.

Joint Chiefs of Staff. *Joint Publication 1: Joint Warfare of the US Armed Forces*. Washington DC: United States Department of Defense, 1991.

———. *Joint Publication 3–0: Joint Operations*. Change 1. Washington DC: Joint Chiefs of Staff, 2018.

———. *Joint Publication 3–0: Joint Operations*. Change 1, 2008. Washington DC: Joint Chiefs of Staff, 2006.

———. *Joint Publication 3–07: Joint Doctrine for Military Operations Other than War*. Washington DC: Joint Chiefs of Staff, 1995.

Jomini, Antoine-Henri, de. *The Art of War*. 1862. Translated by G. H. Mendell and W. P. Craighill. Westport CT: Greenwood, 1971.

Kagan, Donald. *The Archidamian War*. Ithaca NY: Cornell University Press, 1990.

———. *The Fall of the Athenian Empire*. Ithaca NY: Cornell University Press, 1992.

———. *The Peace of Nicias and the Sicilian Expedition*. Ithaca NY: Cornell University Press, 1992.

———. *The Peloponnesian War*. New York: Penguin, 2004.

Karnow, Stanley. *Vietnam: A History*. New York: Penguin, 1986.

Kautilya. *Arthashastra*. Translated by R. Shamasastry. Bangalore: Government Press, 1915.

Keegan, John. *Intelligence in War: The Value—and Limitations—of What the Military Can Learn abut the Enemy*. New York: Vintage, 2004.

———. *The Price of Admiralty: The Evolution of Naval Warfare*. New York: Penguin, 1988.

Kelly, John. *Never Surrender: Winston Churchill and Britain's Decision to Fight Nazi Germany in the Fateful Summer of 1940*. New York: Scribner, 2016.

Ketchum, Robert M. *Saratoga: Turning Point of America's Revolutionary War*. New York: Owl, 1999.

Khaled bin Sultan, with Patrick Seale. *Desert Warrior: A Personal View of the Gulf War by the Joint Forces Commander*. New York: HarperCollins, 1995.

Kissinger, Henry. *Years of Upheaval*. Boston: Little, Brown, 1982.

Kitfield, James. *Prodigal Soldiers: How the Generation of Officers Born of Vietnam Revolutionized the American Style of War*. New York: Simon & Schuster, 1995.

Krause, Michael D., and R. Cody Phillips, eds. *Historical Perspectives of the Operational Art*. Washington DC: Center of Military History, United States Army, 2005.

Langley, Michael. *Inchon Landing: MacArthur's Last Triumph*. New York: Times, 1979.

Latimer, Jon. *Deception in War*. New York: Overlook, 2001.

Layton, Edwin T., with Roger Pineau and John Costello. *"And I Was There": Pearl Harbor and Midway—Breaking the Secrets*. New York: Quill, 1985.

Leckie, Robert. *George Washington's War: The Saga of the American Revolution*. New York: HarperPerenial, 1993.

Lehman, John, *Oceans Ventured: Winning the Cold War at Sea*. New York: W. W. Norton, 2018.

LeMay Center for Doctrine. *Air Force Doctrine Volume 1, Basic Doctrine*. Maxwell AFB AL: LeMay Center for Doctrine, 2015.

Leonhard, Robert R. *The Principles of War for the Information Age*. New York: Ballantine, 1998.

Lewin, Ronald. *Ultra Goes to War: The First Account of World War II's Greatest Secret, Based on Official Documents*. With a new foreword by Max Hastings. London: Grafton, 1988.

Liddell Hart, B. H. *Scipio Africanus: Greater than Napoleon*. London: Da Capo, 1926.

———. *Strategy*. 2nd rev. ed. New York: Meridian, 1991.

Livezey, William E. *Mahan on Sea Power*. 2nd ed. Norman: University of Oklahoma Press, 1986.

Livy (Titus Livius). *The History of Rome in Three Volumes*. Translated by D. Spillan. Halcyon, n.d. Kindle.

Lord, Walter. *The Dawn's Early Light*. New York: Dell, 1973.

Lukacs, John. *Five Days in London: May 1940*. New Haven CT: Yale University Press, 1999.

Lundstrom, John B. *The First Team: Pacific Naval Air Combat from Pearl Harbor to Midway*. Annapolis MD: Naval Institute Press, 2005.

MacDougall, P. L. *The Theory of War: Illustrated by Numerous Examples from Military History*. 2nd ed. London: Longman, Brown, Green, Langmans and Roberts, 1858.

Machiavelli, Niccolò. 1520. *The Art of War*. CreateSpace, 2016. Kindle.

Mahan, A. T. 1890. *The Influence of Sea Power upon History 1660–1783*. 5th ed. New York: Dover, 1987.

Mallick, P. K. *Principles of War: Time for Relook*. New Delhi: KW in association with Centre for Land Warfare Studies, 2009.

Manchester, William, and Paul Reid. *The Last Lion: Winston Spencer Churchill: Defender of the Realm 1940–1965*. New York: Little, Brown, 2012.

Mao Tse-Tung. *On Guerrilla Warfare*. Translated by Samuel B. Griffith II. Urbana: University of Illinois Press, 2000.

Margiotta, Franklin D., ed. *Brassey's Encyclopedia of Military History and Biography*. Washington DC: Brassey's, 2000.

Marshall, S. L. A. *World War I*. New York: American Heritage, 1985.

Massie, Robert K. *Castles of Steel: Britain, Germany, and the Winning of the Great War at Sea*. New York: Ballantine, 2003.

Mattingly, Garrett. *The Armada*. Boston: Houghton Mifflin, 1959.

McCue, Brian. *U-Boats in the Bay of Biscay: An Essay in Operations Analysis*. Washington DC: National Defense University Press, 1990.

McFeely, William S. *Grant: A Biography*. New York: Norton, 1981.

Mc Ivor, Anthony D., ed. *Rethinking the Principles of War*. Annapolis MD: Naval Institute Press, 2005.

McKay, Brett, and Kate McKay. *The Libraries of Famous Men: George S. Patton*. May 28, 2018. https://www.artofmanliness.com/articles/george-patton-reading-list/.

McKiernan, Patrick L. "Tarawa: The Tide that Failed." In *Assault from the Sea: Essays on the History of Amphibious Warfare*, edited by Merrill L. Bartlett, 210–18. Annapolis MD: Naval Institute Press, 1993.

McPherson, James M. *Battle Cry of Freedom: The Civil War Era*. New York: Oxford University Press, 1988.

Miller, John. *Memoirs of General Miller in the Service of the Republic of Peru*, Vol. 1. 2nd ed. Translated by Don Jose Maria de Torrijos. London: Longman, Rees, Orme, Brown, and Green, 1829.

Miller, Nathan. *Broadsides: The Age of Fighting Sail, 1775–1815*. New York: John Wiley & Sons, 2000.

Mitchell, Joseph B., and Edward S. Creasy. *Twenty Decisive Battles of the World*. Old Saybrook CT: Konecky & Konecky, 1964.

Morgan, John G., Anthony D. Mc Ivor, and the Secretary of the Navy's Action Team. "Rethinking the Principles of War." *U.S. Naval Institute Proceedings* 129, no. 10 (2003): 34–38.

Morison, Samuel Eliot. *The Struggle for Guadalcanal: August 1942—February 1943*, Vol. 5 of *History of United States Naval Operations in World War II*. Annapolis MD: Naval Institute Press, 2010.

———. *The Two-Ocean War: A Short History of the United States Navy in World War II*. New York: Ballantine, 1972.

Morse, Philip M., and George E. Kimball. *Methods of Operations Research*. 1st rev. ed. Cambridge MA, Massachusetts: MIT Press, 1951.

Mulvenon, James, and David Finkelstein, ed. *China's Revolution in Doctrinal Affairs: Emerging Trends in the Operational Art of the Chinese People's Liberation Army*. Alexandria VA: CNA, 2005.

Murray, Williamson, and Wayne Wei-siang Hsieh. *A Savage War: A Military History of the Civil War*. Princeton NJ: Princeton Unversity Press, 2016.

Murray, Williamson, and Allan R. Millett. *A War to be Won: Fighting the Second World War*. Cambridge MA: Belknap, 2000.

Myerson, Adam. "Calm after Desert Storm (Interview with Dick Cheney)." *Policy Review* (Hoover Institution). (Summer 1993).

Nagl, John A. *Learning to Eat Soup with a Knife: Counterinsurgency Lessons from Malaya and Vietnam*. Chicago: University of Chicago Press, 2005.

National Commission on Terrorist Attacks upon the United States. *The 9/11 Commission Report: Final Report of the National Commission on Terrorist Attacks Upon the United States*. New York: W. W. Norton, 2004.

NATO Standardization Office. *Allied Joint Doctrine: NATO Standard AJP-01*. Ed. E, version 1. NATO Standardization Office, 2017.

Naylor, Sean. *Not a Good Day to Die: The Untold Story of Operation Anaconda*. New York: Berkley Caliber, 2006.

New Zealand Defence Force. *New Zealand Defence Doctrine (NZDDP-D)*. 4th ed. Wellington: New Zealand Defence Force, 2017.

Nimitz, C. W. "Letter of Instructions." May 28, 1942. http://www.midway1942.com /docs/usn_doc_24.shtml.

100 Most Popular Cognitive Biases. OK Publishing, 2018. Kindle.

Paret, Peter, with the collaboration of Gordon A. Craig and Felix Gilbert, eds. *Makers of Modern Strategy: From Machiavelli to the Nuclear Age*. Princeton NJ: Princeton University Press, 1986.

Parshall, Jonathan B., and Anthony P. Tully. *Shattered Sword: The Untold Story of the Battle of Midway*. Washington DC: Potomac, 2005.

Phillips, Thomas R., ed. *Roots of Strategy: The 5 Greatest Military Classics of All Time*. Harrisburg PA: Stackpole, 1985.

Plutarch. *Lives of the Noble Grecians and Romans*. n.d. Kindle.

———. *Lives of the Noble Romans*. Edited by Edmund Fuller. Translated by John Dryden and revised by Arthur Hugh Clough. New York: Dell, 1966.

Pokrant, Marvin. *Desert Shield at Sea: What the Navy Really Did*. Westport CT: Greenwood, 1999.

———. *Desert Storm at Sea: What the Navy Really Did*. Westport CT: Greenwood, 1999.

Polybius. *The Rise of the Roman Empire*. Translated by Ian Scott-Kilvert. New York: Penguin, 1979.

Pope, Dudley. *The Battle of the River Platte*. New York: Avon, 1990.

Potter, E. B. *Nimitz*. Annapolis MD: Naval Institute Press, 2008.

Prange, Gordon W., with Donald M. Goldstein and Katherine V. Dillon. *Pearl Harbor: The Verdict of History*. New York: Penguin, 1991.

Prokopios. *The Wars of Justinian*. Translated by H. B. Dewing. Indianapolis IN: Hackett, 2014. Adobe PDF eBook.

Reznichenko, V. G., ed. *Tactics: A Soviet View (Soviet Military Thought No. 21)*. Translated by CIS Multilingual Section, Translation Bureau, Secretary of State Department, Ottawa, Canada. Moscow: Voennoye Izdatel'stvo, 1984. (Translation published under the auspices of the U.S. Air Force, 1987.)

Rothenburg, Gunther E. "Moltke, Schlieffen, and the Doctrine of Strategic Envelopment." In *Makers of Modern Strategy: From Machiavelli to the Nuclear Age*, edited by Peter Paret with the collaboration of Gordon A. Craig and Felix Gilbert, 296–325. Princeton NJ: Princeton University Press, 1986.

Royal Australian Air Force. *The Air Power Manual: Australian Air Publication AAP 1000-D*. Sixth Edition, reprinted with corrections May 2014. Canberra: Air Power Development Centre, Department of Defence, 2013.

Royal Australian Navy. *Australian Maritime Doctrine: RAN Doctrine-1*. 2nd ed. Canberra: Royal Australian Navy, 2010.

Rubel, Robert C. "Deconstructing Nimitz's Principle of Calculated Risk: Lessons for Today." *Naval War College Review* 68, no. 1 (Winter 2015): 30–45.

Sandberg, Carl. *Abraham Lincoln: The War Years*. Vol. 1. New York: Harcourt, Brace & World, 1939.

Savkin, V. Ye. *The Basic Principles of Operational Art and Tactics (A Soviet View)*. Translated and published under the auspices of the United States Air Force. Washington DC: U.S. Government Printing Office, 1979. (Originally published in Moscow in 1972.)

Schwarzkopf, H. Norman, written with Peter Petre. *The Autobiography: It Doesn't Take a Hero*. New York: Bantam, 1993.

Searle, Alaric. "Inter-Service Debate and the Origins of Strategic Culture: The 'Principles of War' in the British Armed Forces, 1919–1939." *War in History* 21, no. 1 (2014): 4–32.

Sears, Stephen W. *Chancellorsville*. Boston: Houghton Mifflin, 1996.

———. *Gettysburg*. Boston: Houghton Mifflin, 2003.

———. *Landscape Turned Red: The Battle of Antietam*. New York: Popular Library, 1985.

Seward, Desmond. *The Hundred Years War: The English in France, 1337–1453*. New York: Atheneum, 1982.

Slim, William. 1956. *Defeat into Victory: Battling Japan in Burma and India, 1942–1945*. New York: Cooper Square, 2000.

Smith, Rupert. *The Utility of Force: The Art of War in the Modern World*. New York: Vintage, 2008.

Solomon, Jonathan F. "Maritime Deception and Concealment: Concepts for Defeating Wide-Area Oceanic Surveillance-Reconnaissance-Strike Networks," *Naval War College Review* 66, no. 4 (2013): 87–116.

Spector, Ronald H. *Eagle against the Sun: The American War with Japan*. New York: Vintage, 1985.

Steele, James M. *Nimitz Graybook, Command Summary of Fleet Admiral Chester W. Nimitz, USN, 7 December 1941–31 August 1945*. Vol. 1. Newport RI: U.S. Naval War College, 2013.

Sternhill, Charles M., and Alan M. Thorndike. *Antisubmarine Warfare in World War II*. Washington DC: Operations Evaluation Group, Office of the Chief of Naval Operations, Navy Department, 1946.

Stolfi, Russell H. S. "Barbarossa: German Grand Deception and the Achievement of Strategic and Tactical Surprise against the Soviet Union, 1940–1941." In *Strategic Military Deception*, edited by Donald C. Daniel and Katherine L. Herbig, 195–223. New York: Pergamon, 1982.

——. *Soviet Naval Operational Art: The Soviet Approach to Naval War Fighting*. Monterey CA: Naval Postgraduate School, 1988.

Straker, David. "The 36 Stratagems, in Detail." *Changing Minds*, January 26, 2014. http://changingminds.org/disciplines/warfare/36_strategems/36_stratagems.htm.

Strassler, Robert B., ed. *The Landmark Thucydides: A Comprehensive Guide to the Peloponnesian War*. Translated by Richard Crawley. New York: Touchstone, 1998.

Sun Tzu. *The Art of War*. Translated by Samuel B. Griffin. Oxford: Oxford University Press, 1971.

Swartz, Peter M., with Karin Duggan. *U.S. Navy Capstone Strategies and Concepts: Introduction, Background and Analyses*. Alexandria VA: CNA, 2011.

Symonds, Craig L. *Decision at Sea: Five Naval Battles That Shaped American History*. New York: Oxford University Press, 2006.

Tavris, Carol, and Elliot Aronson. *Mistakes Were Made (but Not by Me): Why We Justify Foolish Beliefs, Bad Decisions, and Hurtful Acts*. Orlando FL: Harcourt, 2007.

Tidman, Keith R. *The Operations Evaluation Group: A History of Naval Operations Analysis*. Annapolis MD: Naval Institute Press, 1984.

Tuchman, Barbara W. *A Distant Mirror: The Calamitous 14th Century*. New York: Alfred A. Knopf, 1978.

——. *The March of Folly: From Troy to Vietnam*. New York: Ballantine, 1985.

Tully, Anthony, and Lu Yu. "A Question of Estimates: How Faulty Intelligence Drove Scouting at the Battle of Midway." *Naval War College Review* 68, no. 2 (2015): 85–99.

UK Ministry of Defence. *Joint Doctrine Publication 0–01: UK Defence Doctrine*. 5th Edition. Swindon, Wiltshire: UK Ministry of Defence, 2014.

U.S. Army. *Field Service Regulations 1923*. Washington DC: United States Army, 1924.

U.S. Department of Defense. *Conduct of the Persian Gulf War: Final Report to Congress*. Washington DC: Department of Defense, 1992.

U.S. Department of the Army. *Army Doctrine Reference Publication 3–0: Operations*. Washington DC: Department of the Army, 2016.

——. *Doctrine Primer: Army Doctrine Publication 1–01 (ADP 1–01)*. Washington DC: Department of the Army, 2014.

——. *Field Manual 100–1, The Army*. Washington DC: Department of the Army, 1994.

——. *Field Manual 100–5, Operations*. Washington DC: Department of the Army, 1993.

——. *Field Manual 3–0: Operations*. Washington DC: Department of the Army, 2008.

——. *Field Service Regulations: Operations (FM 100–5)*. Washington DC: Department of the Army, 1949.

——. *The Soviet Army: Operations and Tactics: Field Manual 100-2-1*. Washington DC: Department of the Army, 1984.

U.S. Department of the Navy. *Naval Warfare: Naval Doctrine Publication 1*. Washington DC: Department of the Navy, 2010.

U.S. Marine Corps. *Marine Corps Operations: Marine Corps Doctrinal Publication 1–0 (MCDP 1–0)*. Washington DC: United States Marine Corps, 2001.

———. *Marine Corps Operations: Marine Corps Doctrinal Publication 1–0 (MCDP 1–0)*. Washington DC: U.S. Marine Corps, 2011.

———. *Marine Corps Operations: Marine Corps Doctrinal Publication 1–0 (MCDP 1–0)*. Change 1. Washington DC: U.S. Marine Corps, 2017.

———. *Warfighting: Fleet Marine Force Manual 1*. Washington DC: U.S. Marine Corps, 1989.

———. *Warfighting: Marine Corps Doctrinal Publication 1 (MCDP 1)*. Washington DC: U.S. Marine Corps, 1997.

U.S. Naval Academy. *Naval Operations Analysis*. Annapolis MD: Naval Institute Press, 1972.

U.S. Naval Institute Proceedings 131, no. 10 (October 2005): 38–55.

Van Avery, Christopher E. "12 New Principles of Warfare." *Armed Forces Journal* (July 2007).

van Creveld, Martin. *Supplying War: Logistics from Wallenstein to Patton*. New York: Cambridge University Press, 1980.

van der Vat, Dan, with research by Christine van der Vat. *The Atlantic Campaign: World War II's Great Struggle at Sea*. New York: Harper & Row, 1988.

Vandergriff, Donald E. "How the Germans Defined Auftragstaktik: What Mission Command Is—and—Is Not." Edited by Dave Dilegge. *Small Wars Journal* (Small Wars Foundation): n.d.

Vegetius (Flavius Vegetius Renatus). "The Military Institutions of the Romans." In *Roots of Strategy: The 5 Greatest Military Classics of All Time*, edited by Thomas R. Phillips, 65-175. Harrisburg PS: Stackpole, 1985..

Vego, Milan. *The Battle for Leyte, 1944: Allied and Japanese Plans, Preparations, and Execution*. Annapolis MD: Naval Institute Press, 2006.

———. "Major Convoy Operation to Malta, 10–15 August 1942 (Operation Pedestal)." *Naval War College Review* 63, no. 1 (2010): 107–53.

War Office. *Field Service Regulations Part 1 Operations*. Repr., with amendments in 1912. London: His Majesty's Stationery Office, 1909.

———. *Field Service Regulations Vol. 2 Operations (Provisional)*. London: His Majesty's Stationery Office, 1920.

Warner, Denis, and Peggy Warner with Sadao Seno. *Disaster in the Pacific: New Light on the Battle of Savo Island*. Annapolis MD: Naval Institute Press, 1992.

Weeks, Mark. *Chess "Initiative"*. May 6, 2006. https://web.archive.org/web/20060506002514/http://chess.about.com/od/reference/g/bldefini.htm.

West, Joseph, trans. *Principles of War: A Translation from the Japanese*. 1969. Reprinted by Combat Studies Institute. Fort Leavenworth KS: U.S. Army Command and General Staff College, 2003.

Whaley, Barton. *Practise to Deceive: Learning Curves of Military Deception Planners*. Edited by Susan Stratton Aykroyd. Annapolis MD: Naval Institute Press, 2016.

———. *Turnabout and Deception: Crafting the Double-Cross and The Theory of Outs*. Edited by Susan Stratton Aykroyd. Annapolis MD: Naval Institute Press, 2016.

Williams, Max Ray. "Jomini, Antoine Henri." In *Brassey's Encyclopedia of Military History and Biography*, edited by Franklin D. Margiotta, 553–56. Washington DC: Brassey's, 2000.

Winnefeld, James A., and Dana J. Johnson. *Joint Air Operations: Pursuit of Unity in Command and Control, 1942–1991*. Annapolis MD: Naval Institute Press, 1993.

Yadin, Y. "A Strategical Analysis of Last Year's Battles." In *Strategy* by B. H. Liddell Hart. 2nd rev. ed. Appendix 2: 386–405. New York: Meridian, 1991.

Yang, Andrew N. D. "The Military of the People's Republic of China: Strategy and Implementation." *UNISCI Discussion Papers* (Research Unit on International Security and Cooperation [UNISCI]) 17 (May 2008): 187–201.

Index

cry-wolf syndrome, 209–10, 213
culture: American, 265; enemy, 262, 263–65, 288; Indian, 3
Cumberland (heavy cruiser), 202
cyberwarfare, 35, 156

d'Albret, Charles, 280
Dardanelles, 182, 192–93
Darius III (king), 173
Davis, Jefferson, 254
"Death Ride," 125
debarkation, prevention of, 77–78
Debt of Honor (Clancy), 178
deception, 33, 40, 87, 104, 118, 125, 155, 216, 234, 295, 297; camouflage and, 187; clues to, 210–12; countering, 209, 213–14, 288; economy of force and, 233; enhancing, 205–14; falling for, 180, 211; goal of, 180–81, 181–82, 185, 205, 288; importance of, 2, 3; information and, 210; initiative and, 119; knowledge and, 266; know your enemy and, 206; "message," 270; physical, 188; preconceptions and, 179; psychological factors and, 210; resisting, 242; surprise and, 158–59, 178, 179, 185, 188; susceptibility to, 206; tactical, 194; ultimate, 201–3; using, 204–5, 209–10
deception (principle of war), 179–214, 295, 297; defined, 180; described, 288
decision-making, 131, 155, 158; information and, 152–53; surprise and, 156
decisive time and place, 17, 32, 85–86. *See also* critical place and time
deconfliction, 133–34
defeating piecemeal (principle of war), described, 32
defense: ascendancy of, 109; countering, 159–60; extending, 70; initiative and, 121–22; offensive and, 16, 34, 115–17; surprise in, 158, 170–71; tactical, 122; weakening, 180
Defense Doctrine (United Kingdom), 28
definition of objectives (principle of war), 31
de Gaulle, Charles, 24, 25, 155, 175
Delaware (gunboat), 138
Delaware River crossing, 155
de Ligne, Field Marshal Prince, 5
Demosthenes, 67
depth and reserves (principle of war), 27
de Saxe, Maurice, 6, 7
Desert Storm at Sea (Pokrant), 133

destruction of enemy to full depth (principle of war), described, 38
Dien Bien Phu, 207
Dieppe raid, 96
Dill, John, 17
discipline, 6, 7, 39, 147; advice on, 2; training and, 4, 5, 252, 291
discrimination test, 54, 57–58, 134–35, 205, 248, 250, 251, 260, 261, 283
dispositions, 2, 4, 16, 143, 253, 262, 272, 293
Doctrine Primer (U.S. Army), 46
dominance, 36, 88; achieving, 104
Dönitz, Karl, 170
Doolittle Raid, 70
Dorchester Heights, fortifying, 114
Duchy of Milan, 139–40
Duncan (destroyer), 228, 229
Dunkirk, 247, 282

eccentricity (principle of war), described, 32
École supérieure de guerre, 24
economic use of combat power (principle of war), described, 32
economy of effort, 19, 231, 232, 240
economy of effort (principle of war), 31, 44; defined, 231–32; described, 29
economy of force, 50, 104, 231, 240; American definition of, 233; deception and, 233; practicing, 232, 234, 287; relative advantage and, 104
economy of force (principle of war), xv, 10, 11, 13, 14, 20, 24, 36, 50, 57, 296; candidate, 231–34; defined, 231; described, 18, 25, 233; security and, 232, 234
Edison, Thomas, 246
Edward, Prince ("Black Prince"), 173
Edward III (king), 92
effectiveness, 87, 105, 150, 200, 287; measuring, 61, 62; preservation of, 39
Einstein, Albert, 218
Eisenhower, Dwight D., 149, 250, 255, 279
El Alamein: deception at, 186–87, 209, 212; stalling at, 95
Elements of Military Art and Science (Halleck), 13
Emancipation Proclamation, 78–79
endurance, physical, 18, 55–56, 237
Enigma code, 191, 206
enterprise, 108, 146–47, 151, 159

Enterprise (aircraft carrier), 139
environment, 274–84, 287; constraints/opportunities in, 274; favorable, 99; knowledge of, 2, 104, 253, 275, 284, 295, 298; mobility and, 275; operational, 28, 50, 51; utilizing, 104, 283–84
environment (principle of war), 274–84, 295, 298; described, 289; new, 274–75
environmental factors, 87, 221, 281–83
Epaminondas, 164
Euclid, xi
Eugene of Savoy, Prince, 112
evacuations, 98, 191, 192–93
Ewell, Richard, 254, 295
Exeter (heavy cruiser), 103
expectations, 207, 210; enemy, 206; false, 206; realistic, 153–54, 217, 221–22, 260, 283, 290
exploiting moral-political factors (principle of war), described, 38

F-4 Phantom fighters, 195, 196
F-105 Thunderchief fighter-bombers, 195, 196
Fabian strategy, 124, 125, 245
Fabius Maximus Verrucosus, Quintus, 124
Falklands War, 141, 149
Fallwell, Marshall, xi, 109, 221
Fedayeen, 144, 265
Feuquières, Antoine de Pas, Marquis de, 6
Field Manual 100–1, The Army (U.S. Army), 20
Field Manual 100–5, Field Service Regulation—Operations (U.S. Army), 20
Field Service Regulations (British Army), 16, 17, 18, 19, 47–48
Field Service Regulations (Soviet Union), 36
Field Service Regulations (U.S. Army), 20
First Punic War, 245
First Triumvirate, 101
Fists of Righteous Harmony, 252
Flaminius, Caius, 172, 266; Hannibal and, 114, 171; Lake Trasimene and, 135–36
Fletcher, Jack, 67
flexibility, 19, 40, 51, 247, 252, 290; complex situations and, 241; indecisiveness and, 241–42, 246, 249; new information and, 243; rigidity and, 246; rule of, 24; violation of, 248
flexibility (principle of war), 19, 31, 41, 236, 252; candidate, 240–49; defined, 240–41; described, 290
Foch, Ferdinand, 14–15

fog of war, 9, 153, 215, 217, 226–30, 259–60, 275, 289; friction and, 259–60; unity of command and, 153
Foote, Andrew H., 138
force, 41, 87, 150, 261; capabilities/limitations of, 255–58; concentration of, 98, 234, 240; distribution of, 231, 262; effective, 105, 287, 291; laws of, 24; maximum, 231; physical, 9; position, 259; potentials, 37; status, 259; tactical, 122; unnecessary use of, 238; utilization of, 231
Forrest, Nathan Bedford, 201–2
Fort Donelson, capturing, 138
Fort Henry, 138, 278
Fort Stedman, 123
foundational elements, unity of effort and, 145–48
The Foundations of the Science of War (Fuller), 18–19
Franks, 269–70
Frederic, 8
Frederick the Great, surprise and, 160
Frederick William, Crown Prince, 9
Frederick William III (king), 9
freedom of action, 24, 25, 29, 156; maintaining, 107, 236; security and, 235
freedom of action (principle of war), 14, 24, 31; described, 25
Fremont, John, 78
French Army, 279; counterattacks by, 174–75; Schlieffen Plan and, 69
friction, 9, 24, 221, 229, 259; minimizing impact of, 153, 217, 220–21, 230, 290; reducing occurrence of, 153–54, 221, 250, 253, 260, 274–75, 277, 289–90
Frontinus, Sextus Julius, 3–4
Frost, Robert S., 241, 247
Frunze, M. V., 36
Fuller, J. F. C., 17, 20; principles of war and, 18, *19;* science of war and, 18–19
full use of all assets (principle of war), described, 38

Gallipoli, 126
Garfield, James, 225
Gates, Horatio, 99, 147–48
Generals and Generalship (Wavell), 267
General Signal Procedure, 229
General Tactics (French Army), 25

imagination, 158, 162; surprise and, 178

Imperial General Staff (British), 19

Inchon, surprise at, 163

indecisiveness, flexibility and, 241–42, 246, 249

Independence (aircraft carrier), 151

Indian Air Force, doctrine of, 31

Indian Army: doctrine of, 31; principles of war of, 30–31

The Infantry Attacks (Rommel), 267

The Influence of Sea Power upon History 1660–1783 (Mahan), 15

information, 56, 85, 151, 243, 244, 263–71, 288; chain of command and, 153; channels for, 209, 213; collecting, 270, 272, 273; contradictory, 177; deception and, 180, 210; decision-making and, 152–53; denying, 212, 272; discredited, 61, 208, 209, 211, 272; double-checking, 260; mastery of, 25; multipliers, 272; rationalizing, 177, 209; receptiveness to, 242; redundant, 213; revealing, 180, 203–4; sharing, 148, 155; unexpected, 176

information superiority (principle of war), 31, 262; described, 35

initiative, 21, 24, 33, 87, 104, 295; choices and, 118–19; deception and, 119; defense and, 121–22; defined, 108; described, 31, 108; exploiting, 106, 118, 125, 126, 127, 284; gaining, 106, 107, 109, 113–17, 118, 119, 120, 121, 125, 126, 235, 238, 287; losing, 111, 123; offensive and, 109–13, 113–17, 119, 120, 122–27; security and, 119; short-term, 118, 123–25, 287; strategic, 121; surprise and, 119; threats and, 119; as wasting asset, 126, 127. *See also* sustained initiative

initiative (principle of war), 31, 36. *See also* sustained initiative (principle of war)

initiative and offensive (principle of war), 106; described, 27

innovation, 158, 168, 177, 247

intelligence, 161; artificial, 286; effective, 155; gathering, 272

International Law of Armed Conflict, 22

interworking (principle of war), 129; described, 39

Iraqi forces, 161, 245, 265

island-skipping campaign, 98

Israeli Defense Force, principles of war and, 26

Israelites, Midianites and, 179

Istiqlal (ship), 189

Italian Navy, 271

Jackson, Stonewall, 126, 254; Chancellorsville and, 98, 155

James D. Mooney Quotes (Mooney), 128

Japanese Army, cooperation by, 136

Japanese Board for Study of the Principles of War, 32

Japanese Combined Fleet, 70

Japanese Ground Self-Defense Force, 32

Japanese Imperial General Headquarters, 136

Japanese Naval General Staff, 70

Japanese Navy, 167, 186, 244; advantage for, 127; cooperation by, 136; naval force ratio and, 97

Jellicoe, John, 194, 259

John (king of France), 173

John F. Kennedy (aircraft carrier), 151

Johnson, Lyndon B., 175, 196

Johnston, Joe, 94, 114–15, 254, 278

joint doctrine, 21, 48, 49, 50, 106, 238; U.S. military services and, 46

Joint Force Command, 46, 51–52, 59–60; air power and, 132

joint operations, principles of, 18, 28, 31, 46, 48, 49, 51, 59, 231, 238–39, 240; American, 21, 22

Joint Publications 3–0: Joint Operations (Joint Publication 3–0), revision of, 22

Jomini, Antoine-Henri, 8, 11, 15, 59; on deception, 233; fundamental truths and, 12; on objective, 13; principles of war and, 12

Jones, R. V.: deception by, 200

judgment, xii, 14, 65, 82, 213; good, 53, 56, 57, 108, 209, 234, 251, 252, 291; improving, 32; initiative and, 81; poor, 254; using, 18, 30, 42, 48, 56, 57, 230, 285

Julius Caesar, x, 8, 91, 146; Belgic forces and, 268

Justinian, 276; Belisarius and, 115–16, 269

Just War doctrine, 22

Kautilya, 3, 275

Keith, Lord, 77

Keynes, John Maynard, 243

key-point strikes (principle of war), described, 35

Khaled bin Sultan, 141

Khe Sanh, 207

Kido Butai, 71, 218

Kimmel, Husband, 136

King, Ernest, 79, 305n37

knowledge, ix; deception and, 266; lack of, 261, 283; obtaining, 271–73, 277; self, 295

known unknowns, 243

know your enemy, 2, 3, 55, 104, 253, 271–73, 284, 295, 298; deception and, 206; surprise and, 178

know your enemy (principle of war), 262–73, 295, 298; defined, 262; described, 288; new, 262–71

know yourself, 2, 3, 253, 260, 295, 297

know yourself (principle of war), 253–61, 295, 297–98; defined, 253; described, 288; new, 253–60

Kurita, Takeo, 127

Kuwait, 73, 190, 278; invasion of, 72, 74, 150, 151, 161, 188

Kuznetsov, Nikolai, 184

Lamachus, 93, 94, 136

Langsdorff, Hans, 103, 202, 203

Late Nineteenth Century, concept of war in, 14–15

Layton, Edwin, 213–14

leadership, 73, 74, 75, 175, 252, 291; American, 255; effective, 94; inspired, 28; strict/ uninterrupted, 38; unified, 34

leapfrogging campaigns, 14, 98

Lee, Robert E., x, 61, 85, 110, 113, 116, 122, 261, 297, 298; advantage for, 294; Ambrose Burnside and, 267; Chancellorsville and, 98, 126, 257; Culp Hill and, 296; deception and, 295; economy of force and, xvi; enemy commanders and, 266–67; Fredericksburg and, 281; George McClellan and, 94, 266–67; George Meade and, 116, 117, 294; Gettysburg and, xv, 89, 110, 117, 143, 254, 294, 306n5; initiative and, 123; J. E. B. Stuart and, 254, 295; lost order of, 94; objectives of, xv, 294; operational decision by, 295–96; Pendleton and, 254; Pickett's Charge and, 257; principles of war and, 293; Rappahannock and, 126; security and, xvii; surprise and, xvii, 295; Ulysses S. Grant and, 107; unity of command and, xvi, 143

Lee, Willis, 257

legitimacy, 21, 22, 238; purpose of, 239; undermining, 60

legitimacy (principle of war), 22, 231, 239; described, 21

Leigh, Humphrey, 169

Leigh Light, 169, 170

Lenin, Vladimir, 263

Leonhard, Robert R., 22, 105, 232

Letter of Instructions (Nimitz), 80, 81, 82, 305n44

Lexington (aircraft carrier), 244

Leyte, invasion of, 65, 68, 137

Liddell Hart, Basil, 83, 110, 116, 173; deception and, 233; indirect attack and, 22, 160; psychologically indirect and, 177; relative advantage and, 90

Light Brigade, Valley of Death and, 109

Ligny, 140, 224

limitations, knowledge of, 255–58

limited strategic aims (principle of war), described, 35

Lincoln, Abraham, 107, 116, 296; inflexibility of, 247; proclamation by, 78–79

Lodge, Henry Cabot, 118

Lofti, Colonel: fake version of, 199–200

logistics, 161, 187, 249; importance of, 26, 251, 290; problems with, 84, 250

logistics (principle of war), 41, 249, 250; near-principle, 252, 290. *See also* sustainability (principle of war)

Longstreet, James, xvi, 61, 89, 195, 254, 295; Gettysburg and, 110; Pickett's Charge and, 225, 257

Louis XIV (king), 112

Lucan, Lt. Gen., 109, 224

Lysander, 182, 183

MacArthur, Douglas, 137

MacDougall, Patrick Leonard, xi, 13

Macedonians, 91, 112

Machiavelli, Niccolò, 5–6, 160

Maginot Line, 174, 232

Mago, 278

Magruder, John, 119, 192

Mahan, Alfred Thayer, 15

Mahan, Dennis Hart, 13

maintenance of morale (principle of war), 19, 31; candidate, 251–52; described, 28; as near-principle, 291

maintenance of morale and fighting spirit (principle of war), described, 27 maintenance of the objective (principle of war), 19; described, 17

Malta, 200, 233

Manassas, 247

maneuver, 40, 47, 50, 104, 118, 125; concept of, 236, 237–38; initiative and, 119; movement and, 236; offensive and, 113–15

maneuver (principle of war), xvii, 20, 41, 42, 230, 240, 296; candidate, 236–38; current, 236–38; importance of, 2

Mao Tse-tung, 257; considerations of, 33–34; on guerrillas, 90; principles of war of, 34; revolution and, 258

Marathon, 86

Marcus Livius Salinator, 62, 237

Mariana Islands, amphibious assault on, 67, 117

Marine Air-Ground Task Force, 48, 133

Marine Corps Operations, 48; revision of, 49, 304n19

Maritime Interception Operations, 142

Maritime Strategy, 193, 265, 266, 315n33

Marlborough, duke of, 112

Marshall, George, 255

Marx, Karl, 263

Marxism-Leninism, 33, 37, 257

mass: achieving, 232; combat power and, 86–87; concentration of, 240; relative advantage vs., 83–105; sustaining, 83; at tactical level, 91–92

mass (principle of war), 2, 8, 12, 17, 20, 42, 96, 232, 287, 296; avoiding, 98; concentration and, xv–xvi; current, 83–84, 87–88, 94–95; following, xvi, 91, 93–94; formulation of, 85–86; guerrillas and, 89–90; idea behind, 87–88; problems with, 84–87; revision for, 84–85. *See also* concentration (principle of war)

mastery of information, rule of, 25

Mathematical Principles of Natural Philosophy (Principia) (Newton), 6

Mauz, Henry H., Jr., 141, 189

maxims of war, 5, 6, 7–8, 9, 12, 13, 14

McClellan, George, 192; Antietam and, 276; operational surprise by, 94; Robert E. Lee and, 94, 266–67

Me-109 fighter, 181

Meade, George: Abraham Lincoln and, 116; advantage for, 296–97; African American

troops and, 63; choices for, 294; Gettysburg and, xv, 116, 257; Henry Halleck and, 296, 297–98; Henry Hunt and, 195; know your enemy and, 298; know yourself and, 297; Petersburg and, 63; Robert E. Lee and, 116, 117, 294; surprise and, 297

Mein Kampf (Hitler), 263

Memories of War (de Feuquières), 6

Mencken, Louis, 215

Metcalf, Joseph, 143

Meuse River, crossing, 98, 174, 233

Mexican War, 266

Meyerson, Adam, 305n29

micromanagement, avoiding, 150, 151

Midianites, 201; Israelites and, 179

Midway Island: deception at, 203–4; defending, 79, 82; Japanese at, 70–71, 75, 77, 80, 81, 91–92, 218

MiG-17 fighters, 195

MiG-21 fighters, 195, 196

Mikawa, Gunichi, 127

military art, principles of, 36–38

military history, 37, 90–91, 105, 127, 172, 285; learning from, ix, x, 53

The Military Institutions of the Romans (Vegetius), 4–5

mindset, 263; flexible, 241, 242, 246, 290

misinterpretation, 109, 153, 215, 217

missiles, 200; anti-ship, 190; ballistic, 73, 167; cruise, x, 55, 167; Silkworm, 190

mission and aim (principle of war), 27

Mississippi River, 198, 245

Mississippi Squadron, 138

Missouri (battleship), 190

misunderstandings, 223, 286; minimizing, 215, 217, 230

mobility, 39, 102, 106; as defense, 132; environment and, 275; tactical, 33

mobility (principle of war), 19, 41, 236; described, 18

Moltke, Helmuth von (the Elder), 14, 26, 220

Moltke, Helmuth von (the Younger), 69, 75

momentum, 28, 32, 47, 48, 175, 303n10

Montgomery, Bernard Law, 19, 28; deception by, 186–87; El Alamein and, 95, 96

Mooney, James D., 128

morale, 7, 26, 39, 87, 158, 271, 287; damaging, 245; described, 251; enemy, 262; good,

Roanoke Island, capturing, 138

robust plans, 217–21; near-principle of war, 230, 290

Rochefort, Joseph, 203–4

Rohrbach Bridge, 276

Roman army, 102, 112, 114, 161, 237, 266; Lake Trasimene and, 172

Roman Empire, 61, 92, 135, 245, 247

Roman Legion, 112, 165

Roman Republic, 124, 130

Romans, xi, 6, 92, 101, 112, 124, 146, 162, 165, 197, 237, 247, 263, 275, 277; Carthaginians and, 172; concept of war and, 3–5; Hannibal and, 252; loss for, 102

Roman Senate, 263, 264

Rommel, Erwin, 211, 267; deception of, 186, 188; supply shortage for, 95; weather/climate and, 279

Roosevelt, Franklin D., 78

Roosevelt, Theodore, 15

Rosecrans, William, 185–86, 210, 225; Braxton Bragg and, 186, 267; deception by, 186

Royal Air Force (British), 64; principles of war and, 18

Royal Military College, 13

Royal Navy (British), principles of war and, 18

Rubel, Robert, 305n44

rules, 6, 10, 12; of engagement, 141–42; tactical, 11; of war, 4, 7, 109

Sadat, Anwar, 61, 265

Salt Lake City (cruiser), 226, 227

San Bernardino Strait, 127, 137, 222, 225–26

San Francisco (cruiser), 226, 228, 258

Saratoga Campaign, 99, 147, 148, 277

Savkin, Vasily, 23, 37, 38, 39

Savo Island, 226–27

Scharnhorst (battleship), 200

Scheer, Reinhard, 194

Schlieffen, Alfred von/Schlieffen Plan, 68–69

Schwartzkopf, H. Norman, 141, 143, 150, 259; deception by, 189

Scipio Africanus, x, 161, 164, 206, 276; conditioning by, 182; deception by, 196–98; Hasdrubal and, 165; surprise by, 162, 165–66

Scott, Norman, 226, 227, 228, 229

Searle, Alaric, 11–13, 15

Second Punic War, 124, 161, 237, 266; decep-

tion during, 196; surprise during, 164. *See also* Hannibal; Scipio Africanus

security, 19, 41, 47, 57, 234, 236; compromising, 60, 209; enhancing, 235; false sense of, 182; freedom of action and, 235; initiative and, 119; obtaining, 50–51; operational, 156

security (principle of war), xvii, 3, 8, 14, 17, 20, 21, 27, 31, 36, 41, 44, 231, 288, 296; candidate, 234–36; current, 234–36; described, 18, 28, 38, 234–35; economy of force and, 232, 234

seizing the initiative early (principle of war), described, 35

selection and maintenance of the aim (principle of war), 19, 28, 29, 30, 44, 60; defined, 60; described, 28. *See also* objective (principle of war)

sensors, 158, 270–71; misusing, 180, 200–201

September 11th, 74, 155, 157, 265; surprise and, 175–76

service doctrine, described, 46

Seven Days' Battle, 254

Seven Weeks' War, 220

Seven Years' War, 77

Shahikot Valley, 66, 222, 283

Sherman, William T., 114–15, 121, 308n17; David D. Porter and, 138–39; Joe Johnston and, 115; terrain and, 278, 279

Shiites, Sunnis and, 264

Shokaku (aircraft carrier), 71, 80

Short, Walter, 136

Sicily, 136, 200; Athenians at, 93; expedition to, 66, 96, 182

Sickles, Dan, 297

simplicity, xvi, 219, 230; lack of, 229; purpose of, 215; rule of, 24

simplicity (principle of war), 20, 27, 31, 44, 219, 240, 290, 296; candidate, 215–30; defined, 215

simultaneity of actions (principle of war), 36

Sinai Peninsula, 61, 265

Singapore, 243

situational awareness, 152, 258, 288, 295

Six Day War, surprise attack of, 267–68

Slim, William: on principles of war, 47–48

Solomon Islands, 167, 226, 268

Somme River, 276

Sorge, Richard, 184

Soviet Military Encyclopedia, 37

Soviet Navy, goal of, 265–66

Soviet Union: deception of, 183–85, 206–7; invasion of, 96, 123, 183, 184–85, 264; principles of war in, 36–42

Spanish Armada, 77, 114, 280

Sparta, 93, 122, 164

Spartans, 86, 122, 173, 183; Athenians and, 182; Thermopylae and, 276

speed, 50, 51, 155, 158, 174, 222, 227, 241, 260, 268; importance of, 2

speed (principle of war), 26, 47, 48

Sphacteria, 122, 173

spies, 188, 204; importance of, 2, 3

Spruance, Raymond, 80, 255, 257; criticism of, 117; Marianas and, 67, 117; Midway and, 91–92

Stalin, Joseph: deception of, 183–84, 206–7; permanently operating factors and, 36–37; persuasion by, 264; rationalization by, 209

Stalingrad, 113, 249–50

Stannard, G. J., 147

steadfastness (principle of war), 38, 251

stratagem (principle of war), 27, 28, 288

Strategemata (Frontinus), 4

strategy, 4, 16, 262; enemy, 265–66; playing games of, x; principles of, 18, 38; science of, 12

Strategy (Liddell Hart), 83

Streight, Abel, 201

strength, 58, 288; collecting, 8; moral-political, 37; recognizing, 253; taking advantage of, 263

strict and uninterrupted leadership (principle of war), described, 38

Stuart, J. E. B. (Jeb), 294; Gettysburg and, 143, 254; Peninsular Campaign and, 143; Robert E. Lee and, 254, 295; surveillance by, 295

suboptimization, 62, 149–50

subordinates: bad news/critical views from, 242; cooperation by, 135, 149, 288, 289; knowledge of, 254–55; objectives and, 150; questions/criticism from, 242; training of, 290

Suez Canal, 61, 185, 222, 265

Sunnis, Shiites and, 264

Sun Tzu, 33, 160, 253, 262, 265, 274, 315n9; advantage and, 2; concepts of war and, 1–3; on deception, 179; discipline/morale and, 2; enemy's weakness and, 87; fundamental elements and, 54–55; know your enemy and, 55; principles of war and, 2–3

superiority, 32, 39, 232; air, 73, 137; information, 31, 35, 262; local, 34, 98; technological, 25

supplies, 5, 98, 123, 249; building up, 95, 293; obtaining, 150; receiving, 97; shortage of, 95

surprise, 19, 40, 87, 118, 125, 155, 242, 265, 287, 295, 297; achieving, 28, 97, 160–61, 163, 178, 180; advantage of, 104, 158, 160, 166; deception and, 158–59, 178, 179, 185, 188; decision-making and, 156; defensive operations and, 158, 170–71; initiative and, 119; know your enemy and, 178; operational, 161–63; physical effect of, 159; psychological effect of, 157, 158, 159, 168, 171–76, 176–77, 178; rule of, 25; strategic, 160–61; tactical, 162–63, 164–66; technological, 160, 166–68, 168–70

surprise (principle of war), xvii, 3, 5, 10, 16, 30, 31, 32, 36, 39, 42, 288, 296, 298; American definition of, 155, 157–58; British definition of, 156, 158; current, 155–56, 156–58, 179; deception and, 158–59; described, 17, 28, 35, 38; importance of, 2, 41; moral effect of, 20; new, 158–79, 295, 297; new, defined, 158; new, described, 288; problems with, 156–58

surveillance, 161, 188, 283, 295; aerial, 204, 211

sustainability, 232, 250, 251, 252, 290

sustainability (principle of war), 28, 29, 31, 44, 240; candidate, 249–51; near-principle, 252, 290. *See also* logistics (principle of war)

sustained initiative, 106, 118, 119, 123–25, 127, 287, 289, 294, 297

sustained initiative (principle of war), 117–27, 294, 297; defined, 118; described, 287

Swinton, Ernest, 166

synergy, 31, 51, 134

Syphax, Hasdrubal and, 196, 197

Syracusans, delay by, 93–94

tactics, 4, 16, 24, 26, 102, 262, 269; fleet, 23; guerrilla, 90; ill-advised, 152; introduction of, 158; Soviet, 41; technology and, 55; unexpected, 242

tactics (Soviet principles of), 18, 38, 39, 41, 106, 129

Taginae, 269

tanks, 102–3, 167, 174, 211; development of, 166; massed, 258

Taranto harbor, attack on, 271

Tarentum, 111–12

Index